Smart Grid Telecommunications

Smart Grid Telecommunications

Fundamentals and Technologies in the 5G Era

Alberto Sendin
Javier Matanza
Ramon Ferrús

WILEY

This edition first published 2021
© 2021 John Wiley & Sons, Inc.

All rights reserved. No part of this publication may be reproduced, stored in a retrieval system, or transmitted, in any form or by any means, electronic, mechanical, photocopying, recording, or otherwise, except as permitted by law. Advice on how to obtain permission to reuse material from this title is available at http://www.wiley.com/go/permissions.

The right of Alberto Sendin, Javier Matanza, and Ramon Ferrús to be identified as the editors of this work has been asserted in accordance with law.

Registered Office(s)
John Wiley & Sons, Inc., 111 River Street, Hoboken, NJ 07030, USA
John Wiley & Sons Ltd, The Atrium, Southern Gate, Chichester, West Sussex, PO19 8SQ, UK

Editorial Office
9600 Garsington Road, Oxford, OX4 2DQ, UK

For details of our global editorial offices, customer services, and more information about Wiley products, visit us at www.wiley.com.

Wiley also publishes its books in a variety of electronic formats and by print-on-demand. Some content that appears in standard print versions of this book may not be available in other formats.

Limit of Liability/Disclaimer of Warranty
In view of ongoing research, equipment modifications, changes in governmental regulations, and the constant flow of information relating to the use of experimental reagents, equipment, and devices, the reader is urged to review and evaluate the information provided in the package insert or instructions for each chemical, piece of equipment, reagent, or device for, among other things, any changes in the instructions or indication of usage and for added warnings and precautions. While the publisher and authors have used their best efforts in preparing this work, they make no representations or warranties with respect to the accuracy or completeness of the contents of this work and specifically disclaim all warranties, including without limitation any implied warranties of merchantability or fitness for a particular purpose. No warranty may be created or extended by sales representatives, written sales materials, or promotional statements for this work. The fact that an organization, website, or product is referred to in this work as a citation and/or potential source of further information does not mean that the publisher and authors endorse the information or services the organization, website, or product may provide or recommendations it may make. This work is sold with the understanding that the publisher is not engaged in rendering professional services. The advice and strategies contained herein may not be suitable for your situation. You should consult with a specialist where appropriate. Further, readers should be aware that websites listed in this work may have changed or disappeared between when this work was written and when it is read. Neither the publisher nor authors shall be liable for any loss of profit or any other commercial damages, including but not limited to special, incidental, consequential, or other damages.

Library of Congress Cataloging-in-Publication Data

Names: Sendin, Alberto, author. | Matanza, Javier, author. | Ferrús, Ramon, author.
Title: Smart grid telecommunications : fundamentals and technologies in the 5G era / Alberto Sendin, Javier Matanza, Ramon Ferrús.
Description: Hoboken, NJ : Wiley-IEEE Press, 2021. | Includes index.
Identifiers: LCCN 2021027143 (print) | LCCN 2021027144 (ebook) | ISBN 9781119755371 (hardback) | ISBN 9781119755388 (adobe pdf) | ISBN 9781119755395 (epub)
Subjects: LCSH: Smart power grids. | 5G mobile communication systems. | Telecommunication.
Classification: LCC TK3105 .S3495 2021 (print) | LCC TK3105 (ebook) | DDC 621.31–dc23
LC record available at https://lccn.loc.gov/2021027143
LC ebook record available at https://lccn.loc.gov/2021027144

Cover Design: Wiley
Cover Image: © Image by Alberto Sendin

Set in 9.5/12.5pt STIXTwoText by Straive, Pondicherry, India

10 9 8 7 6 5 4 3 2 1

To my wife Esti, and our children Eneko, Gilem, Eider, and Paule.

Alberto Sendin

To you, mum, wherever you are. And to my patient and loving wife Alicia and our two cheerful kids Lucas and Chloe.

Javier Matanza

A Mar, Carla i Roger. Tos estimo.

Ramon Ferrús

To the reader: we hope you find the book informative and helpful.

The authors

Contents

Author Biographies *xv*
Preface *xvii*
Acronyms *xxi*

1 The Smart Grid: A General Perspective *1*
1.1 Introduction *1*
1.2 Electric Power Systems *1*
1.2.1 Electricity *2*
1.2.1.1 Frequency and Voltage *3*
1.2.2 The Grid *4*
1.2.2.1 The Grid from a Technical Perspective *5*
1.2.2.2 The Grid from a Regulatory Perspective *9*
1.2.3 Grid Operations *12*
1.2.4 The Grid Assets *14*
1.2.4.1 Substations *14*
1.2.4.2 Power Lines *16*
1.3 A Practical Definition of the Smart Grid *18*
1.4 Why Telecommunications Are Instrumental for the Smart Grid *20*
1.5 Challenges of the Smart Grid in Connection with Telecommunications *23*
1.5.1 Customer Engagement Challenges *23*
1.5.1.1 Customers as Smart Electricity Consumers *23*
1.5.1.2 Customers as Energy Generators *24*
1.5.2 Grid Control Challenges *25*
1.6 Challenges of Telecommunications for Smart Grids *26*
1.6.1 Telecommunication Solutions for Smart Grids *26*
1.6.2 Standards for Telecommunications for Smart Grids *27*
1.6.3 Groups of Interest Within Telecommunications for Smart Grids *29*
1.6.4 Locations to be Served with Telecommunications *29*
1.6.5 Telecommunication Services Control *31*
1.6.6 Environmental Conditions *32*
1.6.7 Distributed Intelligence *34*
1.6.8 Resilient Telecommunication Networks and Services *34*
1.6.9 Telecommunications Special Solution for Utilities *35*
References *36*

2	**Telecommunication Networks and Systems Concepts** *41*	
2.1	Introduction *41*	
2.2	Telecommunication Networks, Systems, and Services Definitions *41*	
2.3	Telecommunication Model and Services *42*	
2.3.1	Telecommunication Model *42*	
2.3.2	Analog and Digital Telecommunications *44*	
2.3.3	Types of Telecommunications Services *45*	
2.4	Telecommunication Networks *46*	
2.4.1	Network Topologies *48*	
2.4.2	Transport and Switching/Routing Functions *48*	
2.4.3	Circuit-switched and Packet-switched Networks *50*	
2.4.3.1	Circuit-switched Technologies *51*	
2.4.3.2	Packet-switched Technology *51*	
2.4.3.3	Multilayered Telecommunication Networks *54*	
2.4.4	Telecommunications Networks and Computing *55*	
2.5	Protocol Architectures for Telecommunication Networks *55*	
2.5.1	Why a Protocol Layered Model Is Needed *55*	
2.5.2	The OSI Model *56*	
2.5.3	The TCP/IP Protocol Stack *57*	
2.5.4	User, Control, and Management Planes *59*	
2.6	Transmission Media in Telecommunications for Smart Grids *59*	
2.6.1	Optical Fibers *61*	
2.6.1.1	Optical Fiber Cables for Smart Grids *63*	
2.6.1.2	Optical Fiber Cables Specifications *65*	
2.6.2	Radio Spectrum *67*	
2.6.2.1	Radio Spectrum for Utility Telecommunications *69*	
2.6.2.2	Radio Spectrum Use *70*	
2.7	Electricity Cables *71*	
2.7.1	PLC Use *73*	
	References *75*	
3	**Telecommunication Fundamental Concepts** *79*	
3.1	Introduction *79*	
3.2	Signals *79*	
3.2.1	Analog vs. Digital *79*	
3.2.1.1	Continuous vs. Discrete *79*	
3.2.1.2	Sampling *81*	
3.2.1.3	Quantizing and Coding *81*	
3.2.1.4	Analog and Digital Signals *82*	
3.2.2	Frequency Representation of Signals *83*	
3.2.2.1	The Continuous-time Fourier Transform *83*	
3.2.2.2	The Discrete-Time Fourier Transform *85*	
3.2.3	Bandwidth *88*	
3.3	Transmission and Reception *89*	
3.3.1	Modulation *89*	
3.3.1.1	Example of a Simple Analog Modulation: Double Sideband *91*	
3.3.1.2	Example of a Simple Digital Modulation: Quadrature-Phase Shift Keying *91*	

3.3.2	Channel Impairments	*93*
3.3.2.1	Attenuation	*93*
3.3.2.2	Noise and Interference	*93*
3.3.2.3	Signal Distortion	*94*
3.3.3	Demodulation, Equalization, and Detection	*97*
3.3.3.1	Signal-to-Noise Ratio and Bit Error Rate	*97*
3.3.3.2	Channel Equalization	*98*
3.3.4	Multiplexing	*99*
3.3.5	Channel Coding	*103*
3.3.5.1	A Simple Example of Coding	*104*
3.3.5.2	Interleaving	*106*
3.3.5.3	Advanced Coding Techniques	*106*
3.3.5.4	Channel Coding in Multicarrier Modulations	*107*
3.3.6	Duplexing	*107*
3.3.7	Multiple Access	*108*
3.3.7.1	TDMA/FDMA/CDMA/OFDMA	*108*
3.3.7.2	Multiple Access Methods	*109*
3.3.7.3	Carrier Sense Multiple Access (Collision Avoidance/Collision Detection)	*109*
3.4	Signal Propagation	*110*
3.4.1	Optical Fiber Propagation	*110*
3.4.1.1	Optical Communications Components	*110*
3.4.1.2	Optical Fiber Propagation Phenomena	*111*
3.4.2	Radio Propagation	*112*
3.4.2.1	Antennas	*113*
3.4.2.2	Array Antennas and Beamforming	*113*
3.4.2.3	Free-space Propagation Phenomena	*114*
3.4.3	Link Budget	*115*
	References	*116*
4	**Transport, Switching, and Routing Technologies**	*117*
4.1	Introduction	*117*
4.2	Transport Networks	*117*
4.2.1	Plesiochronous Digital Hierarchy (PDH)	*118*
4.2.2	SDH/SONET	*119*
4.2.3	DWDM	*121*
4.2.4	Optical Transport Network (OTN)	*123*
4.3	Switching and Routing	*124*
4.3.1	Switching Principles	*124*
4.3.1.1	Switching Process	*125*
4.3.1.2	Solving Switching Loops: Spanning Tree Protocol	*126*
4.3.2	Routing Principles	*127*
4.3.2.1	Routing Classification	*127*
4.3.2.2	Routing Metrics	*128*
4.3.2.3	Autonomous Systems	*129*
4.3.2.4	Routing Algorithms	*129*
4.3.2.5	Routing Protocols	*131*
4.3.3	Ethernet	*132*

4.3.3.1	Carrier Ethernet	*133*
4.3.4	Internet Protocol (IP)	*133*
4.3.5	Multiprotocol Label Switching (MPLS)	*134*
4.3.5.1	Multiprotocol Label Switching – Transport Profile (MPLS-TP)	*134*
	References	*135*

5 Smart Grid Applications and Services *137*
5.1 Introduction *137*
5.2 Smart Grid Applications and Their Telecommunication Needs *137*
5.3 Supervisory Control and Data Acquisition *139*
5.3.1 Components *140*
5.3.2 Protocols *141*
5.3.2.1 Central Infrastructure to Field Protocols *142*
5.3.2.2 Central Infrastructure Protocols *143*
5.4 Protection *143*
5.5 Distribution Automation *147*
5.5.1 Distributed Energy Resources Integration *148*
5.5.2 Electric Vehicles Integration *150*
5.5.3 Fault Location, Isolation, and Service Restoration *151*
5.5.4 Indices for Operations Performance *151*
5.6 Substation Automation *153*
5.7 Metering *158*
5.8 Synchrophasors *161*
5.9 Customers *164*
5.9.1 Demand-side Management *165*
5.9.2 Energy Management *166*
5.9.3 Microgrids *168*
5.10 Power Lines *169*
5.10.1 Flexible AC Transmission System *169*
5.10.2 Dynamic Line Rating *169*
5.11 Premises and People *170*
5.11.1 Business Connectivity *170*
5.11.2 Workforce Mobility *171*
5.11.3 Surveillance *172*
 References *174*

6 Optical Fiber and PLC Access Technologies *179*
6.1 Introduction *179*
6.2 Optical Fiber Passive Network Technologies *179*
6.2.1 Mainstream Technologies and Standards *180*
6.2.1.1 PON Technologies Evolution *180*
6.2.1.2 Supported Services and Applicability Scenarios *183*
6.2.1.3 Spectrum *184*
6.2.1.4 System Architecture *184*

6.2.2	Main Capabilities and Features	*186*
6.2.2.1	Time and Wavelength Division Multiplexing	*186*
6.2.2.2	Features Needed in PONs	*187*
6.2.2.3	Dynamic Bandwidth Assignment	*187*
6.2.3	ITU's GPON Family	*188*
6.2.3.1	GPON	*188*
6.2.3.2	XG(S)-PON	*190*
6.2.3.3	NG-PON2	*190*
6.2.4	IEEE's EPON Family	*191*
6.2.4.1	EPON	*191*
6.2.4.2	10G-EPON	*191*
6.3	Power Line Communication Technologies	*191*
6.3.1	Mainstream Technologies and Standards	*192*
6.3.1.1	PLC Technologies Evolution	*192*
6.3.1.2	Supported Services and Applicability Scenarios	*193*
6.3.1.3	Architecture	*194*
6.3.2	Main Capabilities and Features	*196*
6.3.2.1	Common Transceiver Designs in PLC Systems	*196*
6.3.2.2	PLC Signal Coupling	*197*
6.3.3	Narrowband PLC Systems	*198*
6.3.3.1	ITU-T G.9904 (PRIME v1.3)	*198*
6.3.3.2	Future ITU-T G.9904.1 (PRIME v1.4)	*204*
6.3.3.3	ITU-T G.9903 (G3-PLC)	*205*
6.3.3.4	IEEE 1901.2	*209*
6.3.3.5	ITU-T G.9902 (G.hnem)	*210*
6.3.4	Broadband PLC Systems	*211*
6.3.4.1	IEEE 1901	*211*
6.3.4.2	ITU-T G.996x (G.hn)	*214*
6.4	Applicability to Smart Grids	*215*
6.4.1	Passive vs. Active Optical Fiber Networks	*216*
6.4.2	Broadband PLC over Medium Voltage for Secondary Substation Connectivity	*217*
6.4.3	High Data Rate Narrowband PLC over the Low Voltage Grid for Smart Metering	*218*
	References	*220*
7	**Wireless Cellular Technologies**	*225*
7.1	Introduction	*225*
7.2	Mainstream Technologies and Standards	*225*
7.2.1	Cellular Technologies Evolution	*225*
7.2.1.1	1G and 2G. Voice-centric, Circuit-switched Services	*225*
7.2.1.2	3G. Paving the Way for Mobile Data Services	*227*
7.2.1.3	4G. The First Global Standard for Mobile Broadband	*227*
7.2.1.4	5G. Expanding the Applicability Domain of Cellular Technologies	*228*
7.2.2	Supported Services and Applicability Scenarios	*229*
7.2.2.1	Service Categories	*229*

7.2.2.2	Performance Indicators	*229*
7.2.2.3	Commercial Networks and Private Networks	*229*
7.2.3	Spectrum	*231*
7.2.3.1	Spectrum Harmonization. IMT Bands	*231*
7.2.3.2	Frequency Bands Being Prioritized for 5G	*232*
7.2.3.3	Spectrum Exploitation Models	*233*
7.2.4	3GPP Standardization	*235*
7.3	System Architecture	*237*
7.3.1	High-level Architecture of 4G/5G Systems	*237*
7.3.2	Radio Access Network	*240*
7.3.2.1	E-UTRAN	*240*
7.3.2.2	NG-RAN	*246*
7.3.3	Core Network	*250*
7.3.3.1	Evolved Packet Core	*250*
7.3.3.2	5G Core Network	*252*
7.3.3.3	Transitioning from 4G to 5G	*255*
7.3.4	Service Platforms	*256*
7.3.4.1	IMS and Voice Services over 4G/5G	*256*
7.3.4.2	5G Service Frameworks and Application Enablers	*256*
7.3.5	Main System Procedures	*257*
7.3.5.1	Network Registration	*257*
7.3.5.2	Service Request	*258*
7.3.5.3	PDU Session Establishment	*259*
7.3.5.4	Handover	*260*
7.4	Main Capabilities and Features	*261*
7.4.1	LTE Radio Interface	*261*
7.4.1.1	Operating Bands	*262*
7.4.1.2	Time-frequency Resource Grid	*262*
7.4.1.3	Scheduling, Link Adaptation, and Power Control	*264*
7.4.1.4	Fast Retransmissions and Minimum Latency	*265*
7.4.1.5	Multiple-antenna Transmission and Reception	*265*
7.4.1.6	Carrier Aggregation and Dual Connectivity	*266*
7.4.1.7	Physical Signals and Physical Channels	*266*
7.4.1.8	Mapping Between Physical, Transport, and Logical Channels	*269*
7.4.1.9	Radio Access Procedures	*270*
7.4.2	5G NR Interface	*271*
7.4.2.1	Flexible Waveform and Numerologies	*272*
7.4.2.2	Reduced Latency	*274*
7.4.2.3	Bandwidth Parts	*274*
7.4.2.4	Flexible Placement of the Control Channels	*274*
7.4.2.5	Massive MIMO and Beamforming	*276*
7.4.2.6	New Operating Bands	*276*
7.4.3	Edge Computing Support	*276*
7.4.4	QoS Parameters and Characteristics	*278*
7.4.5	Network Slicing	*278*
7.4.6	Operation in Unlicensed Spectrum	*280*

7.4.7	Private Networks *281*
7.5	Applicability to Smart Grids *282*
7.5.1	Smart Metering *285*
7.5.2	Distribution Grid Multiservice Access *287*
	References *289*

8	**Wireless IoT Technologies** *293*
8.1	Introduction *293*
8.2	Mainstream Wireless IoT Technologies for the Smart Grid *293*
8.3	IEEE 802.15.4-based Technologies: Zigbee and Wi-SUN *294*
8.3.1	Scope and Standardization *294*
8.3.1.1	IEEE 802.15.4 Standard *294*
8.3.1.2	Zigbee *296*
8.3.1.3	Wi-SUN *297*
8.3.2	Network and Protocol Stack Architecture *297*
8.3.2.1	Network Components and Topologies *297*
8.3.2.2	Zigbee Network Architecture and Protocol Stack *300*
8.3.2.3	Wi-SUN FAN Network Architecture and Protocol Stack *300*
8.3.3	Main Capabilities and Features *302*
8.3.3.1	IEEE 802.15.4 Physical Layer *302*
8.3.3.2	IEEE 802.15.4 MAC Layer *303*
8.3.3.3	Zigbee Specifics *304*
8.3.3.4	Wi-SUN FAN Specifics *305*
8.4	Unlicensed Spectrum-based LPWAN: LoRaWAN and Sigfox *307*
8.4.1	Scope and Standardization *307*
8.4.2	LoRaWAN *308*
8.4.2.1	Network Architecture and Protocol Stack *308*
8.4.2.2	Protocol Frame Structure *309*
8.4.2.3	Physical Layer *310*
8.4.2.4	MAC Layer *310*
8.4.3	Sigfox *311*
8.4.3.1	Network Architecture and Protocol Stack *311*
8.4.3.2	Protocol Frame Structure *312*
8.4.3.3	Physical Layer *313*
8.4.3.4	MAC Layer *314*
8.5	Cellular IoT: LTE-M and NB-IoT *314*
8.5.1	Scope and Standardization *314*
8.5.2	Network and Protocol Stack Architecture *315*
8.5.2.1	New Network Attach Method and Connectivity Options *315*
8.5.2.2	New Network Entities *316*
8.5.2.3	Control Plane and Data Plane Optimizations *317*
8.5.3	Main Capabilities and Features *317*
8.5.3.1	LTE-M Radio Access *317*
8.5.3.2	NB-IoT Radio Access *322*
8.5.3.3	Operation in Unlicensed Spectrum *325*
8.5.3.4	LTE-M and NB-IoT Roadmap in 5G *326*

8.6	IoT Application and Management Layer Protocols	*327*
8.6.1	CoAP *328*	
8.6.2	MQTT *328*	
8.6.3	OMA LwM2M *329*	
8.7	Applicability to Smart Grids *329*	
8.7.1	Great Britain Smart Metering System *329*	
8.7.2	Unlicensed Spectrum-based LPWAN Technologies for Smart Metering *331*	
	References *333*	

Index *339*

Author Biographies

Dr Alberto Sendin received the M.Sc. Telecommunication Engineering in 1996, the M.A. degree in Management for Business Competitiveness (GECEM) in 2001, and the Ph.D. degree in Telecommunications in 2013, from the University of the Basque Country in Spain. Starting his professional career with Airtel Movil (now Vodafone), since 1998, he works for Iberdrola (one of the biggest electricity utilities) transforming its telecommunication network as Head of Telecommunications in Spain. He is also a Professor with the Comillas Pontifical University, Madrid, Spain, and formerly with the University of Deusto, Bilbao, Spain, teaching telecommunications since 1999 (microwaves, radiocommunication systems, telecommunication systems, project management, telecommunications for smart grids and Industry 4.0). He is the author and coauthor of eight telecommunication books edited by McGraw-Hill, Artech House, and others; three chapters in edited books by John Wiley & Sons and CRC Press; and tens of peer-reviewed academic papers. His interests are with smart grids, radiocommunication systems, wireline telecommunications systems, and Power Line Communications (PLC). He is an active member of the European Utilities Telecom Council (EUTC) and the PRIME Alliance, and through them, he contributes to ITU and 3GPP standardization activities for Smart Grids.

Dr Javier Matanza received the M.Sc. in Telecommunication Engineering from the Polytechnic University of Valencia, Valencia, Spain, in 2008, and the Ph.D. degree from Comillas Pontifical University, Madrid, Spain, in 2013. He is currently a Research Professional with the Institute for Research in Technology focused on telecommunication technologies applied to power systems. In addition, he is a lecturer with the Comillas Pontifical University in the areas of Linear Systems, Communication Theory, Advanced Digital Communications on the Bachelor Degree in Engineering and Telecommunications Technologies, Master in Telecommunications Engineering, and Master of Science in Smart Grids; he is also the academic coordinator of the later. He has coauthored 14+ publications in JCR-indexed peer-reviewed journals, participated in 29+ international conferences, and worked on 22+ R&D projects. His current interests are in powerline communication technologies, signal processing, and communication network simulations.

Dr Ramon Ferrús received the Telecommunication Engineering (B.S. plus M.S.) and Ph.D. degrees from the Universitat Politècnica de Catalunya (UPC), Barcelona, Spain, in 1996 and 2000, respectively, where he is currently a tenured associate professor within the Signal Theory and Communications department. He has taught in a wide range of basic and specialized courses within the bachelor and master degrees' programs in Telecommunications at UPC, covering topics related to digital communications, data transmission systems, radio communications, satellite communications, mobile communications systems, and network management. At research level, his interests include system design, resource optimization, and network and service management in wireless communications, with his latest activities focusing on the realization of network slicing capabilities in 5G radio access networks and the applicability of data analytics and machine learning techniques for network management. Since 2000, he has participated in 10+ research projects within the 6th, 7th, and H2020 Research Framework Programmes of the European Commission, taking responsibilities as work package leader in some of them. He has contributed to ETSI standardization activities and is currently involved in 3GPP standardization for the adaptation of IoT-NB protocols for non-terrestrial networks. He has also participated in numerous national research projects and technology transfer activities for public and private companies. He is coauthor of two books, three book chapters, and 130+ papers published mostly in IEEE journals and conferences.

Preface

Electric power systems and telecommunications are two of the most basic services supporting our society. Their basic infrastructure nature draws similarities between them, including their high investments needs, their critical nature, and the high expectations over their service quality. However, there are also substantial differences such as the different role that regulation plays in both their technology evolution cycles and the service model.

Electric power systems and telecommunications converge in the Smart Grid concept, as the flagship of a more secure, reliable, resilient, and automated grid, offering a platform to allow a more flexible system model where new power systems technologies and stakeholders may interact, to provide an enhanced electricity service within a more efficient and dynamic power system. Telecommunications are both an enabler of the Smart Grid through the provision of connectivity services to the grid assets, and a system stakeholder that will benefit from this enhanced service.

The grid consists of substations connected with power lines. Substations spread through the territory in which electricity generation is transported and distributed to the customers, across different voltage levels. Transmission and Distribution System Operators rely on their Control Centers to manage electricity service delivery, making use of Information and Communication Technologies, to connect their central and distributed systems' applications, with the grid assets.

Telecommunications networks consist of equipment performing transport, switching, and routing functions over telecommunication media such as optical fiber, radio, and different metallic supporting cables, including power lines; they also integrate ancillary elements and information systems, to create highly resilient telecommunication systems. Telecommunication services are delivered as the final product of Telecommunication systems, and all sorts of end users employ them to remotely communicate people and machines.

Telecommunications have always been instrumental in utility operations. The connectivity to the pervasive and widespread utility assets has always been key to coordinate operation crews and monitor, control and automate the performance of individual grid components and the system as an entity. While trying to leverage commercially provided telecommunication services, utilities have developed private telecommunication networks to fulfill requirements intrinsic to the nature of the grid and the electricity service. On the one hand, environmental conditions and functional performance needs in the grid, vary depending on the underlying infrastructure (Generation, Transmission, Distribution, Point of Supply; substations; power lines; etc.), and a part of them cannot be supported over non-adapted telecommunication networks and services. On the other, utilities have access to rights of way, physical assets, their own power lines as a telecommunication transmission medium, and electricity. These elements have allowed utilities to build their own telecommunication networks to fully comply with their real-time mission-critical needs. These infrastructures and networks have also eventually been used to support telecommunication services by the different fixed and mobile telecommunication systems operators. Thus, a symbiotic relationship exists between both infrastructures, especially relevant when discussing Smart Grids.

The trend in the utility sector is that utilities operate hybrid telecommunication networks. With these networks, utilities provide telecommunication services critical for their power system operation. These networks are hybrid in two aspects. First, they are a combination of private and public (commercial) networks and services, to take benefit from the best of the two domains: private networks can be designed and deployed to fully satisfy requirements, and public services are quick to deploy where requirements fit the needs. Second, different technologies, networks, systems, and telecommunication supporting media (wireline and wireless) are combined, to adapt to the different conditions of the Smart Grid services. All these telecommunications should be managed in an integral way, so that Smart Grid services can be successfully and efficiently deployed and operated.

Electric power systems and Telecommunications are complex functional domains with sophisticated elements that require expert knowledge. The operation of such complex systems increasingly depends on a diverse set of technical aspects and technologies that have usually been the realm of experts, each one in its knowledge field. The complexity of power systems and the complexity of Telecommunications meet and get augmented in the Smart Grid, as the intersection of both. The success in the decision-making process when designing Smart Grid solutions depends on understanding the details of the underlying technologies and systems, so that the best combination of available solutions can be chosen within the goals and constraints of each Smart Grid vision.

This book has the ambition to provide a comprehensive view of the Smart Grid scenario, together with the Telecommunications that are needed to make it happen. The reader is accompanied in a journey around power systems and telecommunications, reaching Smart Grids. The first part of the journey shows the key general and technical aspects of electric power systems and their service needs in terms of their evolution into the Smart Grid. The second part of the journey shows the fundamentals of Telecommunications; it starts with the high-level concepts and follows with the lower-level details of how they are implemented, first at the lower layers of the telecommunication systems, and second in the upper layers' data exchange. The journey stops in several stations, covering the telecommunication systems applicable to Smart Grids, namely optical fiber networks, power line communication (PLC) systems, wireless cellular and wireless IoT systems. Each of these stops offers a selection of example use cases, where the specific group of technologies solves one representative Smart Grid need.

The book is organized in eight chapters.

Chapter 1 provides a view of electric power systems and highlights the relevant aspects and main challenges that determine their evolution into the Smart Grid.

Chapter 2 describes the key concepts and organization of telecommunication networks, systems, and services. It also introduces the fundamental aspects of key telecommunication media in Smart Grid (optical fiber, radio, and power lines).

Chapter 3 goes deep, while keeping descriptive, into the telecommunications framework, and clarifies the foundational technology aspects of telecommunications (analog vs. digital, modulation, medium access, propagation, etc.).

Chapter 4 focuses on data exchange, explaining and describing how telecommunication transport, switching, and routing functions work and are implemented in technologies used in Smart Grids.

Chapter 5 provides a comprehensive view of the different Smart Grid domains, and specifically describes the Smart Grid services, together with the protocols and relevant standards applicable to them.

Chapter 6 builds upon the optical fiber and the power lines, so important for utilities, to cover Passive Optical Networks and PLC technologies, with a focus on their role in the network access domain.

Chapter 7 details wireless cellular 4G and 5G technologies. Being the flagship of the pervasive mobile communications service, this chapter addresses their baseline concepts to get an understanding of the complexity and capabilities of these technologies.

Chapter 8 covers wireless IoT technologies, focusing on those that have an impact on Smart Grids. Zigbee, Wi-SUN, LoRaWAN, Sigfox, LTE-M, and NB-IoT will be introduced.

Acronyms

10G-EPON	10 Gbps Ethernet Passive Optical Network
1G	First Generation
2G	Second Generation
3G	Third Generation
3GPP	Third Generation
4G	Fourth Generation
5G	Fifth Generation
5GC	5G Core Network
5GS	5G System
5QI	5G QoS Identifier
6G	Sixth Generation
6LoWPAN	IPv6 Low Power Wireless Personal Area Network
8PSK	8-Phase Shift Keying
AC	Alternating Current
ACB	Access Class Barring
ACK	ACKnowledgement
ADA	Advanced Distribution Automation
ADM	Add–Drop Multiplexer
ADSL	Asymmetric Digital Subscriber Line
AES	Advanced Encryption Standard
AM	Amplitude Modulation
AMF	Access and Mobility Management Function
AMI	Automatic Metering Infrastructure
AMR	Advanced Meter Reading
ANSI	American National Standards Institute
APCI	Application Protocol Control Information
APDU	Application Protocol Data Unit
API	Application Programming Interface
APN	Access Point Name
APON	Asynchronous Transfer Mode Passive Optical Network
APT	Asia Pacific Telecommunity
ARIB	Association of Radio Industries and Business (Japan)
ARP	Address Resolution Protocol
ARQ	Automatic Repeat reQuest
ASAI	Average Service Availability Index

ASDU	Application Service Data Unit
ASIDI	Average System Interruption Duration Index
ASIFI	Average System Interruption Frequency Index
ASK	Amplitude Shift Keying
ASMG	Arab Spectrum Management Group
ATIS	Alliance for Telecommunications Industry Solutions
ATM	Asynchronous Transfer Mode
ATU	African Telecommunications Union
AUSF	AUthentication Server Function
BCCH	Broadcast Control CHannel
BCH	Broadcast CHannel
BEC	Backward-Error Correction
BEMS	Building Energy Management System
BER	Bit Error Rate
BEV	Battery EV
BGP	Border Gateway Protocol
BLE	Bluetooth Low Energy
BN	Base Node
BPL	Broadband Power Line communications
BPON	Broadband Passive Optical Network
bps	bits per second
BPSK	Binary Phase Shift Keying
C&I	Commercial&Industrial
CAIDI	Customer Average Interruption Duration Index
CAIFI	Customer Average Interruption Frequency Index
CAP	Contention Access Period
CAPEX	Capital Expenditure
CAPIF	Common API Framework
CBA	Cost–Benefit Analysis
CBRS	Citizens Broadband Radio Service
CC	Component Carrier
CCA	Clear Channel Assessment
CDMA	Code Division Multiple Access
CE	Coverage Enhancement
CELID	Customers Experiencing Long Interruption Durations
CEMIn	Customers Experiencing Multiple Interruptions (n)
CEMSMIn	Customers Experiencing Multiple Sustained Interruption (n) and Momentary Interruption Events
CEN	European Committee for Standardization (Comite Europeen de Normalisation)
CENELEC	European Committee for Electrotechnical Standardization (Comité Européen de Normalisation Electrotechnique)
CEPT	European Conference of Postal and Telecommunications
CFP	Contention-Free Period
CFS	Contention-Free Slot
CH	Communications Hub
CIGRE	Conseil International des Grands Reseaux Electriques (International Council on Large Electric Systems)

CIoT	Cellular Internet of Things
CISPR	Comité International Spécial des Perturbations Radioélectriques (International Special Committee on Radio Interference)
CITEL	Comisión Interamericana de Telecomunicaciones (Inter-American Telecommunication Commission)
cMTC	critical Machine Type Communications
CN	Core Network
CoAP	Constrained Application Protocol
CORESET	Control Resource Set
COSEM	Companion Specification for Energy Metering
CP	Control Plane
CP	Cyclic Prefix
CPE	Customer Premise Equipment
CP-OFDM	Cyclic Prefix OFDM
CPP	Critical Peak Pricing
CPR	Critical Peak Rebates
CQI	Channel Quality Indicator
CRC	Cyclic Redundancy Check
CRM	Customer Relationship Management
CSD	Circuit Switched Data
CSI	Channel State Information
CSMA/CA	Carrier Sense Multiple Access/Collision Avoidance
CSMA/CD	Carrier Sense Multiple Access/Collision Detection
CSS	Chirp Spread Spectrum
CTAIDI	Customer Total Average Interruption Duration Index
CTS	Clear to Send
CWDM	Coarse Wavelength Division Multiplexing
DA	Distribution Automation
DBA	Dynamic Bandwidth Assignment
DC	Direct Current
DC	Data Concentrator
DCC	Data Communications Company
DCI	Downlink Control Information
DER	Distributed Energy Resources
D-FACTS	Distribution FACTS
DFT	Discrete Fourier Transform
DFT-S-OFDM	Digital Fourier Transform Spread OFDM
DG	Distributed Generation
DLL	Data-Link Layer
DLMS	Device Language Message Specification
DLR	Dynamic Line Rating
DM	Domain Master
DM-RS	Demodulation Reference Signal
DMS	Distribution Management System
DMVPN	Dynamic Multipoint VPN
DN	Data Network
DNN	Data Network Name

DNO	Distribution Network Operator
DNP3	Distributed Network Protocol
DOE	Department of Energy (USA)
DR	Demand Response
DRB	Data Radio Bearer
DRX	Discontinuous Reception
DS	Distributed Storage
DSB	Double Sideband
DSM	Demand-Side Management
DSME	Deterministic Synchronous Multichannel Extension
DSO	Distribution System Operator
DSS	Dynamic Spectrum Sharing
DSSS	Direct Sequence Spread Spectrum
DWDM	Dense Wavelength Division Multiplexing
EAB	Extended Access Barring
ECC	Electronic Communications Committee
ECHONET	Energy Conservation and HOmecare NETwork
ED	End Device
EDGE	Enhanced Data rates for Global Evolution
eDRX	extended Discontinuous Reception
eMBB	enhanced Mobile Broadband
EMC	ElectroMagnetic Compatibility
EMS	Energy Management System
eMTC-U	enhanced Machine Type Communications Unlicensed
EN	European Norm (Standard)
eNB	eNodeB
EN-DC	E-UTRAN NR Dual Connectivity
ENISA	European Union Agency for Network and Information Security
EPA	Enhance Performance Architecture
EPC	Evolved Packet Core
EPON	Ethernet Passive Optical Network
EPRI	Electric Power Research Institute
EPS	Evolved Packet System
ERP	Enterprise Resource Planning
ETSI	European Telecommunications Standards Institute
EU	European Union
E-UTRAN	Enhanced UMTS Terrestrial Radio Access Network
EV	Electric Vehicle
EVCC	Electric Vehicle Communication Controller
EVSE	Electric Vehicle Supply Equipment
FACTS	Flexible Alternating Current Transmission System
FAN	Field Area Network
FCC	Federal Communications Commission
FDD	Frequency Division Duplexing
FDIR	Fault Detection Isolation and Restoration
FDM	Frequency Division Multiplexing
FDMA	Frequency Division Multiple Access

FEC	Forward Error Correction
FER	Frame Error Rate
FFD	Full Function Device
FFT	Fast Fourier Transform
FH	Frequency Hopping
FLISR	Fault Location, Isolation, and Service Restoration
FM	Frequency Modulation
FR	Frequency Range
FSAN	Full Service Access Network
FSK	Frequency Shift Keying
FTTH	Fiber To The Home
FTTx	Fiber To The x
GB	Great Britain
GBR	Guaranteed Bit Rate
GEM	GPON Encapsulation Method
GFP	Generic Framing Procedure
GIS	Geographic Information System
gNB	gNodeB
GOOSE	Generic Object-Oriented Substation Event
GPON	Gigabit Passive Optical Network
GPRS	General Packet Radio Service
GRE	Generic Routing Encapsulation
GSM	Global System for Mobile
GSMA	GSM Association
GSSE	Generic Substation State Event
GTC	GPON Transmission Convergence
GTP	GPRS Tunneling Protocol
GTS	Guaranteed Time Slot
GW	GateWay
HAN	Home Area Network
HARQ	Hybrid Automatic Repeat Request
HD-FDD	Half Duplex Frequency Division Duplexing
HDR	High Data Rate
HEMS	Home Energy Management System
HEV	Hybrid Electric Vehicle
HFC	Hybrid Fiber-Coaxial
HSPA	High Speed Packet Access
HSS	Home Subscriber Server
HVDC	High Voltage Direct Current
IAN	Industrial Area Network
ICCP	Inter-Control Center Protocol
ICE	Internal Combustion Engine
ICI	Inter-Carrier Interference
ICIC	Inter-Cell Interference Coordination
ICT	Information and Communication Technologies
ID	IDentifier
IED	Intelligent Electronic Device

IETF	Internet Engineering Task Force
IGBT	Insulated Gate Bipolar Transistor
IHD	In-Home Display
IIoT	Industrial Internet of Things
IMS	IP Multimedia Subsystem
IMT	International Mobile Telecommunications
IoT	Internet of Things
IP	Internet Protocol
ISDN	Integrated Services Digital Network
ISI	Inter-Symbol Interference
ISM	Industrial, Scientific, and Medical
ISO	International Organization for Standardization
ITU-R	ITU Radiocommunication Sector
kWh	kiloWatt hour
L2	Layer 2
L3	Layer 3
LAA	Licensed Assisted Access
LADN	Local Area Data Network
LAN	Local Area Network
LASER	Light Amplification by Stimulated Emission of Radiation
LBT	Listen Before Talk
LCAS	Link Capacity Adjustment Scheme
LD	Logical Device
LDP	Label Distribution Protocol
LDPC	Low-Density Parity Check
LDR	Low Data Rate
LED	Light Emitting Diode
LLID	Logical Link IDentification
LMR	Land Mobile Radio
LN	Logical Node
LoS	Line-of-Sight
LPWA	Low Power Wide Area
LPWAN	Low Power WAN
LR-WPAN	Low-Rate WPAN
LSP	Label Switched Path
LTE	Long-Term Evolution
LTE-M	Long-Term Evolution for Machines
LwM2M	Lightweight Machine-to-Machine
M2M	Machine-to-Machine
MAC	Medium Access Control
MAIFI	Momentary Average Interruption Frequency Index
MAIFIE	Momentary Average Interruption Event (E) Frequency Index
MAN	Metropolitan Area Network
MAP	Maximum A-Posteriory
MAP	Media Access Plan
MCL	Maximum Coupling Loss
MCS	Modulation and Coding Scheme

MDM	Meter Data Management
MDMS	Meter Data Management System
MED	Major Event Day
MEF	Metro Ethernet Forum
MEMS	Microgrid Energy Management System
MHDS	Multi Hop Delivery Service
MIB	Management Information Base
MIMO	Multiple-Input Multiple-Output
MME	Mobility Management Entity
mMTC	massive Machine Type Communications
MNO	Mobile Network Operator
MPCP	Multi-Point MAC Control Protocol
MPLS	Multi-Protocol Label Switching
MPLS-TP	Multi-Protocol Label Switching-Transport Profile
MQTT	Message Queuing Telemetry Transport
MS-SPRING	Multiplex Section-Shared Protection Ring
MTC	Machine Type Communications
MVNO	Mobile Virtual Network Operator
NAS	Non-Access Stratum
NASPI	North American Synchrophasor Initiative
NAT	Network Address Translation
NB-IoT	NarrowBand IoT
NB-IoT-U	NB-IoT Unlicensed
NF	Network Function
NFV	Network Function Virtualization
NG-PON2	Next Generation-Passive Optical Network 2
NG-RAN	Next Generation Radio Access Network
NHRP	Next Hop Resolution Protocol
NIC	Network Interface Card
NIDD	Non-IP Data Delivery
NIST	National Institute of Standards and Technology (USA)
NMS	Network Management System
NOP	Normally Open Point
NPBCH	Narrowband PBCH
NPCW	Normal Priority Contention Window
NPDCCH	Narrowband PDCCH
NPDSCH	Narrowband PDSCH
NPN	Non-Public Network
NPSS	Narrowband PSS
NPUSCH	Narrowband PUSCH
NR	New Radio
NR-U	New Radio Unlicensed
NS	Network Server
NSA	Non-Stand Alone
NSSS	Narrowband SSS
NTN	Non-Terrestrial Network
NTU	Network Terminal Unit

NWK	Network	
OAM	Operations, Administration, and Maintenance	
OC	Optical Carrier	
OCh	Optical Channel	
ODN	Optical Data Network	
ODN	Optical Distribution Network	
ODU	Optical Data Unit	
OFCOM	Office of Communications (UK)	
OFDM	Orthogonal Frequency Division Multiplexing	
OFDMA	Orthogonal Frequency Division Multiple Access	
OLT	Optical Line Terminal	
ONT	Optical Network Terminal	
ONU	Optical Network Unit	
OPERA	Open PLC European Research Alliance	
OPEX	OPerational EXpenditure	
OSI	Open System Interconnection	
OSPF	Open Shortest Path First	
OTN	Optical Transport Network	
OTU	Optical Transport Unit	
OXC	Optical Cross-Connect	
PAM	Pulse Amplitude Modulation	
PAN	Personal Area Network	
PCC	Point of Common Coupling	
PCCH	Paging Control CHannel	
PCF	Policy and Charging Function	
PCFICH	Physical Control Formal Indicator CHannel	
Pco	Proxy coordinator	
PDC	Phasor Data Concentrator	
PDCCH	Physical Downlink Control CHannel	
PDH	Plesiochronous Digital Hierarchy	
PDN	Packet Data Network	
PDSCH	Physical Downlink Shared CHannel	
PDU	Packet Data Unit	
PEV	Plug-in Electric Vehicle	
P-GW	PDN Gateway	
PHEV	Plug-in Hybrid Electric Vehicle	
PHICH	Physical Hybrid ARQ Indicator CHannel	
PLCP	Physical Layer Convergence Protocol	
PLMN	Public Land Mobile Network	
PM	Phase Modulation	
PMR	Private (or Professional) Mobile Radio	
PMU	Phasor Measurement Unit	
PNI-NPN	Public Network Integrated NPN	
PON	Passive Optical Network	
POTS	Plain Old Telephone System	
PPDR	Public Protection and Disaster Relief	
PRACH	Physical Random Access CHannel	

PRB	Physical Radio Block
PRIME	PoweRline for Intelligent Metering Evolution
PS	Primary Substation
PSK	Phase Shift Keying
PSM	Power Save Mode
PSS	Primary Synchronization Signal
PSTN	Public Switched Telephone Network
PTT	Push-To-Talk
PUCCH	Physical Uplink Control CHannel
PUSCH	Physical Uplink Shared CHannel
QAM	Quadrature Amplitude Modulation
QoS	Quality of Service
QPSK	Quadrature Phase Shift Keying
RACDS	Resilient AC Distribution System
RAN	Radio Access Network
RCC	Regional Commonwealth in the field of Communications
REST	REpresentational State Transfer
RF	Radio Frequency
RFC	Request For Comments
RFD	Reduced Function Device
RIP	Routing Information Protocol
RLC	Radio Link Control
rms	root mean square
ROADM	Reconfigurable Optical Add-Drop Multiplexer
RPL	Routing Protocol for Low Power and Lossy Networks
RR	Radio Regulations
RRC	Radio Resource Control
RRM	Radio Resource Management
RS	Repeating Station
RTP	Real-Time Pricing
RTS	Ready to Send
RTT	Round-Trip Time
RTU	Remote Terminal Unit
SA	Substation Automation
SA	Stand-Alone
SAE	Society of Automotive Engineers
SAIDI	System Average Interruption Duration Index
SAIFI	System Average Interruption Frequency Index
SAR	Segmentation and Reassembling
SBA	Service-Based Architecture
SCADA	Supervisory Control and Data Acquisition
SCEF	Service Capability Exposure Function
SC-FDMA	Single Carrier FDMA
SCL	Substation Configuration Language
SCP	Shared Contention Period
SDH	Synchronous Digital Hierarchy
SDN	Software Defined Networking

SEP	Smart Energy Profile
SF	Spreading Factor
S-GW	Serving Gateway
SIM	Subscriber Identity Module
SLA	Service-Level Agreement
SMS	Short Message Service
SMWAN	Smart Metering WAN
SN	Service Node
SNPN	Standalone NPN
SNR	Signal-to-Noise Ratio
S-NSSAI	Single Network Slice Selection Assistance Information
SONET	Synchronous Optical NETwork
SRB	Signalling Radio Bearer
SRS	Sounding Reference Signal
SS	Secondary Substation
SSB	Synchronization Signal Block
SSS	Secondary Synchronization Signal
SST	Slice/Service Type
STATCOM	STATic COMpensator
STM	Synchronous Transport Module
STP	Spanning Tree Protocol
STS	Synchronous Transport Signal
SV	Sampled Value
TA	Tracking Area
TASE	Telecontrol Application and Service Element
TAU	Tracking Area Update
TB	Transport Block
TBCC	Tail-Biting Convolutional Code
TC	Technical Committee
TCP	Transmission Control Protocol
TDD	Time Division Duplexing
TDM	Time Division Multiplexing
TDMA	Time Division Multiple Access
TETRA	TERrestrial Trunked RAdio
TIA	Telecommunications Industry Association
TR	Technical Report
TS	Technical Specification
TSC	Time-Sensitive Communications
TSCH	Time-Slotted Channel Hopping
TSN	Time-Sensitive Networking
TSP	Telecommunication Service Provider
TTI	Transmission Time Interval
TWDM	Time and Wavelength Division Multiplexing
UCC	Utility Control Center
UCI	Uplink Control Information
UDM	Unified Data Management
UDP	User Datagram Protocol

UHF	Ultrahigh Frequency
UMTS	Universal Mobile Telecommunications System
UNB	Ultra Narrowband
UPF	User Plane Function
URLLC	Ultra Reliable Low-Latency Communications
UTC	Utilities Technology Council
V2X	Vehicle-to-everything
VC	Virtual Container
VCAT	Virtual Concatenation
VHF	Very High Frequency
VoLTE	Voice over LTE
VPN	Virtual Private Network
WADM	Wavelength Add-Drop Multiplexer
WAMS	Wide-Area Measurement/Monitoring System
WAN	Wide Area Network
WDM	Wavelength Division Multiplexing
Wi-Fi	Wireless-Fidelity
WPAN	Wireless Personal Area Network
WRC	World Radiocommunication Conference
xDSL	x Digital Subscriber Line (covering various types of DSL)
XG-PON	X (10) Gbps Passive Optical Network
XGS-PON	X (10) Gbps Symmetrical Passive Optical Network

1

The Smart Grid

A General Perspective

1.1 Introduction

The Smart Grid is a container of the most modern and evolutionary changes in the power system as a consequence of the advent and adoption of new technologies that progressively add new capabilities to the grid and help it to become a more efficient system. Indeed, the Smart Grid is neither a novel nor a static concept; however, it is bound to be disruptive from the perspective of the achievement of its objectives.

The objectives of the Smart Grid have been broad and ambitious since its inception, decades ago. These objectives have been stimulated and become achievable due to the advances in electric grid technologies and the applicability of Information and Communication Technologies (ICTs) to the grid. With regard to the former, there are new technologies that can be added to the different segments of the grid and change the traditional electricity delivery model (e.g., Distributed Energy Resources [DER]). As per the latter, continuous ICTs' innovations have permeated all industries and the Society as a whole, paving the way toward a digital transformation in utilities, specifically in the areas close to the grid operation.

The ambition of the Smart Grid is to integrate all these electric technologies and ICT innovations into a smart system, empowered with new applications and services, and able to operate more efficiently in all its aspects.

This chapter elaborates on a comprehensive definition of what can be understood for a Smart Grid by introducing the basic elements of a power grid and enhancing those with telecommunication technologies. This definition is complemented with the main challenges that Smart Grids, and more specifically, telecommunication technologies applied to Smart Grids, must face in the coming years.

1.2 Electric Power Systems

Electricity is one of the cornerstones of our Society [1], and as such, its generation, transport, and distribution need to be a fully functional and efficient system. All the dependencies that other essential services for the functioning of the economy and society have on electricity (see Figure 1.1), and the extent to which electricity is also affected by all of them, determine the need to have all of these services evolving in an efficient and coordinated way, while stimulating the adoption of new technologies.

Smart Grid Telecommunications: Fundamentals and Technologies in the 5G Era, First Edition. Alberto Sendin, Javier Matanza, and Ramon Ferrús.
© 2021 John Wiley & Sons, Inc. Published 2021 by John Wiley & Sons, Inc.

1 The Smart Grid

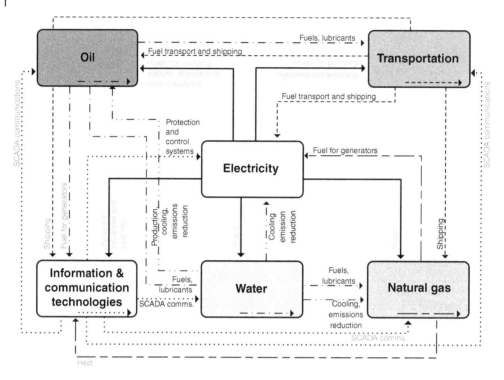

Figure 1.1 Electricity at the core of critical services. Dependencies are based on services provided among them. *Source:* Department of Energy – USA [1].

It is widely recognized that electricity in general, and electrification in particular, are among the major achievements of the twentieth century [2], despite the fact that there are still big parts of the world where electricity is not affordable (electrification today is total in developed countries, while, as reported in [3], there were still 1100 million people in 2016 who live without electricity elsewhere). As A.C. Clarke expressed it in [2], "the harnessing and taming of electricity, first for communications and then for power, is the event that divides our age from all those that have gone before." However, it is also true that electricity supply tends to go unnoticed, as a nearly invisible service attached to our modern way of life, that "is already there."

The electric grid is a complex system reaching every other activity. It is composed of a large number of elements, spread all over where human activity is present; it is controlled to deliver its service in the most reliable and resilient manner. From a purely technical perspective, the grid has evolved improving its associated control capabilities, from its center to the edge across the entire system, and inherits much of the means used in times where remote control was only a wish, and needed electromechanical elements and procedures to minimize manual interventions for incidents resolution. Moreover, and as an inherent characteristic of its nature, much of the electric system is regulated by Governments, meaning that the control over the grid goes beyond the network technical aspects.

1.2.1 Electricity

Electricity is the universal and standard way to transform energy and get it transported everywhere and to everyone. Electricity, as the object of the grid, exhibits a series of properties that justify the complexity of the system behind it.

The size and geographic extension of electric power systems are conditioned by their scope, as a consequence of the purpose of carrying the energy from the places where it is most conveniently produced to the places where it is needed. This is achieved by means of a network (the grid, i.e., Transmission and Distribution) of interconnected elements (e.g., generation systems, power lines, substations) spread over necessarily large geographical areas and integrated to work as a whole.

Electricity, as a product, cannot in practice be stored or shipped in containers. Despite the technological advances in electric batteries and other storage apparatus, handling any amount of energy comparable to a representative percentage of the system's dimension is nowadays still far from feasible (performance aspects, and other limiting factors, may be solved in the future). Thus:

- Electricity must be generated and transmitted to be consumed, involving a necessary real-time dynamic balance between generation and demand.
- Electric power pathways cannot be chosen freely across the network, as it is physics (Kirchhoff's laws) that determines, depending on the impedances in the power lines and the rest of the grid elements, where electricity flows. Thus, the current distribution cannot easily be forced to take any given route, and alternative routes in the grid are highly interdependent.

From an operational perspective, deviations from normal operation may cause the instantaneous reconfiguration of power flows that may have substantial effects on facilities (e.g., substations, power lines, etc.) in the grid and propagate almost instantaneously across the entire system.

Finally, electric power consumption is sensitive to the technical properties of the electricity supply, to the extent that devices may malfunction or simply cease to operate unless the voltage wave is stable over time within certain parameters including shape (sinusoidal), frequency (cycles per second), and value (voltage). The system must have mechanisms to react (detect and respond) instantly to unexpected situations and avoid degradations in service quality.

1.2.1.1 Frequency and Voltage

The frequency of the electricity signal in the different world regions is either 50 or 60 Hz. The waveform adopted by Europe, Asia, Africa, many countries in South America, Australia, and New Zealand for their electricity systems is 50 Hz. North America, some parts of northern South America, Japan, and Taiwan, opted for a frequency of 60 Hz [4, 5].

In contrast, the voltage levels that can be seen in the different parts of the grid span a much larger range of options. A widely accepted, though loosely precise, definition of the voltage levels is:

- Low Voltage (LV), defined as "a set of voltage levels used for the distribution of electricity and whose upper limit is generally accepted to be 1000 V for alternating current" (IEV 601-01-26 [6]).
- High Voltage (HV), defined as either "the set of voltage levels in excess of low voltage" or "the set of upper voltage levels used in power systems for bulk transmission of electricity" (IEV 601-01-27 [6]).
- Medium Voltage (MV), defined as "any set of voltage levels lying between low and high voltage" (IEV 601-01-28 [6]).

The International Electrotechnical Commission (IEC) has standardized three-phase AC rms voltage levels internationally in IEC 60038:2009 within the following ranges:

- Having a highest voltage for equipment exceeding 245 kV: 362 or 420 kV; 420 or 550 kV; 800 kV; 1100 or 1200 kV highest voltages.
- Having a nominal voltage above 35 kV and not exceeding 230 kV: 66 (alternatively, 69) kV; 110 (alternatively, 115) kV or 132 (alternatively, 138) kV; 220 (alternatively, 230) kV nominal voltages.

- Having a nominal voltage above 1 kV and not exceeding 35 kV: 11 (alternatively, 10) kV; 22 (alternatively, 20) kV; 33 (alternatively, 30) kV or 35 kV nominal voltages (there is a separate set of values specific for North American practice).
- Having a nominal voltage between 100 and 1000 V inclusive: 230/400 V is standard for three-phase, four-wire systems (50 or 60 Hz) and also 120/208 V for 60 Hz. For three-wire systems, 230 V between phases is standard for 50 Hz and 240 V for 60 Hz. For single-phase three-wire systems at 60 Hz, 120/240 V is standard. Practically, LV consumers within most 50 Hz regions will eventually be delivered 230 Vac, and 110 Vac in 60 Hz regions.

Thus, it can be said that while LV is clearly below 1 kV, the boundary between HV and MV is commonly placed at 35 kV.

1.2.2 The Grid

The "grid," the power grid, the electric power system, or the electricity supply system, is defined by the IEC as "all installations and plant provided for the purpose of generating, transmitting and distributing electricity."

The power grid is a hierarchical infrastructure comprising a large set of interconnected elements to provide electricity service to its end-customers. The interconnected elements can be grouped in different building blocks, as shown in Figure 1.2. Traditional grids deliver the energy produced by the Generation systems to the Consumption Points, through Transmission and Distribution systems. Generation and consumption are matched in real time.

In a traditional conception of the power system, Generation is conceived as the big power plants where energy transformation into electricity happens. Transmission steps generated voltage levels up, to transport it over long distances with the minimum energy losses. Distribution drives electric energy to all the disperse locations where it is consumed. Finally, Consumption Points are the locations where energy is ultimately delivered.

In practical and intuitive terms, Generation is the block with the big thermal, nuclear, and hydro plants. Transmission grid transports electricity with the costly HV power lines acting as the highways of the energy. Distribution grid is the heterogeneous mix of pervasive electricity assets reaching everywhere (the assets in Transmission and Distribution can be simplified in two, substations and power lines). And Consumption Points conceptually gather the different electricity users and their loads, from commercial and industrial customers to residential ones. Last but not least, Distributed Generation (DG) and/or DER have started to play a relevant role in the power generation closer to end-customers.

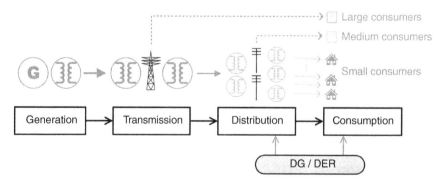

Figure 1.2 Building blocks of traditional electric power systems.

Thus, the power grid is usually understood as a very large network connecting power plants (large or small) to loads, by means of an electric grid that spans countries, with international interconnections. These are referred to as "full power systems," autonomous in their operation.

1.2.2.1 The Grid from a Technical Perspective
Although the grid concept has converged over the past century toward a similar structure and configuration in different world regions and countries, there are however many differences in the details of the infrastructure deployed. These differences can even be found in regions within the same country and even within the same company, exhibiting different implementations of the same concepts depending on a variety of factors. This situation poses difficulties from the equipment standardization and evolution perspective, as it adds complexity to the long lifecycles of grid elements.

1.2.2.1.1 Generation

1.2.2.1.1.1 **Traditional Power Generation** Power plants convert the potential energy of existing resources such as renewable energies (water, wind, solar, etc.) and fuel (coal, oil, natural gas, enriched uranium, etc.) into electric energy.

Traditional centralized power plants generate AC power from synchronous generators. These generators provide in fact three-phase electric power; the voltage source is a combination of three AC voltage sources, i.e., three voltage phasors separated by phase angles of 120°. The frequency of the electricity waveform (i.e. 50 or 60 Hz) is a multiple of the rotation speed of the machine. Voltage is usually no more than 6–40 kV, being determined by the current in the rotating winding (i.e. the rotor) of the generator. The output is taken from the fixed winding (i.e. the stator).

More recent energy sources (e.g., wind turbines and mini hydro units) usually employ asynchronous generators, and the waveform of the generated voltage is not necessarily synchronized with the rotation of the generator. This may create flicker and reactive power quality problems if not properly designed and controlled.

The voltage in the power generation stage is stepped up by a transformer, normally to a much higher voltage. At that HV, the generator connects to the grid in a substation and electricity starts its journey toward the Consumption Points.

1.2.2.1.1.2 **Distributed Generation/Distributed Energy Resources** DG refers to the "utilization of small (0 to 5 MW), modular power generation technologies dispersed throughout a utility's distribution system" [7], i.e., small-size generation that connects into the Distribution part of the system, as opposed to conventional centralized power generation systems. DER (sometimes shortened as DR or Distributed Resources) is literally "a source of electric power that is not directly connected to a bulk power system [... and] includes both generators and energy storage technologies capable of exporting active power to an electric power system" [8].

Although DG, DER, and DR are modern acronyms, the concepts they represent stem from the past [9] and refer to a basic smaller scale generation closer to Consumption Points. The novelty nowadays is that DG/DER/DR intends to be connected to the conventional grid as well and has the potential to be widely adopted across it. This wide adoption is based on their capability to reduce system losses and improve power quality and reliability, thus deferring Transmission and Distribution grid improvements [7].

US EPA (United States Environmental Protection Agency) reinforces the idea and points out the environment factors that the concept, when fully developed, involves [10]: "DG refers to a variety of technologies that generate electricity at or near where it will be used, such as solar panels and combined heat and power. Distributed generation may serve a single structure, such as a home or business, or it may be part of a microgrid (a smaller grid that is also tied into the larger electricity delivery system), such as at a major industrial facility, a military base, or a large college campus. When connected to the electric utility's lower voltage distribution lines, distributed generation can help support delivery of clean, reliable power to additional customers and reduce electricity losses along transmission and distribution lines." Under this definition, solar photovoltaic panels, small wind turbines, natural-gas-fired fuel cells, combined heat and power systems, biomass combustion or cofiring, municipal solid waste incineration, and even Electric Vehicles (EV) may be included.

The role of DG/DER/DR in the electric power system is increasingly relevant, and despite their existing challenges in operational and regulatory terms, "for the many benefits of DG to be realized by electric system planners and operators, electric utilities will have to use more of it" [9]. Moreover, the technical improvements in cost and efficiency terms will make small-size generators come closer to the performance of large power plants.

1.2.2.1.2 Transmission

Power from Generation is connected to the Transmission part of the grid, with transmission lines that carry electric power at various HV levels. The Transmission grid is the backbone of the electric power system covering long distances to connect large and geographically scattered generation plants to demand hubs where Distribution system starts.

A Transmission system corresponds to a web-like structure achieving the back-up of every substation of the grid by all the others. It is a networked, meshed topology connecting generation plants and substations together into a grid that usually is defined at 100 kV or more. The electricity flows over HV transmission lines to a series of substations where the voltage is stepped down by transformers to levels appropriate for Distribution systems.

Transmission power lines are sturdy, durable, and efficient conductors, usually supported by towers. The design of the system needs to be based on mechanical (weight of the conductors, safety distances between conductors, tower and ground depending on the voltage, etc.) and electrical considerations. Transmission lines are typically deployed with three wires along with a ground wire. The conductors are attached to the towers that support them by an assembly of insulators. The towers may support several power lines in the same route, to optimize costs. The system includes sophisticated measurement, protection, and control equipment to prevent its malfunctioning in case of faults (e.g., short-circuits, lightning, dispatch errors, or equipment failure).

Although not common due to higher costs, in congested areas within cities, underground cables are alternatively used for electric energy transmission. The technology to be used is more sophisticated and applies to the lower voltage ranges. These underground Transmission systems are preferable from the environment perspective.

Although most Transmission systems are AC, a mention needs to be made to DC systems. DC transmission systems require expensive converter stations to convert to the regular AC systems. They are used because they present some advantages (namely, no reactive power flows, higher transmission capacity, lower losses, and lower voltage drops for the same voltage and size of the conductors, controllability of the flow, no frequency dependence, reduced stability problems) over AC Transmission systems in applications such as long distances, submarine cables, and the interconnection systems with different security standards or system frequency.

1.2.2.1.3 Distribution

Distribution segment is widely recognized as the most challenging part of the grid due to its ubiquity. Distribution networks are more subject to failures than Transmission networks. They have

HV, MV, and LV levels. Further than the formal definition of voltage levels, HV usually comprises 132 (110 in some places), 66, and 45 kV; MV 30, 20, 15, 10, 6.6 kV, etc.; and LV, levels below 1 kV.

The Distribution network[1] consists of power lines connecting primary substations (PSs) and secondary substations (SSs), the former in charge of transforming voltage from HV to MV and the latter from MV to LV. The parts of the Distribution network with the higher complexity are the MV and LV grids. MV has concentrated the attention of grid infrastructure evolution in technical and technological terms in recent years; on the contrary, the LV has witnessed less evolution. Thus, LV grids present more complex and heterogeneous topologies than MV grids.

MV grid topologies (Figure 1.3) can be classified in three groups, although their operation is radial:

- Radial topology. Radial lines are used to connect PSs with SSs, and the SSs among them. These MV lines (often named "feeders") can be used exclusively for one SS or can reach several of them. Radial topologies show a tree-shaped configuration when they grow in complexity.
- Ring topology. A ring topology is an improved evolution of the radial topology, connecting SSs to other MV lines to create redundancy, and from there to a PS to close the ring. This topology is fault-tolerant and overcomes the weakness of radial topology when one element of the MV line gets disconnected. The elements in the MV circuit need to be maneuvered to reconfigure the grid and connect SSs.
- Networked topology. Networked topology consists of PSs and SSs connected through multiple MV lines to provide a variety of distribution alternatives. In the event of failure, many alternative solutions may be found to reroute electricity.

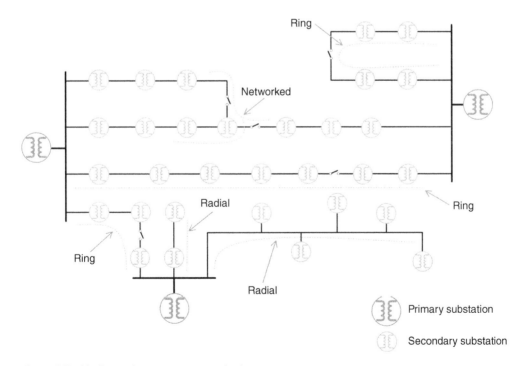

Figure 1.3 Medium voltage common topologies.

---

1 Distribution networks are run by the so-called Distribution Network Operators or DNOs. They are increasingly referred to as DSOs (Distribution System Operators) in their evolution.

1 The Smart Grid

LV topologies are much more diverse than MV's. LV networks may have grown in a not very coordinated way, depending on the extension and specific features of the service area, the type, number, and density of points of supply (loads), country- and utility-specific operational procedures. Each SS typically supplies electricity to one or several LV lines, with one or multiple MV to LV transformers at the same site. LV topology is typically radial, as in Figure 1.4, having multiple branches that connect to extended feeders. LV lines are typically shorter than MV lines. LV Distribution systems can be single-phase or three-phase. In Europe, e.g., they are usually three-phase, 230 V/400 V systems (i.e. each phase has a rms [root mean square] voltage of 230 V and the rms voltage between two phases is 400 V).

1.2.2.1.4 Consumption Points

Customers' concurrent energy consumption patterns drive the needs of the electric power system. Thus, the Consumption Points are extremely relevant from a technical perspective. Traditionally, electric grids have been oversized due to the difficulties to measure, understand, and modify these consumption patterns. However, behind a Consumption Point, customers can be found, and as stakeholders of the system, their contribution to it needs to be considered.

Customers need to receive a reliable and agreed-quality electricity service, as they connect their loads to the grid and must be guaranteed that the supply will be available. These load functioning may demand a service with different requirements depending on the nature of the customer (residential, commercial, or industrial).

Customers must be charged for their use of the system, and utilities have developed technology and processes to determine the consumption of electricity at the grid edge, where Consumption

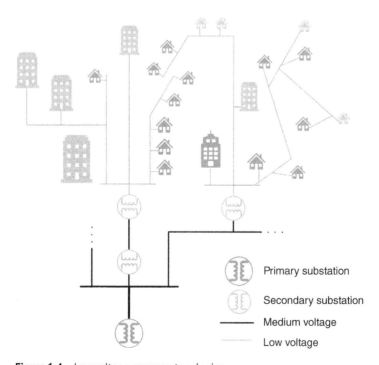

Figure 1.4 Low voltage common topologies.

Points are located. These metering systems (see Chapter 5) have been based on meters installed on the border between the grid and each customer. These meters have traditionally been the only visible connection of customers with the utility.

However, the role of the customer in electric power systems is changing, and it needs to gain visibility within the utility. In a similar manner, the electric power system must be visible for customers to the extent that his active participation in the system is required (see Chapter 5).

1.2.2.2 The Grid from a Regulatory Perspective

Regulation is instrumental to understand electric power systems. In essence, Regulation [4] is the series of principles or rules to control individual and collective human behavior. Regulation, with or without the authority of law and with an *ex ante* intention, is used to control, direct, or manage an activity, organization, or system. From the legal perspective, regulation is a form of secondary legislation issued under the authority of primary legislation, whose effective application it is to secure.

Governments regulate industries to improve their performance, be it to guarantee that no agent goes against the common interests, to steer an industry's performance toward improving "general welfare," to protect consumers, and to protect investors from the State. Regulation is implemented through the design of rules, the structuring of the industry, and the supervision of agents' behavior and industry performance.

The core underlying criterion supporting the regulation of electric power systems is the maximization of social and environmental welfare in the production and consumption of electric power. This involves several fundamental concerns. First, efforts should be devoted to minimizing the costs incurred in providing the service (both investment and recurring costs). Second, the quality of supply must also be satisfactory (including reliability – security in short-term operation, and adequacy for long-term evolution, and "product" quality factors). Third, sustainable development, defined as the one meeting the needs of today without compromising those of the future, is needed. All these concerns may be contradictory and trade-offs should be established.

1.2.2.2.1 Regulatory Models

Regulatory models for electric power systems are different in every country, although they often deal with two dimensions that classically determine how the model is implemented:

- Which activities, of all the activities needed in the electric power system, need to be separated from the others? These activities are to be unbundled.
- Which activities can be performed on a competitive basis? These activities can be deregulated or regulated in a different way.

These two dimensions create the different models. In one extreme, if none of the activities are unbundled or deregulated, we have the vertically integrated monopoly. In the other extreme, the activities (including Generation, supply – Transmission and Distribution – and energy trading) are said to be vertically disaggregated and are performed competitively.

In a vertically integrated monopoly, electric utilities are awarded a territory to supply electricity to. Thus, the utilities own and operate all the generation and network assets in those territories, and they plan and implement the expansion of production and network capacity under the guidelines and authorization of the regulatory authority, within a remuneration for the utility based on the incurred cost of service (including a rate of return for invested capital), an agreed price for end-users within a satisfactory reliability level, and with the allowed environmental impact.

In an organization of the electric power systems where activities are vertically disaggregated, the competition is introduced through wholesale electricity markets, which are open to all generators (incumbents and new entrants) and to all consumer entities. In a wholesale electricity market, the electricity market price is fixed by competing forces and serves as a reference for medium- and long-term contracts of different types. The agents trading on such markets are generators, different categories of supplier companies representing consumer interests, or acting as intermediaries. In this context, consumers are clients free to choose the supplier based on the available commercial propositions.

The beginning of the power sector (nineteenth and the early twentieth century) was driven by private initiative and competition, as electricity was not yet a pervasive must for the Society. Soon, in most countries, the situation was superseded by strong governmental intervention in the form of public ownership or utilities as regulated monopolies. States in most countries assumed a heavy planning and intervention, being the sole regulator. The situation has remained like this until the 1990s, when unbundling and regulation of network activities has happened, to separate them from the business part that can be performed in a competitive way.

Nevertheless, even in highly deregulated environments, Transmission and Distribution grids are subject to the existence of business-relevant regulation. These networks, part of the whole system, are considered natural monopolies, as they do not have characteristics that allow the provision of their services under a market-based regulatory regime. Consequently, these networks have enormous market power, and this is the reason why these grid-associated activities must at least be wholly independent of other competitive businesses within the electric power system (i.e., generation and retailing – offering electricity to end-users). Thus, their remuneration must be regulated, and they must be obliged to provide open and nondiscriminatory access to their facilities, allowing the rest of the competitive framework to happen. This highly regulated nature of Transmission and Distribution activity is extremely important, as networks are one of the most important elements in Smart Grids.

To understand the regulatory factors affecting this part of the system (the more directly connected to the evolution toward Smart Grids), we will focus on the monopolistic vertically integrated electric power system. Regulation must ensure that companies reach efficiency objectives, i.e., an optimal trade-off between efficiency and service quality:

- It needs to control the revenues from electricity sales so that the company can earn. Such revenues must be sufficient to enable the utility to make necessary investments while earning an adequate return on the capital invested and to cover its operating costs.
- It needs to ensure that collected revenues are neither detrimental to consumer interests nor to the competitiveness of a country's industry, as electricity is an essential service.

There are two traditional approaches to solve this problem:

- Cost-of-service (or rate-of-return) regulation. It allows utilities to recover all incurred costs, i.e., investment costs (a fair return on investment included) plus operation costs. The mechanism requires that utilities present their investment plans to run their business within the established framework and taking into account future growth needs and estimated future operation costs. Then, a regulatory commission may accept, reject, or propose to amend the proposals, until an agreement is reached.

- Incentive-based regulation. It is in fact an extension of the previous one, intending to improve it through a cost control (the idea is that both Society – through better prices – and the utility may benefit). Under the assumption that price paid do not necessarily mirror costs at any given time, explicit monetary incentives are established for the regulated companies allowing them to make a profit when they are able to lower costs. Prices are fixed for a certain period, and the utility may keep all or part of the cost reductions. Target revenues may also include incentives associated to other service characteristics, such as quality of service improvement and environmental aspects.

1.2.2.2.2 Transmission and Distribution Regulation

Electric network infrastructures require large investments spread through the territories they serve and are bound to the physical space where they are located. From an intuitive perspective, it would most likely not be efficient to have two competing electricity grids in the same area providing the same service. From the analysis of real practices, the widest accepted reality is that Transmission and Distribution activities are natural monopolies and, as such, are closely regulated.

If technical details are analyzed, the conclusions become evident. Transmission and Distribution parts of the grid are characterized by different aspects leading to the efficiency of a single infrastructure to provide the service. First, the duplication of Transmission or Distribution grids is not affordable in terms of cost, land use, and environmental impact. Second, in the case of the Transmission grid, the unit cost of transmitting electricity (cost of 1 km of line per unit of transmission capacity) declines abruptly with a line's total transmission capacity, involving that a single element performs better than two. Third, the essentiality of the service in a territory, the operation of the network as a whole, the prevention of market control scenarios, and the technical design constraints advocate for this monopoly type of regulation.

However, Transmission and Distribution grids are different among them. Each one has a different impact and complexity within an electric power system and do pose a different burden and risk to the utility operating it. Thus, the way in which they are regulated should be different:

- Transmission networks have a huge impact in wholesale markets, and regulation should focus on their proper performance.
- Distribution networks are the origin of over 90% of end-consumers outages, and regulation should bear in mind the quality of supply.
- While Transmission infrastructure is relatively standard and can be audited and controlled easily, Distribution grid differs greatly among territories and its assets are widespread. Thus, Distribution requires much more simplified procedures to calculate remuneration (compared to Transmission), typically incentive-based.

The economic characteristics of Transmission networks are associated to investments. This statement is comprehensive of operation and maintenance costs, as they are proportional to the investment. Relatively speaking, the economic weight of Transmission networks with respect to all activities involved in the electric power system is significantly lower than Generation and Distribution, and it is typically between 5 and 10% of the total cost of electricity.

As with Transmission network, Distribution networks' costs include the investment needed to strengthen the existing grid and build new facilities to expand it. The cost components also include the obvious operation and maintenance activity, energy losses costs, and those of auxiliary activities such as meter reading and billing for network services. It is important to mention that Distribution network charges are generally separated into connection charges and use-of-system charges. The former is a single payment when a new connection (or upgrade in the existing connection) is required, and the latter are the periodic payments made by network users to cover the total cost of the regulated service.

All in all, Transmission and Distribution electric utilities have the need and the opportunity to invest heavily in their networks, as long as that is justified, and approved by Regulators. Then, they can translate these investments in long-term recognized costs that will be part of the service tariffs.

A final note on this cannot miss the point that the emergence of the new developments in power systems (DG/DER/DR-including Distributed Storage or DS, EVs, Demand-Side Management [DSM], etc.) is challenging the technical *status quo* of electricity business structure and may ultimately motivate the appearance of new regulatory models that may adapt better to the new situation. However, present-day grid evolutionary needs must be addressed within the current scenario. A smart regulatory design and implementation will be required to guide financial resources to support Smart Grids.

1.2.3 Grid Operations

Electric power system management is a complex undertaking covering technical, economic, regulatory, social, business, and environmental factors. The management of a power system combines investment planning and system operation and maintenance to ultimately deliver the electricity supply. These processes have both short-term and long-term components across different grid segments.

Investment planning is a process projecting itself anywhere from 2 to 15 or more years. The process involves determining which and when new generation and network facilities are to be installed. Factors taken into account are demand growth forecasts, technical alternatives and costs, budgetary limitations, strategic considerations of generation resources, grid reliability criteria, environmental constraints, etc.

Power system operation and maintenance are performed under the assumption that production and consumption have to be always in balance: a mismatch between supply and demand in a large system cannot happen, as the overall dynamic balance would be compromised and the supply of electricity across a significant amount of the grid potentially lost. At the same time, system parameters, namely, voltage and frequency, must remain within predefined operation thresholds in the short and long terms. Maintenance of power systems is not conceptually different to the practices of any other business: reactive maintenance is needed when an unplanned failure occurs; preventive (proactive) maintenance is planned to minimize future system failures.

Power system operations and investment planning meet in system planning. System planning is part of the grid operations, and its consequence is the identification of the investment needed. System planning, as a process, intends to understand and forecast load location and needs to adapt the grid consequently. System planning considers the generic elements mentioned earlier and considers a short- and a long-term scope. Long term requires a system model and considers load evolution and associated changes needed in the existing system (e.g., new power lines or substations, refurbishment of existing infrastructure: feeders, transmission capacity) along with relevant constraints (e.g., budget) to develop different prospective scenarios and adapt to them. Short-term planning implies detailed analysis of the existing infrastructure, both Transmission and Distribution segments. Different topological and performance data feed studies to analyze voltage drops (to identify weak points in the grid), sectionalizing options (to minimize outages), conductor adequacy (based on power required), etc. It will ultimately affect the evolution of all system components.

Protection and control are electric power system functions that are transversal to all grid management areas. System operations depend on a combination of automated or semiautomated control and actions requiring direct human intervention. Operations are assisted by electromechanical grid elements and recently enhanced with the support of ICTs:

- Protection. This function ensures the safety of the system, its elements, and the people. Protection schemes must act in real time when there is a condition that might cause personal injuries or equipment damage. However, protection cannot avoid disturbances in the system. Faulty conditions (faults) are detected and located based not only on grid voltage and current measurements but also on some other parameters. A fundamental part of the operation of a power system is to quickly detect and clear faults, rapidly and selectively disconnecting faulty equipment and automatically reclosing for supply recovery in case of transient failures.
- Control. Power system operators manage their grids from Utility Control Centers (UCCs). Different UCCs exist, dealing with different grid domains (grid segment and/or territory). Most routine operations of a well-designed system should not require any human intervention; however, a number of manual operations are needed. Power system data are constantly and automatically collected for the required analysis of system performance for planning and contingency analysis.

Producing the power needed in the system is the task for the Generation segment in a traditional electric power system. In traditional monopolistic environments, vertically integrated utilities knew when, where, and how much electricity was going to be needed and scheduling of energy production was relatively easy. More recent market-based decentralized approaches have increased the complexity of the task for the sake of the system efficiency. Short-term markets (bidding mechanisms) and bilateral agreements between producers and consumers need to be coordinated. Eventually, energy producers offer their production capacity and get it awarded in, with a price assigned per MW to the different generation sources.

Technical aspects of the different conventional generation sources are taken into consideration together with the economic mechanisms. Their different constraints (e.g., costs, speed of startup and shutdown, capacity factor, forecast of day availability) determine unit commitments [11]. These are assignments of a production rate and temporal slot some time in advance of the real need. To cope with unexpected contingencies, the reserves (i.e., the "back-up" generation units) are planned as well.

Frequency stability is a key aspect of power system operation. It starts in the Generation segment. Traditional production of electricity (hydro, fuel, nuclear) involves mechanical elements (e.g., water, steam, or gas flowing through a turbine) and has an effect on the rotational speed of the turbine that consequently determines the exact frequency of the electricity signal. If rotational speed is higher, frequency is higher too. Loads also affect system frequency: when load is heavy, the turbine will tend to rotate more slowly, and the output frequency will be lower. This effect needs to be compensated on the generator side (with the mechanical resources) to keep the frequency as close as possible to the 50 or 60 Hz nominal value.

Voltage levels are the other key aspect that needs to be controlled. Loads in the system exhibit a reactive behavior and, if the consumption of reactive power is excessive, the generated output power will not be efficiently used. The grid (Transmission and Distribution segments) takes care of compensating loads to keep reactive consumption low and maximizes the real power flowing in the system. As a last resource, generators may need to act.

Transmission systems support the electricity transported in the power line. These systems need to be highly reliable, resilient, and able to dynamically adapt to physical limitations and tolerances of the cables (e.g., thermal) to minimize system losses. Control of the reactive part of the load is done with Volt-Ampere Reactive (VAR) regulation elements (inductors, capacitors, and semiconductor switches) deployed in the grid; other losses are highly dependent on weather conditions (humidity and temperature). Geomagnetical-induced currents must also be taken into account [12], as they might cause damage specially in long conductors (more common in North America than in Europe). Transmission is highly coordinated with Generation, as network capacity expansion must be coordinated with any new generation plant in the grid, and the hubs it must reach depend on where the electricity is needed.

Distribution systems are the most significant in terms of territory coverage and have a fundamental role in supply availability and quality control. Distribution operations take care of the control and the voltage regulation (i.e., LV levels within limits in both the unloaded and full-load conditions), power factor (i.e., the reactive part of the load), harmonics (degradation of the waveform frequency, more common due to the growing presence of solid-state switching devices in the grid), and voltage unbalance among phases in multiphase systems. Literature [4] specifically identifies concepts such as supply outages (supply interruptions of different duration), voltage drops (dips in supply voltage), overvoltage (voltage increases caused by network events), voltage wave harmonics (deviations from the fundamental frequency), and flicker (low-frequency fluctuations in voltage amplitude frequency).

1.2.4 The Grid Assets

Grid assets are the elements of the electric power system. This section will focus on the network part of it, i.e., Transmission and Distribution segments.

All the network elements are important not only because they are the elements making the grid work but also because they are the elements to be controlled, the ones needing telecommunication services, and the ones that can be used to support the means to develop telecommunication networks (e.g., optical fiber cables on the power lines).

1.2.4.1 Substations

Substations are the grid nodes where voltage level is transformed. PSs transform voltage from HV to MV and SSs from MV to LV. Both PSs and SSs use transformers for this purpose, and as voltage is reduced, current is increased. Voltage transformation takes place in several stages and at several substations in sequence. These substations are connected through power lines.

Other than the transformer as the main component of a substation, substations consist of a complex set of circuit breaking, and control equipment arranged to direct the flow of electric power. Substations perform several functions, namely, safety (isolating parts of the system in case an electric fault occurs), operation (minimization of energy losses; separation of parts of the network when maintenance or network upgrades are needed), and interconnection (if different electric networks with diverse voltages need to be connected).

PSs (Figure 1.5) are designed for several specific functions such as regulating voltage to compensate for system changes, switching transmission and distribution circuits, providing lightning protection, measuring power quality and other parameters, hosting communication, protection, and control devices, controlling reactive power, providing automatic disconnection of circuits experiencing faults, etc. They contain elements such as line termination structures, switchgear, transformer(s), elements for surge protection and grounding, capacitors and voltage regulators, and electronic systems for protection, control, metering, etc. These elements are usually connected to each other through conductor buses or cables within the facility. PSs are quite homogeneous in

Figure 1.5 Primary substation examples.

its design and external appearance. Most of them are big outside compounds where power lines and metallic structures can be easily identified.

SSs (Figure 1.6) are located at the edge of MV networks, close to end-users, where electricity is transformed to LV for Consumption Points. LV grids are deployed to reach the customers, and therefore, SSs are located close to the end-users. In Europe, SSs normally supply an LV area corresponding to a radius of some hundred meters around the SS; North and Central American Distribution networks consist of an MV network from which numerous (small) MV/LV transformers supply one or several consumers, by direct service cable (or line) from the SS transformer location (thus SSs may be just a number of small transformers on top of a pole).

The number of SSs in the grid exceeds the number of PSs with a factor than can be in excess of 100. Due to their relevance in the evolution into the Smart Grid, it is worth describing the structure of an SS, from its MV input to its LV output:

- MV lines. These are the power lines carrying voltages and currents that originate from PSs.
- Switchgear or MV panels. These are the interfaces between MV lines and the transformer or transformers. The switchgear protects the transformer and allows the interconnection of MV feeders. Switchgear maybe open-air or encapsulated air-insulated (old versions) or encapsulated with a gaseous dielectric medium (usually sulfur hexafluoride, SF6).
- Transformer. The device that steps down the voltage from MV levels to LV.
- LV panel. The element located beyond the transformer, connected to its secondary winding, and distributing the electric power in a number of LV feeders that will eventually reach Consumption Points. It is usually a large single panel or frame composed of four horizontal bars where the LV

Figure 1.6 Different types of secondary substations.

feeders are connected (for the three phases and the neutral). The LV switchboard usually is fitted with switches, overcurrent, and other protective elements.
- LV feeders. These are the power lines supporting specific voltages and currents that deliver electric energy to customers. Usually several LV lines come out of a single SS, which then provide electric service to buildings and premises around.

SSs are more diverse in its nature, dimensions, and external appearance than PSs. The main reason lies in the need to adapt to the physical surroundings and specificities of the Consumption Points. SSs can thus be located indoor (in shelters, integrated in building spaces, or underground) or outdoor (with overhead transformers on poles or in compact surface cabinets – padmount transformers – or similar).

1.2.4.2 Power Lines

Cables, as the main component of power lines, are very important elements in terms of their numbers and wide-spread installation. To have a reference of the dimension, organizations representing Transmission and Distribution system operators ENTSO-E [13] and E.DSO [14] in Europe report almost 0.5 million km of HV cables and 8 million km of MV and LV cables.

Power lines connect substations, Generation and Consumption Points, and are also used inside substations and the consumer installations. The cables that are suspended on towers, or ducted underground, need to manage the voltages and currents (i.e., power, as voltage multiplied by current equals power), in an effective way. Thus, conductor material, type, size, and current rating characteristics are key factors in determining the choice of the proper cable for transmission lines, distribution lines, transformers, service wires, etc. [15].

Heating of conductors determines the current that can flow through. This heating depends on the resistance of the conductor (Joule effect), being the resistance per km a constant: the larger the diameter of the conductor or the better the conductivity of the metal used, the lower the resistance. Conductors are rated to support the current that causes them to heat up to a predetermined amount of degrees above ambient temperature, acceptable for the grid design of the utility. Current ratings depend on weather conditions and operational conditions to be determined by system designers.

Utilities use different conductor materials for different applications, being copper, aluminum, and steel the primary conductor materials used. Conductivity, durability, weight, strength, and cost are the factors that, combined, determine their use in the power system. Copper and aluminum are the best conductors (copper better than aluminum), but aluminum is lighter and more rust-resistant than copper, while not as durable. Steel is the worst conductor, but its strength justifies its use in the core of aluminum conductors. The conductors can be found as a solid structure or stranded to create a more flexible structure than can also combine different metals to achieve the needed results. In general, solid conductors show better conductivity, while stranded conductors present improved mechanical flexibility and durability.

Power lines are more than individual cables, and cables are more than the conductors that carry electric power. Power lines can present different single cables running in parallel or can bundle them together if it is convenient for laying costs. Thus, conductors (single conductor or several conductor or cores) are fitted in cable structures that will manage to deliver the service needed, be it laid on overhead structures (towers or poles of different materials) or underground (ducted, preferably).

Underground networks are increasingly favored over overhead ones. Over recent years, Europe [16] has seen an increasing trend to replace existing overhead distribution lines with underground cabling, drivers being the higher reliability and safety of supply of underground solutions, their higher acceptance among citizens due to a reduced environmental impact, and the continuous decrease in the cost factor compared to overhead. In Europe, 41% of the MV grid and 55% of the LV grid are reported underground.

HV transmission grid lines consist of a combination of steel and aluminum conductors usually running overhead, except in some cases close to the urban areas. The towers supporting the weight of the conductors need to keep safety distances between them, with the metallic structure of the tower, and the ground [4]. To reduce the so-called corona discharge (rupture of the insulation capacity of the air around the conductors due to high electrical fields), each phase of the line is generally divided into two, three, or more conductors. Inductance depends largely on the relative geometric position of the three phases on the tower. Capacitance among the conductors and with the earth also exists and determines their capacitance to ground. The inductive effect is dominant under heavy load situations (reactive energy consumption), while the capacitive effect is prevalent during light load periods (generation of reactive energy). HV underground systems are rare as their very short distance between the line and the ground requires the installation of heavy-duty insulators. These lines have higher capacitive effect than overhead lines.

Overhead MV conductors can be classified in two major categories: homogeneous and non-homogeneous. The first category includes copper, AAC, and AAAC, and the second category includes ACSR, ACAR, ACSS, or AACSR. Wires can be insulated or not insulated. The bare-wired (un-insulated) is the most common type of line used in overhead power lines. Covered wires (for improved reliability) are AAC, AAAC, or ACSR conductors covered with PE or XLPE (see Table 1.1).

Underground MV lines are considered safer and statistically more reliable than overhead ones. However, installation costs are significantly higher, although their maintenance cost is lower. Underground MV cables are basically a metallic conductor surrounded by an insulation system

Table 1.1 Typical components of overhead medium voltage conductors.

Acronym	Name
AAC	All-aluminum conductor
AAAC	All-aluminum alloy conductor
ACSR	Aluminum conductor steel reinforced
ACAR	Aluminum conductor alloy reinforced
ACSS	Aluminum conductor steel supported
AACSR	Aluminum alloy conductor steel reinforced
PE	Polyethylene
XLPE	Cross-linked polyethylene
PVC	Polyvinyl chloride
EPR	Ethylene propylene rubber

and a protective system. Its structure consists of the following elements: the inner conductor, the conductor shield, the insulation, the insulation shield, the neutral or shield, and the jacket. Underground cables can be single-core or three-core (thus three phases can be grouped together) and can have additional armored layers over the insulation to provide the cable with additional mechanical protection.

LV lines carry electricity from SSs to individual LV customers. In European countries, the LV grid is usually a larger infrastructure than the MV grid ([16] reports that LV is more than 63% of the distribution grid), while in North and Central American practice utilities' LV grids are practically nonexistent.

In Europe, the output from a transformer in an SS is connected to LV panels via a switch or simply through isolating links. LV panels usually have from 4 to 12 LV-way, three-phase, four-wire distribution fuse boards or circuit-breaker boards and connect electricity to LV feeders reaching out to customers (link boxes may be found along the way, as far as an LV distribution cabinet located at the entrance of buildings and houses, with fuses that often delimit the edge of the utility grid), where electricity meters are located. In the USA, e.g., the common structure is that the distribution is effectively carried out at MV, such that the MV grid is, in fact, a three-phase four-wire system from which single-phase distribution networks (phase and neutral conductors) supply numerous single-phase transformers. These transformers are center-tapped in their secondary windings to produce LV single-phase three-wire supplies that usually reach customer meters through overhead lines.

LV overhead lines may use either bare conductors (usually aluminum or copper) supported on glass/ceramic insulators or an aerial bundled cable system to be laid outdoor on poles or wall-mounted.

LV underground lines are often found in medium- to large-sized towns and cities, inside utility tunnels, laid in ducts or tubes, or directly buried in trenches. These cables show a typical structure of conductor insulated with similar materials to those discussed for MV (the metallic screen is not mandatory) and are protected by an outer PVC jacket.

1.3 A Practical Definition of the Smart Grid

Electric power systems have evolved and adapted to the growing needs of electrification both in terms of reach and of power consumption increase. From the first isolated and single-purpose grids to the nowadays interconnected power systems, the grid has enlarged its capability and has achieved the high marks that allow the rest of the society assume the supply of energy as a commodity.

However, the adoption of changes in power systems is not as dynamic as in other domains, industries, or services. Indeed, a fast transformational pace is not a key characteristic of utilities both because technology cycles in utility industry take longer than in other industries, due to the substantial investments required by many of their infrastructures, the regulation and the endurance expected in electricity service.

All in all, the evolution of the grid over the last decades, probably since the 1980s, has been accelerated, but more strongly since the term Smart Grid was coined. However, there is no standardized or globally accepted definition of it; instead, Smart Grids are defined differently around the world, in different world regions, in different utilities, by different regulators, etc., to reflect local requirements and goals [17]. Moreover, from the initial references to the Smart Grid concept in the 1990s (when Smart Grid was not the commonly agreed expression yet – see [18–20]), the successive examples, implementations, and instances of Smart Grids have shown such a divergence as the one included in the wide scope of the ideas behind the concept.

Grid modernization [21] has been an overarching concept in the Smart Grid evolution. Although the idea collates a great variety of grid evolutionary material aspects, grid assets refresh, adoption of new grid-edge technologies [22], new technologies in energy storage and microgrids domains, and large shares of renewable energy (i.e., DER) have emerged as principal components. All these elements aim at a change toward a more resilient, responsive, and interactive grid [21], to improve the reliability of the system (network and services).

For this purpose, these technologies must be properly integrated in the systemic, operational, and regulatory framework of utility business. Therefore, referring to utility business, legislative and regulatory actions have been taking place to allow these business changes to be introduced. Indeed, utility business, regulatory framework, and associated utility rates are in constant revision to adapt to the new reality. In this new context of energy as an enabler of our Society progress, and with the environmental concern as a major one, the role of consumers comes also into perspective. Consumers overcome their role as passive objects of the electric system and appear as active pieces of the overall service experience taking an active role in system-wide performance (helping to shape the system requirements and operations through their active participation producing and storing energy or adapting consumption patterns). The active participation of the different stakeholders (customer being a central element) in the system will achieve higher levels of energy efficiency across the value chain.

Future power grids may not be equal to those of today. However, taking into consideration the history behind power systems, we can state that power grids will not be radically different in neither the short nor the medium term, and the changes will happen in an evolutionary way. Thus, existing grid infrastructure will play a key role, and its integration with the new grid technology is both a must and a key in the process of leveraging existing assets.

And it is here that ICTs, also commonly referred to as *digital technologies* [23], come into play. The advances in electronics, computation, and telecommunications gathered around the ICTs are continuously impacting different aspects of our Society.

Utility industry is a very special (neither minor, nor simple) example of adoption of ICTs. Although some in utility industry write the equation "Smart Grid = Grid + ICT," this is an excessive simplification. While it is true that ICTs integration is one of the most prevalent ideas behind the quest for "Smart" in the Grid, there are many other components that are instrumental. What it is also true is that the incorporation of ICTs to human activities does not by itself improve them, and this is why the role of ICTs for the grid in its evolution toward a smarter grid must be understood in terms of "enablement" of what needs to happen in the grid, with the new ICT capabilities helping both existing and new grid components to seamlessly integrate with the grid as an enhanced system.

Some high-level definitions (Table 1.2) have had a higher impact driving actions enabling the Smart Grid. Within the idealistic vision drawn by them, sits the ambition of a "better" electric grid, superior to the existing one, characterized or involving the following elements:

- Resilient electric power system.
- Grid infrastructure modernization.
- Power quality assurance.
- Efficiency in the power delivery system and in the customers' consumption.
- Reduced environmental impact of electricity production and delivery.
- Combination of bulk power generation with DG resources.
- Storage as technology increasingly available in the grid edge.
- Automation of operational processes.
- Increased number of sensors and controls in the electricity system.
- Monitoring and control of critical and non-critical components of the power system.

No two conventional grids are the same today. Thus, even with common objectives, and despite the efforts in standardization and proper frameworks definition, Smart Grid implementations in utilities will be different one from the other.

1.4 Why Telecommunications Are Instrumental for the Smart Grid

Out of the four high-level definitions of Smart Grids in Table 1.2, two of them refer to ICTs when they mention "digital technologies." It is encouraging that high-level definitions come that close to our topic of interest.

Technical references on Smart Grids show a more precise idea of the relevance of ICTs and Communications in particular. Franz et al. [29] highlight the "convergence of the electricity system with ICT technologies" aspect of the Smart Grid; [30] is more comprehensive with the purpose and signals the difference between information technologies and communications ("two way exchange of electricity information") in its application to the Smart Grid: "A Smart Grid refers to a next-generation network that integrates information technology (Smart) into the existing power grid (Grid) to optimize energy efficiency through a two-way exchange of electricity information between suppliers and consumers in real time." This approach is also expressed in [20] that highlights the enabling aspects of the different components of the ICTs, in connection with the elements of the grid, and the ultimate goal of getting smarter grids ("A Smart Grid is the use of sensors, communications, computational ability and control in some form to enhance the overall functionality of the electric power delivery system"), and in [18], that is more explicit in what the improvement of the grid should be ("The IntelliGrid vision links electricity with communications and computer control to create a highly automated, responsive and resilient power delivery system").

The importance of the communication's part of the ICTs is highlighted in [31] ("Technologically, the Smart Grid can be viewed as a superposition of a communication network on the electric grid"). And the explicit reference of the telecommunications connectivity pervasiveness comes from [32] in its *Grid 2030* as "a fully automated power delivery network that monitors and controls every customer and node, ensuring a two-way flow of electricity and information between the power plant and the appliance, and all points in between."

Table 1.2 Main definitions of the Smart Grid.

Body	Definition
The Smart Grid European Technology Platform [24]	A smart grid is an electricity network that can intelligently integrate the actions of all users connected to it (generators, consumers, and those that do both) to efficiently deliver sustainable, economic, and secure electricity supply.
The U.S. Department of Energy [25]	A smart grid uses digital technology to modernize the electric system – from large generation, through the delivery systems to electricity consumption – and is defined by seven enabling performance-based functionalities [17]: • Customer participation. • Integration of all generation and storage options. • New markets and operations. • Power quality for the twenty-first century. • Asset optimization and operational efficiency. • Self-healing from disturbances. • Resiliency against attacks and disasters.
The International Energy Agency (IEA) [26]	A smart grid is an electricity network that uses digital and other advanced technologies to monitor and manage the transport of electricity from all generation sources to meet the varying electricity demands of end-users. Smart grids coordinate the needs and capabilities of all generators, grid operators, end-users, and electricity market stakeholders to operate all parts of the system as efficiently as possible, minimizing costs and environmental impacts while maximizing system reliability, resilience, and stability [27].
The World Economic Forum [28]	Key characteristics of the Smart Grid [17]: • Self-healing and resilient. • Integrating advanced and low-carbon technologies. • Asset optimization and operational efficiency. • Inclusion. • Heightened power quality. • Market empowerment.

The reference to ICT within the "digital technologies" concept is needed when referring to the wider aspects of the Smart Grid. ICT definition is a broad concept intended to cover all technologies (hardware and software) that manage and process information data and transmit them through telecommunication networks. ICT is certainly used extensively, but a dissection of its different parts is needed if we want to understand its impact to Smart Grids, and specifically the areas that can limit its realization.

The two main components of ICT are the information and the communications. "Information" is again a broad term defined by IEC as "knowledge concerning objects, such as facts, events, things, processes, or ideas, including concepts, that within a certain context has a particular meaning." The idea is better identified when the definition of Information Technology (IT) equipment is analyzed: "equipment designed for the purpose of (a) receiving data from an external source [...]; (b) performing some processing functions on the received data (such as computation, data transformation or recording, filing, sorting, storage, transfer of data); (c) providing a data output [...]."

"Communication" is the other component of ICTs. Communication is, according to the definition of the IEC, the "information transfer according to agreed conventions." The "agreed conventions" piece refers to the protocols or language that is used, and the "information transfer" is what traditionally is implicit in the term "Telecommunications." Indeed, "telecommunications" are defined as "any transmission, emission or reception of signs, signals, writing images and sounds or intelligence of any nature by wire, radio, optical or other electromagnetic systems" in [33] and implies connectivity provided over different physical media, allowing a variety of information sources.

ICTs, understood as computing and electronic elements coupled with telecommunication networks, have historically been appreciated and used by the power utility industry at the core of their operations. Distant grid elements have evolved from being monitored and operated locally to a centralized control performed from UCCs. Central and/or remote intelligence has progressively taken control of most elements of the grid in its different parts. Automatic collection of distributed information allows performing grid simulation, and operation and maintenance activities effectively, as it would not be possible without ICTs to analyze thousands of complex parameters without manual intervention. Automated systems have an instrumental role in utility operations, take complex decisions, and execute actions over remote grid assets based on data coming from many distributed grid components. Both the infrastructure (grid) and algorithms (intelligence) are fundamental for the Smart Grid, and the "glue" that integrates them is ICT [34].

All Smart Grid strategies and visions are founded upon the availability of telecommunications connectivity. Most Smart Grid applications, in the different segments of the electric power system, rely upon the availability of a telecommunications network for interconnection of their components [35]. Some of these segments bring less difficulties, e.g., when investment allowance is granted, or the distributed nature and number of the assets involved are low, or the telecommunication connectivity is already available and does not involve any special requirements. However, when any of those circumstances, or several of them, do not happen, the difficulties may cripple Smart Grid adoption despite all efforts in areas that are not related to telecommunications. Remarkably, Distribution and Transmission segments are (in this order) the most challenging fields. With no doubt, we cannot consider any of those two segments in a monolithic way, as they consist of many different components. Their various parts are intrinsically disparate and present distinctive challenges. Thus, in each case, we will need to see which connectivity is needed depending on the part of the grid needing "smartness," and for which Smart Grid application or service.

Thus, telecommunication connectivity is more important than ever for utilities. Although utilities have historically used telecommunications to protect and control their grids, the challenge of extending telecommunication access to potentially millions of geographically dispersed end-points over large service areas, is inherent to the Smart Grid and remains unsolved even considering the sole telecommunications market. On one hand, telecommunication markets (TSPs, equipment vendors, etc.) tend to favor profitable population segments and concentrate their network efforts where return on investment can be maximized (thus leading to terms such as "telecommunications gap" – term coined in "Maitland Report" [36] to describe the different telephone access density in the different parts of the world, or the more recent "digital divide," that copes both with access to information in terms of information technology, and in terms or communications connectivity after this foundational document of modern telecommunications development). On the other hand, standard residential, commercial, or industrial telecommunication services supported by existing networks and equipment do not by default comply with the type of need and service-level guarantees the utilities have in their operational environments and complex processes [37], tailored to maximize the safety and resiliency of the electric power system as a whole.

The challenge of telecommunications connectivity is to grow today's utilities' existing telecommunication networks by possibly several orders of magnitude and in very diverse circumstances.

Until the advent of the Smart Grid, telecommunications connectivity needs in utilities were limited to some of their assets. As new elements are brought to the grid, changing the way the grid must be operated, a pervasive control and monitoring is needed, and telecommunications connectivity becomes a bottleneck, and eventually the weak-link of all the Smart Grid strategy. Due to their varied nature, location, and requirements of the Smart Grid assets, telecommunications connectivity for their Smart Grid services will not be systematically and cost-effectively provided over one single technology. The telecommunication network solution will be a hybrid one, combination of a mix of private and public/commercial telecommunication solutions. The optimal blend will be different for each utility due to historical, economical, technical, market, or strategic reasons [34].

The history of utilities cannot be understood without the telecommunication networks and services supporting their operations. The future of utilities with Smart Grid will reinforce this reality.

1.5 Challenges of the Smart Grid in Connection with Telecommunications

There are some Smart Grid challenges tightly connected to the use of telecommunications technologies and services. They can be grouped in two broad categories *Customer Engagement* and *Grid Control*.

1.5.1 Customer Engagement Challenges

The Consumption Point is now transformed into a customer, with changing needs and capabilities, in contrast to its view as a plain electricity service subscriber. It is important not only that, as a customer, it demands a quality service but also that the customer has a potential to contribute to the electric power system in various forms.

1.5.1.1 Customers as Smart Electricity Consumers

The customer is the entity driving the consumption patterns and electricity demands that, when aggregated across all the different types of customers (residential, industrial, etc.), define the power system load curve (see Figure 1.7).

The major concern of electric power system operators, apart from the hourly consumption prediction to manage generation sources in real time, is the general reduction of the curve peaks, and the possibility to control the load (consumption) at the moments where the system may not be prepared to cope with it.

If the consumption pattern can be influenced, the total electricity demand can be flattened, while keeping total energy consumed the same. This effect implies that the system does not need to be dimensioned to cope with the worst-case condition of electricity demand. On the other hand, the system operator needs to have tools available to control the loads present in the network (i.e., to be capable of reducing the number of them connected or, to curtail their consumption) in a near-real-time manner.

The customer needs to be convinced of taking a more active role to yield part of his freedom to consume to the system operator (for a certain incentive), for the system to be optimized. To favor such involvement, customers need to be aware of the offered possibilities, and this is when they need to perceive the usefulness of their contribution and need to have easy-to-use access to the mechanism that enable this participation in the system. Customers will then only be convinced if

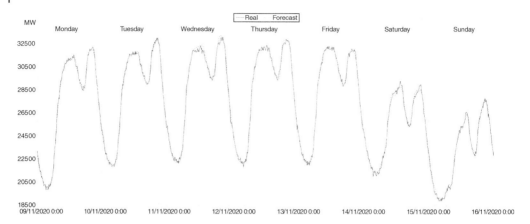

Figure 1.7 Sample weekly aggregated electricity demand curve. Consumption in the Spanish peninsula. *Source:* Real-time Spanish Electricity Demand [38].

they are able to easily see the result of their effort, and, if this effort is not cumbersome and facilitated by the utility through the use of processes and tools.

DSM is the key concept to get the customer to participate in the electric power system. DSM includes all the activities performed by the utilities to "influence" the customer demand to balance instantaneous grid electricity supply with the demand. DSM groups together a set of activities including pure energy efficiency initiatives, where Demand Response (DR) is the most important one (see Chapter 5).

1.5.1.2 Customers as Energy Generators

Customers have now a set of technologies (DER [39]) that allow them to participate as an agent that has the possibility of producing part of the energy they need, and even help the grid, making any excess of generation available for the system.

There are multiple DER elements, including DG but also Energy Storage (ES), and not forgetting Electric Vehicles (EVs), that play a role for the system (positive, as "batteries on wheels"; challenging, as "moving loads") further than their direct environmental impact (reduction in fossil energy sources consumption).

DG includes wind power, solar power, geothermal power, biomass, and fuel cells. They can be manageable or unmanageable from the grid perspective, depending on several factors such as their installed capacity, the availability of the energy they produce, and the connectivity and intelligence of their ancillary elements for grid integration. Notwithstanding, when these elements are available and connected at end-user premises of small producers and communities, the consumer has the dual role of consumer and producer (i.e., a prosumer). However, the overall traditional concept of the energy power system protection and control needs to adapt to this new reality.

EVs, on their side, have a very interesting role to play. EVs contribute to the overall energy waste reduction, system efficiency, and fossil fuels and emission reduction and are seen as moving batteries to help the system. This can be done by means of the storage of certain types of energy produced at times where the consumption cannot take advantage of them [40] and be used to charge EVs and have it delivered when and where it is needed.

1.5.2 Grid Control Challenges

Within the Transmission segment, challenges are those inherent to its role in the system. Despite the evolution of the electric power system technologies, Transmission has not been affected by technical disruptions that impose a new reinvention of its nature. However, their traditional functions can be pretty much enhanced.

The need to keep energy loss at a minimum justifies the need to understand and control the different power line parameters involved in ampacity (i.e., maximum value of electric current – Amperes) calculations. Grid stability is another concept of great importance, as the connection of bulk-generation at different parts of the network can cause instability if the voltage phase difference between the power signals is too different at the two ends of the line. Last but not least, harmonics of the fundamental frequency (50 or 60 Hz) have to be controlled, as they would be propagated along the grid, impacting overall quality of the electricity wave.

There are other needs related with the physical aspects of the cables, leading to keep the infrastructure available and extending its useful life [41]. There is a growing need to easily detect the origin of faults in the power lines and be able to identify and communicate its exact location. These needs can be supported with modern telecommunications-related technology. These technologies can also help to control strain in the cables when they are being laid out, specifically in cables under special conditions (e.g., submarine routes).

Distribution grid challenges are the ones that have been driving the evolution of the Smart Grid concept in the network control side. The paradigm of a static unidirectional network delivering electricity from the core to the edge has changed into an entangled mix of new technologies and assets, with varied properties, that needs to be organized. The new idea of a "platform" [42], rather than a "grid" that can serve the different Distribution grid stakeholders, has become popular due to its use in some major initiatives.

Automation tries to drive the grid toward becoming a self-healing entity. Automation in utilities consists of Substation Automation (SA) and Distribution Automation (DA). Automation process started at substation level, mainly PSs (hence SA), and moved out toward the edge (hence DA), increasing its ambition to reach every corner of the grid with that kind of automatic signals driving electricity through the most appropriate routes (like information is routed in telecommunications networks, differences aside). The aid of ICTs and telecommunications in grid automation is highlighted in [43] with the term ADA (Advanced Distribution Automation). The ultimate objective of the self-healing grid is to improve the service availability introducing efficiencies in the grid operation (less manual and field interventions).

The appearance of DER alters the traditional equilibrium of things in the Distribution grid. E.g., DG and EVs come with some caveats that change the landscape of traditional grid operation. On the more general side, operational procedures need to be modified as they can no longer assume that energy flow is not unidirectional (restoration manoeuvers may need power to be completely for the safety of people and assets). On the other, a greater visibility and control of the elements of the grid, including those of the new technologies close or in the Consumption Points, to enable the operational procedures.

Now the Smart Grid in the Distribution segment needs to be seen as the superposition of the grid with the new grid technologies, and a telecommunications network providing connectivity to any corner of the grid to measure, control, and remotely operate the different IEDs (Intelligent Electronic Devices) attached to the grid assets [35]. A different proportion of central and distributed intelligence will use this connectivity to run algorithms able to make autonomous decisions and perform sophisticated actions that will be monitored from UCCs.

1.6 Challenges of Telecommunications for Smart Grids

Telecommunication networks, systems, and the services they provide are defined to fulfill a set of technical requirements within the state-of-the-art solutions and the affordable economic conditions.

Thus, not any telecommunications solution can fulfill the need of any service. However, from a pragmatic perspective, some needs and/or some industries may trim their requirements to the capabilities of existing telecommunications solutions.

From a practical Smart Grid implementation perspective, one of the main decisions when selecting telecommunication solutions is the option of deploying private telecommunication networks, as an alternative to using commercial telecommunication services from TSPs. It is often not a question or selecting one or the other, but how to select the best combination of both, leveraging the full control that private networks offer, with the use of TSP services that may be already available in third-party deployed networks (TSPs'). In both cases, but probably more relevant when the services are provided by TSPs, there are two important aspects that need to be taken into account. The first one is the national regulation of the conditions under which any telecommunication solution is licensed in a given territory; the second, pretty much affected by the first one, is the network design and its implementation on field, together with the operation and maintenance aspects.

Thus, the service adequacy of a telecommunications solution needs to be evaluated considering the integration of several aspects, at least, the solution conception based on the requirements to be fulfilled, and its practical implementation within the constraints of network design decisions, regulation and business plans. As a consequence, the assessment of how the different Telecommunication solutions fit Smart Grid requirements must evaluate many different alternatives under a non-evident set of considerations.

1.6.1 Telecommunication Solutions for Smart Grids

Smart Grid infrastructure refresh cycles are much slower than telecommunication's. In general terms, we could say that the useful lifetime of a grid technology is always higher than any telecommunication solution that could be used in it, with factors ranging from 1.2 to 2.5 (see examples in [44]). If a telecommunications solution is to be found for a specific Smart Grid need, it is not just the requirements, or the basic technology supporting it that deserves attention, but the way in which the telecommunication solutions are to be sustained over time and how to better approach this.

The consideration clearly exceeds technical aspects, but probably economic aspects as well, as the feasibility of modifying infrastructure aspects on the grid must be limited in order not to affect grid reliability or to interfere with normal grid operation. Thus, substations cannot be frequently stopped, and customers cannot be visited without a good reason.

There is an inherent risk of making wrong choices when deciding on the telecommunication solutions for Smart Grids:

- The technology assessment period (including analysis, proof of concept, service adjustment, and tender process) may take so long that a relevant percentage of the product or service lifecycle is gone.
- The network or service deployment rhythm, as it is encompassed with the grid infrastructure delivery, may take so long that a relevant percentage of the product or service lifecycle is gone.

- If the selection of the technology involves a service provided by a TSP, the availability of the service or its service conditions may vary over time, and it may invalidate products developed to adapt to this service.

Utilities usually take some measures to adapt to these risks:

- Private telecommunication network solutions are chosen, even if there are commercially available alternatives, to avoid that their investments or costs are decided outside their industry.
- Access to niche telecommunication solutions and product vendors. Apart from the grid knowledge of many of these vendors, they offer their solutions with long-term availability commitments.
- Integration of telecommunication solutions and even technology inside grid elements. The use of assets that are intrinsically part of the grid (optical fiber cables, PLC, etc.) guarantee that they evolve as the grid infrastructure does.
- Integration of replaceable telecommunication modules within the devices to be connected, or as separate boxes connected through simple standard interfaces, to minimize disruptions due to external changes.

All these precautions come with a price. Private telecommunications network solutions require an expertise that needs to be found within the utility, contracted as a service, or existing within niche TSPs. Niche solutions have the handicap that they are not usually best of breed, as they are not based on the ultimate knowledge and resources from market leaders; moreover, as the niche market may be comparatively small, the attraction of competitors is limited, so that it may end up with prices (costs) that are not competitive. Integration of telecommunication solutions within the grid requires that utilities open their grids to telecommunication experts so that the field they need to explore is available to develop and test new ideas; this might not be easy being the grid a place where safety and reliability of service is a key aspect.

Thus, there are some considerations to observe when designing the Smart Grid evolution with Telecommunications:

- Smart grids telecommunication solutions must be always understood as a moving target; i.e., any solution, while in the process of being delivered, must be managed observing its next evolutionary step, as due to the longer technology cycles in Smart Grids, telecommunication solutions will change faster that Smart Grid ones.
- Telecommunications solutions must come to the Smart Grid infrastructure in the form of technological waves that will be applied to the existing grids, in such a way that along the life cycle of a grid asset, several evolutionary telecommunication solutions will be applied.
- Requirements imposed by the Smart Grid must be realistic but challenging, to find a solution within the existing solutions, while simultaneously providing requirements to the next wave of telecommunication solutions for Smart Grids.
- Telecommunication solutions must be scalable, meaning that networks and network end-points will be able to increase in extension and number (respectively) without general network degradation.
- No single solution will ever exist for the "smarter grid." All Smart Grid evolutions will be a consequence of complex decisions, based on regulatory and business aspects, electric and telecommunications technical, and strategy considerations that will configure unique solutions.

1.6.2 Standards for Telecommunications for Smart Grids

The initiatives to standardize ICT-related Smart Grid solutions started to crystallize in the 2010s. The overarching idea was to define some reference architectures, identify existing standard solutions, identify gaps in them for Smart Grid purposes, and both adopt and adapt existing standards to eventually create new ones where appropriate.

Different efforts have taken place. The reference models and compilations of solutions are useful but lack specific examples on how to use them to provide real solutions. Their value comes from the identification of applicable standards and the process to produce new ones.

Standards ([45]) are instrumental to achieve interoperability, and within their different scopes, they must allow for the interconnection of standard-compliant equipment from different manufacturers. Standards have many origins; solutions that end-up being a standard not always start in standardization bodies. The most extreme case is de-facto standards that, starting purely in industry, may end as a formal standard solution.

IEC, ITU, IEEE, ETSI, CEN, CENELEC and ANSI are probably the main standardization bodies that have engaged with Smart Grids.

IEC's (International Electrotechnical Commission) influence domain is within electrotechnologies (i.e., electrical, electronics, and related technologies) and is probably one of the best options for Smart Grid standards to grow. IEC members are the National Committees (one country, one vote) that appoint delegates and experts to produce consensus-based standards. The IEC identified hundreds of Smart Grids standards [46].

ITU (International Telecommunication Union) is the United Nations' agency specializing in ICT and is organized in three sectors, namely, Radiocommunications (ITU-R), Standardization (ITU-T), and Development (ITU-D). A major role of ITU is on radiocommunications (the World Radiocommunication Conference – WRC – being the major reference point), coordinating spectrum allocation at a worldwide level. ITU-T created a focus group for Smart Grid activities, now within Study Group 15 [47]. Many relevant Smart Grid-applicable ITU recommendations will be mentioned throughout the different chapters.

IEEE (Institute of Electrical and Electronics Engineers) is a well-known technical professional organization serving professionals involved in all aspects of the electrical, electronic, and computing fields and related areas of science and technology. IEEE is organized in "societies," and the ones related to Smart Grid activities are the IEEE Communications Society and the IEEE Power & Energy Society. The IEEE references over 100 standards related to the Smart Grid, and many of them will be mentioned along this book.

ETSI (European Telecommunications Standards Institute) [48] is the European Union (EU) ICT-related recognized body, that jointly with CEN (Comité Européen de Normalisation, or European Committee for Standardization) [49] and CENELEC (Comité Européen de Normalisation Electrotechnique, or European Committee for Electrotechnical Standardization) [50] are the European Standard Organizations. The ETSI works closely with the National Standards Organizations (NSOs) in the European countries, to an extent that all European Standards (ENs) become national standards of the different European member states. ETSI works very close to the EU institutions, and in the Smart Grid domain, the European Commission issued M/490, M/441 and M/468 mandates to CEN, CENELEC, and ETSI to develop standards for Smart Grids, Smart Metering, and EV charging.

ANSI (American National Standards Institute) [51] works also very closely with USA institutions, to facilitate the standardization and conformity assessment in the United States, for the better performance of the internal market. ANSI develops accreditation services to assess the competence of organizations certifying products and personnel and provides a framework for American National Standards (ANSs) to be developed out of common agreements. The ANSI label may be used when the organization producing the standard meets ANSI requirements; this is the case with the IEEE.

Last but not least, 3GPP (3rd Generation Partnership Project) [52] unites a set of telecommunications standard development organizations (ARIB, ATIS, CCSA, ETSI, TSDSI, TTA, TTC) in a stable environment to produce the so-called 3GPP technologies' specifications. These specifications cover cellular telecommunications technologies for mobile telecommunications.

1.6.3 Groups of Interest Within Telecommunications for Smart Grids

The impossibility of defining one Smart Grid solution that fits all, draws the attention to interest groups and associations that gather utilities and Smart Grid stakeholders. Some relevant ones are NIST, CIGRE, EPRI, and UTC.

The NIST (National Institute of Standards and Technology) [53] is a nonregulatory federal agency and one of the USA's oldest physical science laboratories now within the U.S. Department of Commerce. In the Smart Grid domain, the NIST specifically established the Smart Grid Interoperability Panel (SGIP) to accelerate standards harmonization and advance into the implementation and interoperability of Smart Grid devices and systems. The SGIP evolved in 2013 as a non-profit private-public partnership organization, SGIP 2.0, and in 2017 merged with Smart Electric Power Alliance (SEPA).

The CIGRE (Conseil International des Grands Reseaux Electriques, or International Council on Large Electric Systems) is an international non-profit association whose objective is to promote collaboration with experts from all around the world by sharing knowledge and joining forces to improve electric power systems today and in the future. CIGRE [54] works with experts in Study Committees (SCs) overseen by the Technical Committee. The SC D2 focuses on ICT applied to "digital networks"[2] from HV to the Distribution grid (smart meters, IoT, big data, EMS, etc.), communication solutions for information exchange in the smart delivery of electric energy [55].

The EPRI (Electric Power Research Institute) [56] is an independent, non-profit organization, bringing together scientists and engineers and experts from academia and industry. EPRI addresses the challenges in all the aspects of the electricity domain. In the Smart Grid domain, the Intelligrid concept as the architecture for the Smart Grid of the future is one of the best-known contributions.

The UTC (Utilities Technology Council) is a global trade association with the purpose of creating a favorable business, regulatory, and technological environment for companies that own, manage, or provide critical telecommunication systems in support of their core business (e.g., utilities including electricity utilities). Although UTC's initial focus was getting radio spectrum allocations for power utilities, it is now focused on ICT solutions together with its members and associated groups (EUTC – Europe – [57], UTCAL – Latin America – [58], etc.) in different world regions.

1.6.4 Locations to be Served with Telecommunications

All electric power grid assets potentially need telecommunication services, of one kind or another. Relying only on commercial telecommunication networks to connect these assets spread over a wide extension of territory and sometimes placed in remote (e.g. solar wind farms, transmission power lines) or hard-to-reach locations (SSs, meters, street cabinets, fuse boxes, etc.; they tend to be placed underground or inside metallic enclosures) is not feasible in many occasions simply due to coverage limitations, setting aside other considerations related to service control and assurance.

---

2 Some utilities use the term "digital networks" to refer to their new grid technologies and components capable of taking advantage of ICTs.

Commercial telecommunication networks are mainly driven by the large consumer market and, even with the universal service obligations [59] established for the telecommunications market, coverage level is dissimilar across the territory and not all citizens get access to the same portfolio and performance of telecommunication services (bandwidth, throughput, latency, etc. are usually better in urban and suburban scenarios than in rural). Coverage obligations when licences conditions are applied usually focus more on people coverage than territory, as may seem logical thinking on revenues. Moreover, even considering satellite-based solutions offered in the market for the case of areas lacking appropriate terrestrial infrastructures (e.g. rural areas or remote locations), those solutions are not always applicable due to either technical limitations (e.g. lack of "visibility" of the satellite in valleys) or cost considerations (satellite access is currently far more expensive than terrestrial-based solutions).

Another aspect that needs to be considered is the feasibility of the service end-point locations to host the necessary telecommunication equipment. It is not only that a telecommunications device to provide the service is needed (with its special characteristics) but also that there needs to be enough space to install it and an appropriate power supply (in HV and MV networks, the existence of the typical LV AC or DC power is not readily available). The room aspect might become crucial, as there has been a historical trend to reduce the available space in grid assets. Even if this is something to be solved in future grid assets evolution, it is a constraint today with legacy infrastructure.

The HV segment of the electric power system do not usually has problems in terms of physical space available within premises. There are usually suitable shelters (e.g., substation control building) that have been pre-conditioned to host electronics. The situation might be different when it applies to grid elements in the compound exterior area that may be close to grid component to be monitored, measured, or controlled. Providing power supply to these elements may require special transformers close to them or internal wiring through the substation. When it comes to cables, the situation is different, as if there is a need to incorporate some sort of monitoring along the power lines, the availability of power can be restricted to the inductive feeding possibilities.

The MV and LV grid segments are often less prepared to host electronic devices in general, and telecommunication equipment in particular. Power supply circumstances, however, are different in MV and LV, as in LV there is always AC power available by default. Thus, a combination of strategies is needed both to adapt physical spaces and telecommunication devices, where costs will be a matter of how repetitive the solutions will be and how many premises to adapt.

The variety of MV substation types is wide and any possible classification is non-standard. A broad classification follows population density of the area where they are installed: urban areas usually present underground and in-building type of substations (e.g., ground level and basement); suburban areas share the in-building type with the urban areas and include the shelter-type and the pad-mounted transformers. In contrast, rural areas prevalently show pole-mounted transformers. In terms of power lines, underground prevail in urban and suburban, where it gets mixed with on-wall and overhead mounting, typical of rural.

In-building and shelter-type SSs are not so much a constraint in term of space as pad-mounted constructions. Due to the lack of space in pad-mounted and pole-mounted SSs, there is usually the need of an outdoor enclosure. In all SSs cases, the access to LV AC power is readily available in the secondary winding of the transformer (DC power supply and battery solutions might be a concern).

However, the situation of the accessibility to LV power is more difficult in the parts of the MV grid where no SSs exist. In these places, the only standard possibility to feed electronic devices with a non-negligible power need lays in the installation of a special transformer that increases the cost of the solution and may oblige the utility to adapt and reinforce the pole to which the element is attached. Inductive solutions may exist but are not suited for telecommunication devices needing to work in permanent operation.

The LV grid is still a challenge for telecommunications, as it has not been until very recent times when utilities have realized that, to improve the service quality, the LV grid must start to be monitored and controlled. With the likely exception of the meter locations construction, street-cabinets and fuse boxes were never probably understood prone to host telecommunications. Meter rooms, as the result of the effort to avoid the installation of meters inside homes, are presented as built-in wall enclosures or conditioned rooms for meters, depending on the house-type. Thus, meter rooms and such spaces may have the capability (and the utilities the right) to host telecommunication equipment, as derived from the need to access meters for remote reading purposes. LV grid present a more amiable situation, as LV power supply is always available and there are easy and cheap ways to connect to the cables with piercing devices.

1.6.5 Telecommunication Services Control

The impact of telecommunication services on Smart Grid operations is a key aspect of the enablement of telecommunications for the grid. Telecommunications are not a simple add-on over the grid, but a transformation vector. Whenever grid operations rely on telecommunications, telecommunications become as critical to the utility as any other resource.

Telecommunications private network design and/or third-party TSP service selection must be performed to allow the provision of the Smart Grid needed services within the utility license conditions. When the utility operates a private network for this purpose, the performance responsibility is within its decision range. However, when third-party TSP services are involved, Service-Level Agreements (SLA) must be clearly defined to govern service delivery conditions.

ITU-T E.860 provides a framework for SLAs, defined as "a formal agreement between two or more entities that is reached after a negotiating activity with the scope to assess service characteristics, responsibilities and priorities of every part. An SLA may include statements about performance, tariffing and billing, service delivery and compensations." SLAs are based in some formal definitions that must be well known to both parties, especially when referred to the service characteristics that need to be expressed in telecommunication technical terms. However, they must also collate (see ITU-T G.1000 and its reference to ETSI ETR 003) all different service functions (sales, service management, technical quality, billing processes, and network/service management by the customer) and quality criteria with each of them (speed, accuracy, availability, reliability, security, simplicity, and flexibility), to be consistent with the objective.

Smart Grid needs must be transformed into telecommunication technical parameter requirements that measure network performance. Network performance is defined as in ITU-T Recommendation E.800, as "the ability of a network or network portion to provide the functions related to communications between users." Performance is critical for applications to deliver their expected benefits and must be properly planned and controlled in networks that share resources among different users.

Two of the most important performance parameters in telecommunications are throughput and latency. These general parameters are included in the network performance objectives of different networks (e.g., ITU-T I.350 and Y.1540 for digital and IP-based networks):

- Throughput is the maximum data rate where no packet is discarded by a network. Throughput is usually measured as an average quantity, with control over the peak limits.
- Latency is the time it takes for a data packet to get from one point to another. Latency is usually defined as a requirement below a certain limit.

Service performance is assessed in terms of its Quality of Service (QoS). ISO 8402 defines QoS as "the totality of characteristics of an entity that bear on its ability to satisfy stated and implied needs." ITU-T E.800 defines it as "the collective effect of service performance which determine the degree of satisfaction of a user of the service." Beyond QoS, QoE (Quality of Experience as in ITU-T E.804) enters into the subjective domain of the service users' perception that cannot be neglected as it reflects previous experiences and expectations. Thus, SLA, QoS, and QoE go beyond the pure technical perspective.

Eventually, telecommunication systems must absorb all these service requirements and integrate them in their complex network structures. These networks will not only deliver them but also control and manage its performance in different circumstances. ITU-T Study Group 12 "Performance, QoS and QoE" [60] is leading this work, as in ITU-T E.804 "QoS aspects for popular services in mobile networks" and ITU-T Y.3106 "Quality of service functional requirements for the IMT-2020 network."

1.6.6 Environmental Conditions

The harsh conditions of most grid sites are one of the most important aspect that need to be taken into account when designing a telecommunications network for Smart Grids or when using telecommunications services provided by a TSP. It needs to be born in mind that most electric grid premises are neither set up in the way a telecommunications site would be, nor service points are typical residential houses.

On the contrary, there are many aspects that need to be considered to design telecommunication products for Smart Grids. We will refer to them as non-functional requirement ("functional" refers to their function within the telecommunications network), and they include electrical, Electromagnetic Compatibility (EMC), and environmental requirements. Most of them prevent a cost-effective introduction of telecommunication elements in the Smart Grid.

IEC and IEEE are the two main bodies that tackle the specific requirements for devices to be installed at substations and similar locations.

IEC has several series of standards setting requirements as a reference for Smart Grid-related equipment in substations. Technical Committee TC 57 "Power Systems Management and Associated Information Exchange," dealing with "power systems control equipment and systems including EMS (Energy Management Systems), SCADA (Supervisory Control And Data Acquisition), distribution automation, teleprotection, and associated information exchange for real-time and non-real-time information, used in the planning, operation and maintenance of power systems," is the most relevant in this context, while TC 95 "Measuring relays and protection equipment" has a certain role for some device types in the grid (their non-functional requirements might also be considered to the extent where there is not a better reference):

- IEC 60870 "Telecontrol Equipment and Systems", within TC57. This series has a broad scope in terms of monitoring and control, not just in the substation:
  - IEC 60870-2-1 focuses on electromagnetic compatibility.
  - IEC 60870-2-2 focuses on environmental conditions (climatic, mechanical, and other nonelectrical influences) and partially supersedes IEC 60870-2-1.
- IEC 61850 "Communication Networks and Systems for Power Utility Automation," also within TC 57, is much more recent than the IEC 60870 series and focuses in substations and power plants:
  - IEC 61850-3 focuses on environmental aspects of utility communication and automation IEDs and systems.

- IEC 60255 "Measuring relays and protection equipment" and "Electrical Relays" in TC 95 gives requirements specifically for protection devices.

In IEEE, the Power and Energy Society's Substations Committee produced IEEE 1613 to specify "standard service conditions, standard ratings, environmental performance requirements and testing requirements for communications networking devices installed in electric power substations." It complements IEC 61850-3. This standard has broadened its scope with IEEE 1613.1 (in collaboration with the Transmission and Distribution Committee), to cover other devices installed in all electric power facilities, not just substations, specifically applicable for devices used in DA and DG. Interestingly, it explicitly mentions device testing and performance requirements for communications via Radio Frequency (RF), Power Line Communications (PLC), Broadband over Power Line (BPL), or Ethernet cable.

In terms of electrical requirements, devices that are to be part of a Smart Grid deployment usually have to respect specific constraints in terms of voltage and frequency levels (admitted tolerances), power supply redundancy, battery lifetime, etc. These constraints are normally specific to the type of equipment and its location. IEC and IEEE differ in the way they fix some of these values.

A group of very relevant requirements refers to EMC. EMC collates different groups of aspects: radiated and conducted emissions, immunity to radiated and conducted disturbances, insulation, electrostatic discharge immunity, electrical fast transient/burst immunity, surge immunity, voltage-dips/interruption immunity, etc. For conducted or radiated emission limits CISPR 32 is usually considered. Immunity requirements may be taken from tests proposed in IEC 61000-4 series.

The rest of non-functional requirements fall in the environmental category. This group refers to climatic and mechanical (vibration, shock, seismic) conditions, to be taken into account in the product lifecycle (storage, transportation, and in normal-use regime) according to their different environmental conditions (weather-protected, temperature control; stationary use, mobile use, portable use, etc.). There is a very complete reference in Europe in ETSI 300 019 series. The tests for the different classes are based on IEC 60068-2 series "Environmental Testing – Part 2: Tests." Alternatively, a much simpler reference is IEEE 1613.1. IK codes (IEC 62262) and IP codes (IEC 60529) for protection against ingress of solid foreign objects and against ingress of water with harmful effects need also to be taken into consideration, although these are not very different from other fields. Recommended limits for all these aspects can be found in [34].

There is a very important aspect that influences the relationship between telecommunications and the substation environment, and it has to do with earthing/grounding aspects [61]. Substations protect all elements inside them from unexpected HV events such as power faults and lightning strikes (anything that may cause a discharge of large amounts of electrical energy into its surroundings) by providing an equal-potential zone with its ground grid. The substation ground grid connects all metal parts together to create an equipotential zone, so that everything within the compound is at the same potential. While this creates a safety mechanism for the substation, it will affect any externally connected metallic cable, if it is not properly engineered. This situation has been widely studied for PSs in IEEE 367-2012, IEEE 487-2015, and IEEE 1590-2009, and although HV insulators have been developed to allow for the safe connection of, e.g., copper pairs reaching the substation from TSPs premises, the installation of optical fiber in dielectric cables, or radio-based solutions, is preferred to connect substations. The situation is similar in SSs; although distances within SSs are much smaller, the MV ground reference is different to the LV ground reference of the neutral wire, and this imposes extra insulation requirements in devices connected to the MV and LV ground references, simultaneously.

1.6.7 Distributed Intelligence

There has been a recurrent discussion in utility industry around the IT and Telecommunication aspects of ICTs, when it comes to the availability of both in grid asset sites.

Computing power and telecommunications have been at large scarce resources. This fact, connected to the evolution of some traditional control system applications (i.e., SCADA), has configured different historical trends when deciding if system intelligence had to be distributed or centralized.

If we take the IT component, processing power has been expensive and bulky in the old times. At the same time, SCADAs started at power plants and PSs to automate operations locally (see Chapter 5) and that long-distance Telecommunications were not readily accessible in many remote areas. These circumstances created an electric power system control structure that installed part of its computing power in distributed premises (substations).

However, the expansion of telecommunication capabilities, combined with the size reduction of computing power and the enhanced automation capabilities of evolving SCADAs, pushed a wave of SCADA system concentration that leveraged the cost savings produced by synergies in the SCADA operations. Thus, distributed intelligence started to diminish.

SCADA evolution was parallel to remote meter reading systems. The situation of these systems changed the perspective of the telecommunications access to substations. From the need to access a few PSs of early SCADAs, to the need to access many SSs where high-throughput connectivity was not affordable, changed again the approach. At this point, the idea of "data concentrators" (see Chapter 5) appeared, as a way to downsize central MDM systems with a certain distributed intelligence.

Those two parallel stories of Smart Grid systems embryos, different in the use of ICT components, have reached our days, and numerous initiatives in Smart Grids show the influence of both.

In the telecommunications domain, the availability of high-throughput connectivity is common today, and telecommunication bottlenecks are not to be expected in urban, suburban, or populated rural areas. Thus, the distributed intelligence concept is no longer generally a must. On the contrary, the general availability of communications to access all SSs leaves the decision on where to place computing power to utilities.

However, telecommunications themselves, and for their purpose of adapting to the various market needs, are witnessing two trends or types of systems in their evolution:

- Internet of Things (IoT)-type of connectivity, as a connectivity intended to provide some sort of low-end telecommunication capabilities to any location.
- Telecommunications network intelligence closer to the network end-points, helping to provide low latency services that might be demanded by some users.

Thus, another level of complexity is added to the Smart Grid distributed intelligence discussion that should consider these new trends of telecommunications market.

1.6.8 Resilient Telecommunication Networks and Services

Two different concepts mix when referring to resiliency in telecommunications: reliability and resiliency. Before we refer to them, we need first to focus our attention on the availability concept, prevalent in the measurement of the service conditions.

Availability is defined in ITU-T E.800 as the ability "of an item to be in the state to perform a required function at a given instant of time or at any instant of time within a given time interval, assuming that the external resources, if required, are provided." When the network is available, its performance can be observed and measured. Many ITU-T and ITU-R recommendations define availability objectives. Few references, however, exist for Smart Grid-related telecommunication services, although it can be broadly assumed that the strictest Smart Grid services need 99.999% availability (this means 315.36 seconds of unavailability during the year).

To have networks and services in the "available" state, reliability concept appears. It is defined in ITU-T E.800 as "the probability that an item can perform a required function under stated conditions for a given time interval." Reliability is different to resiliency. Resiliency can be defined as "the ability of an entity, system, network or service to resist the effects of a disruptive event, or to recover from the effects of a disruptive event to a normal or near-normal state" [62]. Thus, while reliability focuses on preventing that the system fails, resilience focuses on recovering when it fails.

Resilience in telecommunications implies carrier-grade (carrier-class) network elements (equipment with redundancy in its vital parts – such as power supply), reliable design of network links (with the proper availability assurance), protection of paths through redundant routes that will not fail simultaneously, licensed spectrum usage (legally protected not be interfered), battery backup, etc. While these elements can be properly controlled in private networks, it can just be controlled through SLAs in commercial services.

Finally, security or cybersecurity aspects need to be mentioned. Security aspects must be considered even if only related to the availability of networks, as attacks can turn any network or service into either a non-performing one or a non-available one (e.g., denial-of-service [63]). As an example of this importance, North American Electric Reliability Corporation (NERC) has developed and enforced regulations that demonstrate the importance of security in electric utilities. Security is not just an add-on of telecommunication networks, but a by-design aspect that needs to be present at any stage of the telecommunication service and its network components (technology definition, products implementation, deployment, operations, and delivery).

1.6.9 Telecommunications Special Solution for Utilities

The deployment of Smart Grids involves a transformation of the electric grid infrastructure, specifically Distribution grid infrastructure. This infrastructure is closely connected to the grid. The fact that the electric infrastructure has many aspects in common with the infrastructure aspects of telecommunication networks opens the door for synergies. In particular, there are several elements that are available for utilities, and in some cases, exclusive to them, that are used to deploy telecommunications infrastructures for the Smart Grid:

- Rights of way. Rights of way have been necessary to deploy substations and power lines and their associated sites, towers, poles, and ducts. These infrastructure elements can host other cables (not only power lines) for telecommunication purposes. In fact, there is an abundance of examples of this carriers' carrier business model that has been used for the expansion of existing telecommunication networks [64]. Legislative bodies around the world have also granted the use of these existing rights of way to TSPs.
- Optical fiber cables, unique to utilities. To take advantage of towers, ducts, and cables where optical fiber can be deployed in close proximity to HV power lines, special cables have been developed and are today of wide-spread use (see Chapter 2). HV and LV solutions exist to carry optical fiber close to or inside power cables as well.
- Towers and poles are used and can host radio base station to gain height and improve coverage within existing assets. This possibility minimizes costs and quickens deployments of radio (wireless) solutions.

- Power Line Communications (PLC) are the way to support the transmission of telecommunication signals inside electric power cables. This technology is as diverse and versatile as radio (wireless) and has helped utilities through all ages to communicate distant points among them, for voice communication purposes, low (64 kbps) and high (200 Mbps) data rate needs.
- Surface waves working around all sorts of overhead electricity cables, both naked or jacketed [65]. An application example of this technology is being developed and promoted by [66] and its future is uncertain.

References

1 Department of Energy – USA (2017). *Transforming the Nation's Electricity System: The Second Installment of the Quadrennial Energy Review (DOE/EPSA-0008)* [Online]. https://www.energy.gov/sites/prod/files/2017/02/f34/Quadrennial%20Energy%20Review--Second%20Installment%20%28Full%20Report%29.pdf (accessed 4 October 2020).
2 Constable, G. and Somerville, B. (2003). A Century of Innovation: Twenty Engineering Achievements that Transformed Our Lives. Washington, DC: The National Academies Press.
3 (2017). Universal energy access by 2030 is now within reach thanks to growing political will and falling costs – news. *Energy Access Outlook 2017* (19 October 2017). https://www.iea.org/news/universal-energy-access-by-2030-is-now-within-reach-thanks-to-growing-political-will-and-falling-costs (accessed 24 October 2020).
4 Pérez-Arriaga, I.J. (ed.) (2013). Regulation of the Power Sector. London: Springer-Verlag.
5 IEC (2020). *World Plugs: List View by Frequency*. https://www.iec.ch/worldplugs/list_byfrequency.htm (accessed 24 October 2020).
6 IEC 60050 (2020). International electrotechnical vocabulary – welcome. *Electropedia: The World's Online Electrotechnical Vocabulary*. http://www.electropedia.org/ (accessed 18 October 2020).
7 Goodman, F.R. (1998). *Integration of Distributed Resources in Electric Utility Systems: Current Interconnection Practice and Unified Approach*. Electric Power Research Institute – EPRI, Technical TR-111489. https://www.epri.com/research/products/TR-111489 (accessed 24 October 2020).
8 IEEE Std 1547-2018 (2018). *Standard for Interconnection and Interoperability of Distributed Energy Resources with Associated Electric Power Systems Interfaces*. IEEE [online]. https://standards.ieee.org/standard/1547-2018.html (accessed 8 October 2020).
9 (2007). The Potential Benefits of Distributed Generation and the Rate-Related Issues that may Impede Its Expansion. Report Pursuant to Section 1817 of the Energy Policy Act of 2005 (June 2007) [Online]. https://www.ferc.gov/sites/default/files/2020-04/1817_study_sep_07.pdf (accessed 4 October 2020).
10 US EPA (2015). Distributed generation of electricity and its environmental impacts. *US EPA* (4 August 2015). https://www.epa.gov/energy/distributed-generation-electricity-and-its-environmental-impacts (accessed 4 October 2020).
11 Kassakian, J.G. and Schmalensee, R. (eds.) (2011). The Future of the Electric Grid: An Interdisciplinary MIT Study. Massachusetts Institute of Technology.
12 Bush, S.F. (2014). Smart Grid: Communication-Enabled Intelligence for the Electric Power Grid. Wiley: Chichester, UK.
13 Statistical Factsheet 2018 (2020). *Brussels – Belgium* (June 2019) [Online]. https://eepublicdownloads.blob.core.windows.net/public-cdn-container/clean-documents/Publications/Statistics/Factsheet/entsoe_sfs2018_web.pdf (accessed 4 October 2020).

14 E.DSO in Numbers (2017). *Brussels – Belgium* [Online]. https://www.edsoforsmartgrids.eu/wp-content/uploads/EDSO-in-Numbers-2017.pdf (accessed 4 October 2020).

15 Blume, S.W. (2017). Electric Power System Basics for the Nonelectrical Professional, 2e. Wiley-IEEE Press.

16 (2014) *An Introduction to Medium and Low Voltage Cables in Distribution Networks as Support of Smart Grids*. Brussels – Belgium (June 2014) [Online]. https://www.europacable.eu/wp-content/uploads/2017/07/Introduction-to-Distribution-Networks-2014-06-16.pdf (accessed 4 October 2020).

17 Madrigal, M., Uluski, R., and Mensan Gaba, K. (2017). Practical Guidance for Defining a Smart Grid Modernization Strategy: The Case of Distribution (Revised Edition). The World Bank.

18 Haase, P. (2005). INTELLIGRID: a smart network of power. *EPRI J.* 2005 (Fall): 26–32.

19 Carvallo, A. and Cooper, J. (2015). The Advanced Smart Grid: Edge Power Driving Sustainability, 2e. Boston: Artech House.

20 Gellings, C.W. (2009). The Smart Grid: Enabling Energy Efficiency and Demand Response. The Fairmont Press Inc.

21 A. Proudlove, B. Lips, and D. Sarkisian (2020). North Carolina Clean Energy Technology Center, The 50 States of Grid Modernization: Q1 2020. Quarterly Report (April 2020) [Online]. https://nccleantech.ncsu.edu/wp-content/uploads/2020/04/Q12020_gridmod_exec_final.pdf (accessed 4 October 2020).

22 World Economic Forum (2017). *The Future of Electricity. New Technologies Transforming the Grid Edge* (March 2017) [Online]. http://www3.weforum.org/docs/WEF_Future_of_Electricity_2017.pdf (accessed 24 October 2020).

23 World Telecommunication/ICT Indicators Database (2020). https://www.itu.int/en/ITU-D/Statistics/Pages/publications/wtid.aspx (accessed 24 October 2020).

24 E.DSO (2020). European technology platform for smartGrids. *About Us*. https://www.edsoforsmartgrids.eu/policy/eu-steering-initiatives/smart-grids-european-technology-platform/ (accessed 3 October 2020).

25 US Department of Energy (2020). *About Us*. https://www.energy.gov/about-us (accessed 3 October 2020).

26 International Energy Agency (IEA) (2020). *About*. https://www.iea.org/about (accessed 3 October 2020).

27 International Energy Agency – IEA (2011). Technology Roadmap: Smart Grids. OECD Publishing.

28 World Economic Forum (2020). *Our Mission*. https://www.weforum.org/about/world-economic-forum/ (accessed 3 October 2020).

29 Franz, O., Wissner, M., Büllingen, F. et al. (2006). Potenziale der Informations- und Kommunikations-Technologien zur Optimierung der Energieversorgung und des Energieverbrauchs (eEnergy). Bad Honnef: WIK.

30 Park, M.-G., Cho, S.-B., Chung, K.-H. et al. (2014). Electricity market design for the incorporation of various demand-side resources in the Jeju smart grid test-bed. *J. Electr. Eng. Technol.* 9 (6): 1851–1863. https://doi.org/10.5370/JEET.2014.9.6.1851.

31 Goel, S., Bush, S.F., and Bakken, D. (eds.) (2013). IEEE Vision for Smart Grid Communications: 2030 and Beyond. IEEE.

32 GRID 2030 (2003). *A National Vision for Electricity's Second 100 Years* (July 2003) [Online]. https://www.energy.gov/oe/downloads/grid-2030-national-vision-electricity-s-second-100-years (accessed 7 October 2020).

33 Radio Regulations (2020). http://www.itu.int/pub/R-REG-RR (accessed 18 October 2020).

34 Sendin, A., Sanchez-Fornie, M.A., Berganza, I. et al. (2016). Telecommunication Networks for the Smart Grid. Artech House.

35 (2020). *Electricity System Development: A Focus on Smart Grids Overview of Activities and Players in Smart Grids* [Online]. https://www.unece.org/fileadmin/DAM/energy/se/pdfs/eneff/eneff_h.news/Smart.Grids.Overview.pdf (accessed 7 October 2020).

36 Maitland, D. (2020). *Report of the Independent Commission For World Wide Telecommunications Development*. ITU – Independent Commission for World Wide Telecommunications Development (December 1984) [Online]. http://search.itu.int/history/HistoryDigitalCollectionDocLibrary/12.5.70.en.100.pdf (accessed 8 October 2020).

37 Forge, S., Horvitz, R., and Blackman, C. (2014). Is Commercial Cellular Suitable for Mission Critical Broadband?: Study on Use of Commercial Mobile Networks and Equipment for "Mission-Critical" High-Speed Broadband Communications in Specific Sectors: Final Report. Luxembourg: Publications Office.

38 (2020). *Real-time Spanish Electricity Demand*. https://demanda.ree.es/visiona/peninsula/demanda/total/ (accessed 29 November 2020).

39 Rocky Mountain Institute (2014). *The Economics of Grid Defection* [Online]. https://rmi.org/insight/economics-grid-defection/ (accessed 8 October 2020).

40 *Smart Grids Explained – YouTube – EENERGY*. https://youtu.be/5cIy-5c1DdE.

41 Cherukupalli, S.E. and Anders, G.J. (2020). Distributed Fiber Optic Sensing and Dynamic Rating of Power Cables. Wiley.

42 REV Connect (2020). *Accelerating Innovative Energy Partnerships in New York State*. https://nyrevconnect.com/ (accessed 24 October 2020).

43 T. Godfrey (2020). *Guidebook for Advanced Distribution Automation Communications*. Electric Power Research Institute – EPRI, Technical Report 3002003021 (November 2014) [Online]. https://www.epri.com/research/products/000000003002003021 (accessed 8 October 2020).

44 Taxation Ruling TR 2020/3 (2020) [Online]. https://www.ato.gov.au/law/view/document?docid=TXR/TR20203/NAT/ATO/00001 (accessed 8 October 2020).

45 ISO/IEC (2020). *ISO/IEC Guide 2:2004 Standardization and Related Activities – General Vocabulary Ed. 8* [Online]. https://isotc.iso.org/livelink/livelink/Open/8389141 (accessed 29 November 2020).

46 IEC (2020). *Smart Grid Standards Map*. http://smartgridstandardsmap.com/ (accessed 24 October 2020).

47 Study Group 15 at a Glance (2020). *ITU-T Study Group 15 – Networks, Technologies and Infrastructures for Transport, Access and Home*. https://www.itu.int/en/ITU-T/about/groups/Pages/sg15.aspx (accessed 24 October 2020).

48 ETSI (2020). Standards, mission, vision, direct member participation. *ETSI*. https://www.etsi.org/about (accessed 24 October 2020).

49 European Committee for Standardization – CEN (2020). *Who We Are*. https://www.cen.eu/about/Pages/default.aspx (accessed 24 October 2020).

50 CENELEC (2020). *About CENELEC – Who We Are*. https://www.cenelec.eu/aboutcenelec/whoweare/index.html (accessed 24 October 2020).

51 American National Standards Institute – ANSI (2020). *ANSI Introduction*. https://www.ansi.org:443/about/introduction (accessed 24 October 2020).

52 (2020). *About 3GPP*. https://www.3gpp.org/about-3gpp (accessed 29 November 2020).

53 National Institute of Standards and Technology (2015). *About NIST* (5 January 2015). https://www.nist.gov/about-nist (accessed 24 October 2020).

54 Conseil International des Grands Reseaux Electriques (2020). *Introducing CIGRE*. https://www.cigre.org/GB/about/introducing-cigre.asp (accessed 24 October 2020).

55 CIGRE Study Committee SC D2 (2020). *CIGRE SC D2 > Home. Information Systems and Telecommunication*. http://d2.cigre.org/home.asp (accessed 24 October 2020).

56 Electric Power Research Institute (2020). *EPRI Home*. https://www.epri.com/about (accessed 24 October 2020).

57 EUTC (2020). *Welcome to European Utility Technology Council*. https://eutc.org/ (accessed 29 November 2020).

58 (2020). *Utilities Telecom & Technology Council America Latina*. http://www.utcamericalatina.org/ (accessed 29 November 2020).

59 OCDE (2020). *Universal Service Obligations*. DAF/COMP(2010)13 (April 2010) [Online]. http://www.oecd.org/regreform/sectors/45036202.pdf (accessed 8 October 2020).

60 ITU-T Study Group 12 – Performance, QoS and QoE (2020). *Study Group 12 at a Glance*. https://www.itu.int/en/ITU-T/about/groups/Pages/sg12.aspx (accessed 8 October 2020).

61 Blume, S.W. (2011). High Voltage Protection for Telecommunications. Piscataway, NJ: IEEE Press.

62 (2011). *Enabling and Managing End-to-End Resilience*. Heraklion – Greece [Online]. https://www.enisa.europa.eu/activities/identity-and-trust/library/deliverables/e2eres/at_download/fullReport (accessed: 8 October 2020).

63 CISA (2020). *Understanding Denial-of-Service Attacks*. https://us-cert.cisa.gov/ncas/tips/ST04-015 (accessed 24 October 2020).

64 (2020). *Navigant Research Report Shows Adopting a Fiber Optic Backbone Is a Strategic Imperative for Utilities' Long-term Success*. https://guidehouseinsights.com/news-and-views/navigant-research-report-shows-adopting-a-fiber-optic-backbone-is-a-strategic-imperative-for-utiliti (accessed 18 October 2020).

65 Bennet, R., Barnickel, D.J., Barzegar, F. et al. (2013). Millimeter-wave surface-wave communications. US 8897697 B1, 6 November 2013.

66 History of the JRC (2020). https://www.jrc.co.uk/about-jrc/history-of-the-jrc (accessed 29 December 2020).

2

Telecommunication Networks and Systems Concepts

2.1 Introduction

Telecommunications, understood as communications (information transfer according to agreed conventions) by optical, radio, wire, or other electromagnetic means, have evolved and adapted to the different historical technology evolutions. Understanding their specificities, and their evolutionary prospective features, is key to select the most appropriate combination of them.

Telecommunication networks and systems components sit on top of these basic telecommunications supporting media. The way these elements are organized to interoperate and span all territories is crucial to make scalable, reliable, secure and fit-for-purpose communications.

As a result of this, the different communication services can be provided, as needed, for the different users of telecommunication networks. The nature of the telecommunication systems will determine the types of services that can be delivered and to which extent the different network types must be combined and interconnected to achieve the planned result. This combination of resources must follow a coherent model to allow networks' end-to-end management.

There are some important underlying concepts in telecommunications that explain how these scalable and interoperable telecommunication systems can achieve their goals. They cover transport and switching/routing, circuits and packets, layered models, etc. and they are described in this chapter.

This chapter introduces and describes the organization and differences between the telecommunication networks, systems, and services. It also includes a section on the fundamental aspects of telecommunication media key for the Smart Grid, namely optical fiber, radio, and power lines (electricity cables).

2.2 Telecommunication Networks, Systems, and Services Definitions

Telecommunication services delivery is the goal of telecommunication networks or systems.

There is no formal difference between the network or system terms, when referring to telecommunications. The United States generally refers to Telecommunication System, while elsewhere, the term Telecommunication Network is more often used [1, 2].

However, modern telecommunications can be better understood if we take advantage of the definition of "system" in plain English, to establish a difference that is generally understood in the telecommunications industry. Thus, *Telecommunications Network* can be used to refer to the basic traditional telecommunication equipment, interconnected, to perform its communication basic function, and the term *Telecommunication System* can refer to Telecommunication Networks that are accompanied by non-telecommunication elements (IT elements such as databases and information systems) to complement them with additional intelligence and capabilities. Other than that, the two terms will be used throughout this book.

The formal definition of a Telecommunication Network has evolved. There is a set of definitions that focus on the infrastructure perspective, highlighting the elements inside it. From the "set of nodes and links that provides connections between two or more defined points to accommodate telecommunication between them" [3], to the "entirety of equipment (comprising any combination of the following: network cable, telecommunication terminal equipment, and telecommunication system or installation) that are indispensable to ensure normal intended operation of the telecommunication system" [4]. This evolution shows the complexity that Telecommunication Systems have acquired, evolving from the times where telecommunications meant voice and television, to everything merging into "data."

However, there are other definitions that, focusing on the user's perspective, define the Telecommunication Network as the "entirety of means of providing telecommunication services between a number of locations where installations provide access to [. . .] services" [2]. This definition brings the attention to the widely used Telecommunication Service concept, defined as "an offering by one or more network operators to satisfy a specific telecommunication requirement" [2].

A *Telecommunication Service* is then any communication capability offered to a user by a Telecommunication Network or System. There are many different types of Telecommunication Services and no harmonized classification for them. Telecommunication Services can be provided with different types of Telecommunication Networks or Systems, combining different supporting means, telecommunication technologies, and other ancillary elements. SLAs, QoS, and QoE are inherent to the provision and operation of Telecommunication Services and key in their definition. As it will be shown in this book, the different Telecommunication Services will offer different added values that will make them fit for Smart Grids.

2.3 Telecommunication Model and Services

2.3.1 Telecommunication Model

The fundamental purpose of telecommunications is the exchange of data between two parties. Figure 2.1 shows several examples of the communication model for this purpose. The first one is an analog voice communication between two individuals; the second one is the communication between a workstation and a server over a public analog telephone network; and the third one is the same, but through a digital network.

The key elements of the model are as follows:

- Source. It is the device that generates the data to be transmitted. Examples in the classical telecommunication context are telephones and personal computers; in the Smart Grid domain (see Chapter 5), they may include meters, Remote Terminal Units (RTUs), protection relays, etc.
- Transmitter. The data generated by a source system are not transmitted directly in their original form: a transmitter transforms and encodes the information in such a way that electromagnetic signals are produced to be connected to some sort of transmission system. An example in the

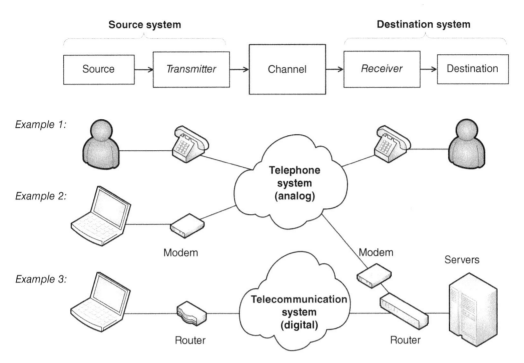

Figure 2.1 Generic telecommunications model.

classical telecommunications context is a modem that takes a digital bit stream from an attached device (e.g. a personal computer) and transforms it into an analog signal that can be handled by the telephone network. An example in the Smart Grid domain is the case of an RTU that needs to send its digital data to the SCADA (see Chapter 5).
- Channel. With an on-purpose simplification of Telecommunication Systems, a channel can be a single transmission line over any transmission media (e.g. optical fiber, radio – wireless, power line, etc.) or most commonly a complex set of connected telecommunication equipment (performing Transport, Switching and or Routing functions – see Section 2.4.2) and transmission media (see Section 2.6), configuring a complete Telecommunication Network providing the channels.
- Receiver. It is the device that accepts the signal from the channel and converts it into the signal that can be handled by the destination device. For example, a modem will accept an analog signal coming from a network or transmission line and convert it into a digital bit stream for the computer.
- Destination. It is the device that takes the incoming data from the receiver.

Thus, the communication of distant points needs at least a media connecting them and the transmitters and receivers to make it work. When a multiplicity of such telecommunications need to be provided for different end users, the opportunity of a Telecommunication System appears.

The art of sending and receiving signals is called Transmission, defined as "the transfer of information from one point to one or more other points by means of signals" [5], or as "the propagation of a signal, message, or other form of intelligence by any means such as optical fiber, wire, or visual means" [6].

Transmission media used in telecommunications are classified into physical (optical fibers, power lines, and other metallic conductors) and nonphysical (free space). Physical telecommunication media are typically bundled in cables, which provide the required physical conditions to be deployed and cover the long distances required. Nonphysical transmission happens "over the air," using the free space in what is known as radio (wireless).

Last but not least, Traffic concept must be introduced. Traffic in a transmission medium or in a network has to do with the amount of information being transmitted, and it is measured with the traffic intensity parameter. This concept is similar to its common use in road transportation. One of the main challenges when designing a telecommunication system is network dimensioning (i.e. the calculation of the network nodes capacity and the links between them). To determine the needed resources for each service traffic, traffic engineering calculation techniques are used. Traffic intensity, as defined in ITU-T E.600, is measured in Erlang (E) in circuit-switched networks. In data networks, common measurements are based on transmission data rates (e.g. average data rate, in bits per second) and the total data volume (e.g. the amount of Bytes transferred in a day).

2.3.2 Analog and Digital Telecommunications

Analog and digital concepts are often not explained, as it is taken for granted that everyone knows their meaning. However, it is important to understand them, as the impact of digital in telecommunications was disruptive.

The formal definition of analog and digital data in [6] highlights that "analog data implies continuity, as contrasted to digital data, that is concerned with discrete states." More specifically, when dealing with digital telecommunications, as a mean to convey data, when the end points use a telecommunication mean to transmit and receive the data, they make a digital interpretation of the transmitted and received signals as in Figure 2.2. Thus, the information content of the digital signal is concerned with the discrete states of the signal, such as the presence or absence of a voltage, the presence or absence of a light, etc. The details of this will be covered in Chapter 3.

Digital telecommunications have many advantages [7, 8] over analog:

- Improved reception quality, as noise accumulation problems are solved with digital regenerative repeaters.
- Error detection and correction capability through signal processing.
- Increased efficiency of telecommunication channels use (with multilevel digital modulations, higher data rates can be achieved).
- Adaptation to integrated circuits, to manage digital states of solid-state technology.
- Compatibility with network signaling needs and native digital network terminals (i.e. computers).

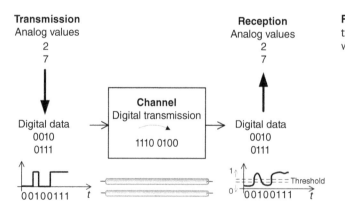

Figure 2.2 Digital telecommunications in an analog world.

Thus, all new telecommunication systems today are digital. However, there are still legacy telecommunication networks that have not completed their digital transformation due to the big impact of replacing or improving the existing infrastructure, and the disruptive effect on end-user devices (e.g. vast parts of the public telephone system, radio, and television broadcasting).

2.3.3 Types of Telecommunications Services

Telecommunication services are provided at different layers (see Section 2.4.3.3) and with different purposes and characteristics.

Services can be classified according to the information inside the messages. Thus, services can be audio, video, or data. Although digitalization has achieved that both audio and video can be transformed into data, there are still analog systems used to transport non-digitalized audio or video. In these telecommunication systems, and in digital ones, the service characteristics needed to carry audio or video (digitalized or not) are different from those of other types of non-real-time data.

Services can also be classified as unidirectional or bidirectional. While most services imply the transfer of information in both directions, there are services such as broadcast services (radio and TV) that only send information from the source and do not expect any information in return. The term [2] *transmission channel* is used to express this as the "one direction" nature of the transmission channel, while the term *telecommunication circuit* alludes to the "transmission in both directions" (bidirectional), as it consists of two channels, one in each direction.

There is another way to refer to unidirectional and bidirectional communications. Although in the past, literature referred to simplex, half-duplex, and full-duplex communications, today [2] refers to simplex and duplex, respectively, to name unidirectional (in any direction) and bidirectional (simultaneous) communications.

Services can also be classified as point-to-point or point-to-multipoint. The former are the classical communications between one source and one destination, while the latter intend to communicate from a source to a multiplicity of destinations. Point-to-multipoint usually assumes that the destinations are in fixed locations [2], while if the destinations are expected to be moving in an area, they are referred to as point-to-area [2].

Services can also be classified according to the nature of the resources that are reserved in telecommunication networks for end-to-end communications. In digital telecommunication networks, we can find circuit-switched data transmission services (also present in analog world) and packet-switched data transmission services. While circuit switching [2] reserves "transmission channels or telecommunication circuits to provide a connection for exclusive use of the users for the duration of a call or service," packet switching [2] takes advantage of the possibility to cut information in packets that are sent to its destination through channels that are not reserved, but shared. Thus "a packet occupies a channel only for its duration of transmission; at other times the channel is available for the transmission of other packets belonging to the same message or to other messages."

Last but not least, telecommunication regulators monitor and control the telecommunication services offered in their countries, both with the objective of developing the telecommunications infrastructure to support the nation's objectives (digitalization to support all economic, political, and social activities), and to control the costs and final prices of such services and the supporting infrastructure with the proper competition framework. Services are then typically classified as retail or wholesale. Retail telecommunication services are typically offered in residential and enterprise markets. Wholesale services are offered among TSPs and to specific industries.

Retail and wholesale services can be additionally classified as fixed or mobile. They are fixed if delivered with telecommunication systems oriented to offer services in fixed locations (e.g. PSTN, xDSL, HFC, or FTTH). They are mobile if they are provided by telecommunication systems supporting the mobility of their users, even if these services are in static locations (e.g. 3GPP technologies).

Retail and wholesale services are also classified based on their use. Thus, they can be telephony services or data transmission-oriented. Data transmission services, when offered in retail residential markets, are usually intended to provide Internet access; however, if they are offered in enterprise markets, they can be used to interconnect company locations and create private enterprise networks. Data transmission services, especially for wholesale markets, can be based on Time Division Multiplexing (TDM), Ethernet, or IP (see Chapter 4). For TDM, different structures will address transmission throughputs within well-defined multiples. In Ethernet and IP cases, a higher degree of throughput modularity exists. TDM services are increasingly being abandoned in favor of Ethernet and IP services. In the context of wholesale services, point-to-point circuits are typically referred to as leased lines.

Although these are the regular telecommunication services offered in retail and wholesale markets, there are others that are common in TSPs' services offerings. SMSs (Short Messages Services) in mobile communications, closed users' group telephony for enterprise segments, satellite-provided services, radio and television services are commonly found in the telecommunications market.

After all this classification, however, the reader needs to bear in mind that many terms in telecommunications, as in other disciplines, are sometimes used loosely. Thus, the terms *Telecommunication Channel*, *Telecommunication Circuit*, and even *Telecommunication Line* [6] are used indistinctively to refer to any connectivity between end points, be it logical or physical, and independently of the nature of the communication or the network supporting it.

2.4 Telecommunication Networks

Telecommunication networks spread over territories to connect end users among them, to provide telecommunication services. Telecommunication networks are hierarchically organized for this purpose and define different network areas and abstraction entities.

The terminology used to refer to this organization of telecommunications is not homogeneous across literature; however, we can refer to two different groups of terminology that, with different origins and implications, are used to refer to the distributed telecommunication network organization.

The traditional world of telecommunication networks was organized around Backbone and Access networks. Backbone was used to refer to the long-distance point-to-point connections of elements that gave way to the Access segment, i.e. the access to end users individually.

With the complexity growth of these networks, the term Backbone has evolved into Core [9, 10] and the meaning of Backbone has remained applicable to the "spine" connectivity of core network elements, with the Core meaning not only interconnection over very distant locations but also the part of the network where common system elements were located. The growth in size of telecommunication networks forced the appearance of an aggregation zone between Access and Core, usually named as Aggregation (or Distribution, at a metropolitan – "Metro") level.

Last but not least, there are some radio-specific terms that can be found in literature, and that are often used loosely. One example of such terms is *Backhaul*, meaning the connection (point-to-point) of Access/Aggregation with the Core of the system [11]. Formally, backhaul communication is the "transport of aggregate communication signals from base stations to the core network" [12, 13]. For the sake of completeness, there is a recent term named *Fronthaul* that is associated with the appearance of "control" system elements also in the Aggregation part.

The influence of computer networks gave origin to another parallel way to refer to the organization of the networks using the terminology of "Area Networks." First, the terms Local Area Network (LAN) and Wide Area Network (WAN) were used (WAN was used to interconnect LANs). With the growth in the size of computer systems, the Metropolitan Area Network (MAN) concept emerged to connect LANs in the same urban area, and subsequently, WAN connected MANs [6, 14–16].

Starting from these three acronyms, others were also popularized at a later stage. Some of them closer to the customer (PAN [Personal Area Network], HAN [Home Area Network], etc.); and some others complementing previous definitions but referring to wireless (radio) solutions for these computer network domains with the addition of the letter "W" in combination with some of the acronyms (the most common being WPAN and WLAN) [17].

Specific to Smart Grids, but with a US and wireless (radio) technologies flavor, two more acronyms have been coined. They are the Neighborhood Area Network (NAN) and Field Area Network (FAN):

- NAN is often associated to Smart Metering, referring to the segment connecting utility premises with the utility WAN.
- FAN refers to the Transmission and, more often, Distribution network interconnection, connecting substations, and the different grid elements for DA purposes.

To finalize this acronym definition, Building Area Network (BAN) and Industrial Area Network (IAN) are mentioned in [18], but with little Smart Grid impact.

A general perspective of how all these definitions match the Smart Grid context is provided in Figure 2.3.

There is a nontechnical classification of telecommunications networks that needs to be mentioned in the Smart Grid context. Telecommunication networks can be classified as private or public.

Public telecommunications networks are those that are authorized to be used to provide services to the "general public." They offer their services between defined network termination points and can interconnect among them. Public is the attribute indicating this, and it does not indicate any

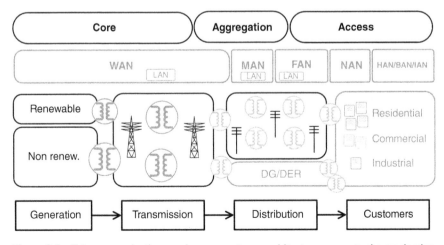

Figure 2.3 Telecommunications and power systems architecture conceptual organization. *Source:* Based on [19].

ownership aspect. Public networks are often also referred to as commercial networks, and examples of them are those providing voice and data (e.g. Internet) connection services.

Private telecommunication networks are those that serve just a predetermined set of users. Again, private does not indicate ownership aspects, but that the services are for self-use to cope with the needs of specific organizations and not for resale. In these networks, the provision of telecommunications services or portions of it includes the construction, maintenance, or operation of the system. Examples of private networks are those created by private companies (including utilities), state enterprises, or government entities.

2.4.1 Network Topologies

Network topology refers to how various telecommunication equipment and interconnecting media on a network are physically or logically arranged in relation to each other. These topologies are sometimes inherent to the transmission media when thinking in physical interconnections terms or can be decided in other occasions.

Physical network topology refers to the physical connections and interconnections among telecommunication devices (nodes) through the interconnection media (e.g. cables). Logical network topology is an abstract creation referred to the conceptual understanding of how data moves through it.

Each topology has advantages and disadvantages, and final arrangements depend on the targets of the different networks and standards. Following are the most common topologies (Figure 2.4):

- Point-to-point. It consists of the connection of one node facing another one. The repetition of these structure provides Line configurations, which are typical of radio-relays to reach distances not covered with a single hop.
- Bus. It consists of the connection of a multiplicity of nodes to a shared media. This topology is generally used when the media cannot be used to separately interconnect individual nodes.
- Star. It consists of a central node that is connected to the rest of the nodes, directly, but to each of them individually. This topology is also referred to as hub-and-spoke.
- Tree. This topology configures hierarchical node interconnections, from the root toward the branches. From the root-node, different nodes depending on it will be subsequently acting as secondary root-nodes for nodes deeper in the topology.
- Ring. This topology is configured through a repetition of point-to-point connections, where the last node is additionally connected with the first one. Ring structures provide resilient network structures that can cope with single failures in network elements or connections.
- Mesh. This topology consists of an eclectic structure of elements interconnected among them with a mix of point-to-point, star, tree, and ring topologies. It is often used in networks where the location of nodes cannot be anticipated in planned network design.
- Full-mesh or fully connected. This topology fully connects each and every node with all the rest of the nodes. This topology can be found in data centers where a very high redundancy is needed.

2.4.2 Transport and Switching/Routing Functions

Any modern telecommunication system combines transport and switching/routing functions in its network devices. In some cases, these functions are implemented in different network devices or can be found integrated in multipurpose equipment:

- Transport function is present to carry the signal between two network nodes, in a point-to-point or line fashion to cover large distances. Furthermore, ring structures usually exist to create resilient transport networks.

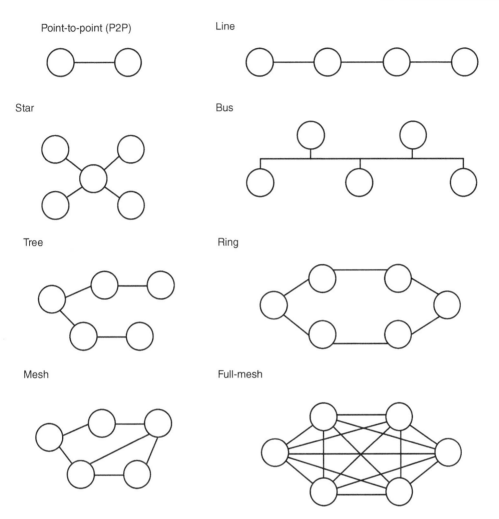

Figure 2.4 Network topology types.

- Switching/Routing functions are needed when the information must go through one or several network nodes between the transmitter and the receiver, and different paths can dynamically be taken. Network nodes with these functions "switch" or "route" the information selecting the path to reach the final destination.

Transport typically defines fixed routes that are manually and *ex-ante* configured in the network. All data involved in a transmission of information between end points follow the same route.

Switching/Routing is used when the origin and the destination are known, but the data packets can follow different routes to reach that destination. Each packet can potentially follow a different route, given that the packet reaches its destination, even if not in the same order as it was sent. Switching/Routing connects end points based on devices' addresses, from the origin to the destination (usually on top of the Transport function).

Transport and Switching/Routing functions can be compared with road transportation. A car driving through a road connecting departing point A and a destination B can drive following different routes, and when reaching a crossroad, must select the appropriate path to reach

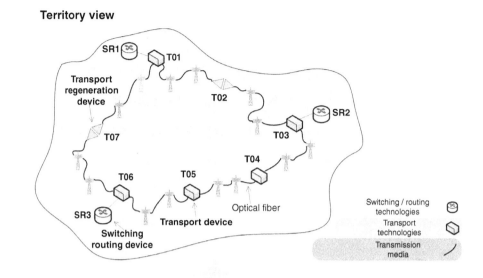

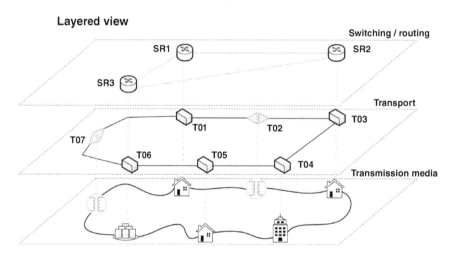

Figure 2.5 Transport and Switching/Routing are complementary.

destination. Each road segment "transports" him to the next road node or junction where he needs to switch (select its route) to the next road segment in its path toward destination.

Transport and Switching/Routing are complementary technologies (see Figure 2.5) that are usually implemented with different networks and devices. However, they are sometimes integrated in the same network element with the advent of digital processing capabilities [20, 21]. Keeping them as different networks maximizes the flexibility in delivering telecommunication services.

2.4.3 Circuit-switched and Packet-switched Networks

In order to effectively connect end points within a telecommunication network, these end points must be provided paths or connection routes through which they will be exchanging data. This can be achieved with circuit switching or packet switching methods.

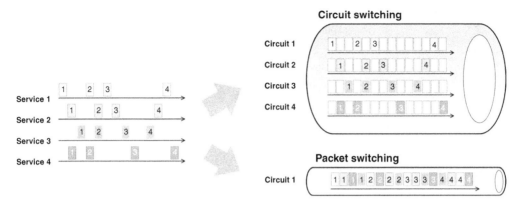

Figure 2.6 Traffic efficiency in circuit or packet-based networks.

Circuit switching was the original way to achieve this connectivity. As an example, the telephone system uses circuits to connect exchanges where telephones are attached. This circuit-based connection is static, as the fixed telephone service had very clear and static service (coverage) targets and was highly deterministic and predictable. In circuit-switching, several logical connections or circuits are established over the physical links of the network nodes, so that data flows through those dedicated circuits in a very fast way. In circuit switching, data is not usually presented as packets, but fitting into specific structures.

Packet switching was an improvement to better accommodate data traffic as packets (not just voice, or even not voice) in telecommunication networks, in a more traffic-efficient way (see Figure 2.6). Rather than defining fixed and rigid circuits, packets move from node to node until they reach their destination. Packet switching may come with a higher latency, as packets need to be analyzed one way or the other, at every switching network node.

Telecommunication devices switching circuits or packets are different. Circuit switches are position-based, and bits arriving in a certain position are switched to a different position based on a combination of physical aspects (interface, time, wavelength, etc.). Packet switches are label-based, and they use information in packet headers (labels) to decide how to switch the packet.

2.4.3.1 Circuit-switched Technologies

Circuit switching (Figure 2.7) was the dominant technology in voice communications since the invention of the telephone. However, permanently established circuits may not be very efficient in other domains such as computer communications, since the resources for the connection are being reserved even if there is no current communication on it.

2.4.3.2 Packet-switched Technology

Packet switching came to solve the inefficiency of permanent circuits for most data traffic. Packet switching has its origin in LANs and evolved to be eventually used in WANs. Due to the adaptation to different environments and needs, they have incorporated auxiliary functions to their basic core of identifying the next hop in the path of a packet toward its destination and today include other features such as packet prioritization, aggregation and disaggregation, etc.

Figure 2.8 shows how packet-based traffic is organized to connect end devices.

52 | *2 Telecommunication Networks and Systems Concepts*

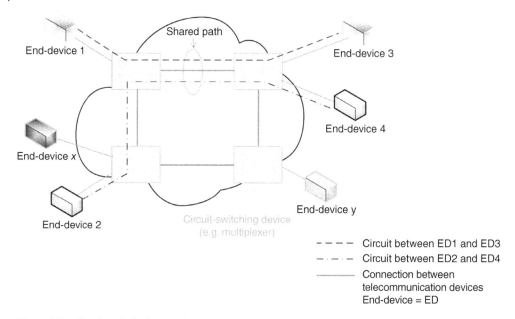

Figure 2.7 Circuit-switched network.

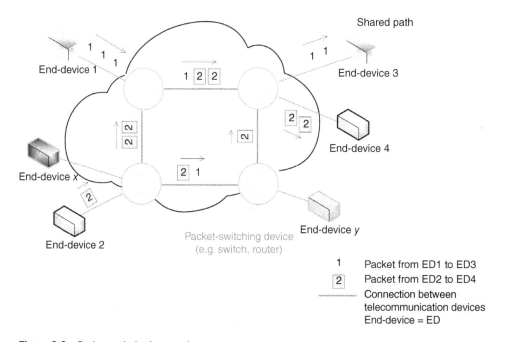

Figure 2.8 Packet-switched network.

Given the difficulties inherent in finding the most efficient way to compute optimal paths for message delivery, different strategies and algorithms exist (see Chapter 4). The minimum elements packet-based technologies include are as follows:

- Packet Data Unit (PDU) formats, with header, payload, and checksum (see Section 2.5.2).
- Control PDUs, to implement the information that needs to be exchanged to make the protocol work.
- Segmentation and reassembly procedures, to create, e.g. shorter length packets, and enable rearrangement with the proper order at the destination end point.
- Acknowledgment procedures, e.g. to inform the source that the packet has been correctly received.
- Error control and recovery procedures, to detect and correct bit errors in the information conveyed.

Packet switching can be classified as connection-oriented and connectionless [22]:

- Connection-oriented work mode or "virtual circuit" approach, when circuits are created to carry data. In connection-oriented packet technologies, a "virtual path" is created among the network elements involved in the path between origin and destination prior to data transmission. The path does not reserve network resources permanently (it is created and dissolved before and after the transmission), and all packets follow the established path.
- Connectionless work mode (or datagram approach) does not pre-establish any route or resource, and all packets follow the best possible path at any point in time. Packets reach the destination at different times and possibly out of sequence, so that they need to be rearranged.

Connection-oriented mode involves some phases:

- Connection establishment. Before the information is transmitted, a physical end-to-end circuit must exist between both parties. This means that all the nodes in between the transmitter and receiver must synchronize to provide this physical connectivity.
- Data transfer. Once the connection or circuit is established, information transmission can start.
- Connection release. After the data has been transferred, all the resources used in the connection establishment phase need to be released to be used for another transmission.

Connectionless packet switching divides the message to be sent into pieces (packets) and includes control and labeling information for the packet to be addressed to the receiving party. Thus:

- Network efficiency is higher since the same physical link can be dynamically shared between two or more transmissions. Packets are stacked in a queue and are transmitted as soon as resources are available. On the negative side, the extra control information of packets adds individual packet overhead.
- Transmission times are optimized since nodes with higher speeds can communicate with each other at higher data rates.
- Traffic increase makes the network react differently in connection-oriented and connectionless networks. The former may block (no more "circuits" can be established). The latter may manage congestion. On the contrary, the latter affects latency and may prevent the operation of real-time traffic.
- Priorities can be implemented, so that if a node has several packets queued for transmission, it can transmit those with the highest priority first. These packets will thus experience a lower delay than those of lower priority.

2.4.3.3 Multilayered Telecommunication Networks

Telecommunication networks are organized in layers. These layers are a mix of functions, technologies, media, and/or services and do not follow strict rules. The layers referred to in this section are different to the layers of the models covered in Section 2.5.1, from a conceptual perspective.

Layers intend to convey a logical perspective of the different domains to which telecommunication elements belong, while at the same time they provide a sense of physical location and interconnection.

Figure 2.9 shows an example of how telecommunication networks are arranged. Every TSP or private company will decide which are the layers of the network that better suit its objectives and the means and technologies for each layer. Then, it will select which topology is best for each layer and segment of the network, and how they are integrated:

- At physical level, the most appropriate and available transmission media will be selected to achieve the highest degree of connectivity and redundancy in network points.
- At transport level, different technologies to cover service needs and networks segments will be chosen and interconnected. Both legacy and future-looking networks and services may affect how the network evolves and transitions to adapt to technology evolution.
- At switching and routing level, the different network domains, and the number of services, and its performance needs will make different technologies to be combined in different ways to allow an optimal network design that favors scalability and service delivery control.

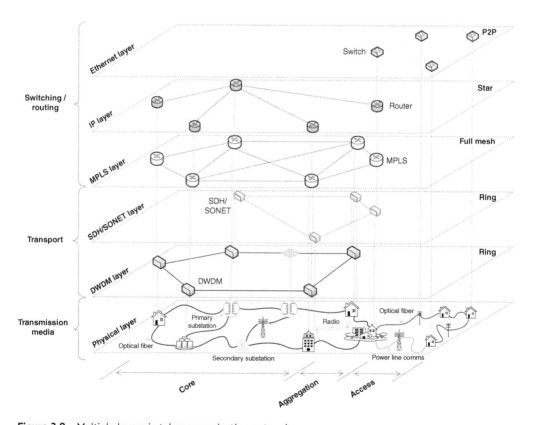

Figure 2.9 Multiple layers in telecommunication networks.

2.4.4 Telecommunications Networks and Computing

A final reference needs to be made to some concepts that are relatively new to telecommunication systems but that, coming from the computer science field, are gaining relevance with regard to achieving the low latencies that some services need, and complementing the IoT world. Cloud, Fog, and Edge computing appear as the way to create an application processing layer, with computing power and storage capacity in different locations, evolving from a purely centralized architecture to a distributed close-to-the-edge one, to host the computing needed by certain applications in connection with network services.

From a historical perspective, Cloud computing started as "a model for enabling ubiquitous, convenient, on-demand network access to a shared pool of configurable computing resources (e.g., networks, servers, storage, applications and services) that can be rapidly provisioned and released with minimal management effort or service provider interaction" [23]. Cloud replaces private non-scalable computing resources and became a major market trend. However, the centralization of computing resources does not allow the low latencies needed in some services (e.g. video analytics, augmented reality, etc., for which 4G and specially 5G 3GPP specifications try to offer services). Stemming on this need and probably due to the loss of market space that TSPs have suffered in Cloud-based services, some computing initiatives take the computing power further down to the telecommunications network edge.

Edge computing comes as the answer from TSPs to provide the computing resources needed in the network edge, to cope with low and ultralow latency services defined in their network specifications to serve specific industry needs, e.g. industry (factory) automation. Within Edge computing initiative, MEC (Multi-access Edge Computing) sticks out as a relevant one in the TSP space, especially among mobile cellular communications, within 3GPP technologies. MEC [24] started with its "M" meaning mobile. MEC offers "cloud-computing capabilities and an IT service environment at the edge of the network. This environment is characterized by ultra-low latency and high bandwidth as well as real-time access to radio network information that can be leveraged by applications."

Right between the Cloud and the Edge [17], and seemingly agnostic to wireline or wireless technologies, Fog computing [25] addresses a "resource paradigm that resides between smart end-devices and traditional cloud or data centers" [23]. Fog computing defines itself as a hierarchically organized computing resources architecture, intended to minimize the request–response time from/to supported applications, to provide local computing resources for the end devices and, when needed, network connectivity to centralized services.

Fog computing includes the reference to the Mist computing concept as well, where "mist" refers to specialized-dedicated nodes that exhibit low computational resources and are intended for geographically dispersed locations, low-latency computational resources even closer to the smart end devices they serve (often sharing the same locality). In Fog computing vocabulary, the Edge computing is not to be confused with the Edge computing concept in MEC, as for Fog computing, it just refers to the network layer of the end devices and their users.

2.5 Protocol Architectures for Telecommunication Networks

2.5.1 Why a Protocol Layered Model Is Needed

The organization of the telecommunication elements involved in the communication among different network elements as described in Section 2.4 is not the only challenge in end-to-end communications fulfillment.

The procedures involved in the exchange of data between devices connected with such elements need to be considered and included in a comprehensive model to achieve seamless scalable and interoperable end-to-end communications.

An example may be used to describe the type of procedures that need to be included in this model, using the case of a file transfer between two computers, or a field device with a central server. The example considers that there is a physically direct path between both entities, through the network. Then, the following actions should be in place:

- The transmitting party needs to provide the telecommunication network with the identification of the desired destination party.
- The transmitting party must be sure that the destination is not only ready to receive data (i.e. the receiver circuit must not be busy with other tasks) but also to capture the file.
- The format of the file needs to be understood by both parties.

To make the complete process work, there must be a high degree of cooperation between the devices involved in the two communication ends. The way how this is orchestrated is by following the "divide and conquer" principle: these tasks are divided into subtasks, each of which is performed in separated modules (software or hardware).

The way a protocol architecture is typically analyzed is by arranging the modules in a vertical stack with a variable number of layers. Each layer in the stack performs the subset of interrelated tasks that are necessary to communicate with the other system, at that same layer.

In general, the most basic functions are left to the layer immediately below, so that the upper layer is not involved in the details of these functions. Most importantly, each layer provides a set of *service points* to the layer immediately above it. Ideally, the layers are defined in such a way that changes in one layer do not require changes in the others.

2.5.2 The OSI Model

In 1984, the ISO (International Organization for Standardization) published ISO 7498, which defines the layers or levels that should form a protocol architecture for communication between devices. This structure is known as the OSI reference model and can be found in ISO/IEC 7498-1:1994, "Information technology – Open Systems Interconnection – Basic Reference Model: The Basic Model."

ISO defined the set of layers, as well as the services to be performed by each one of them. The division groups the functions that were conceptually similar in a way that each layer is small enough, to avoid processing overloads. The resulting reference model has seven layers (Figure 2.10). These layers provide services among the adjacent ones, and the way to exchange them is through primitives and parameters (the typical primitives are request, indication, response, and confirm). The main functions of each layer are described below [26]:

- Physical Layer. It is the physical interface between devices. It defines aspects that go from the mechanical description of the interface connectors to the modulation and media access to be used (see Chapter 3). This layer includes the telecommunication supporting means in Sections 2.6 and 2.7.
- Data-Link Layer (DLL). While previous layer takes care of the mechanisms for transmitting bits, the data link makes this transmission reliable. The main function for this layer is the detection and correction of errors. The layer above the DLL can be totally confident that the information delivered to it is completely reliable and error-free. DLL basic functions can be divided in two groups: multiple access and resource sharing, and traffic control.

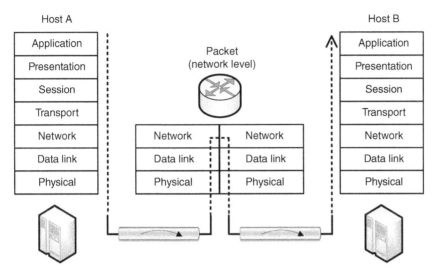

Figure 2.10 Representation of the OSI layered model.

- Network Layer. It frees the upper layers from the need to know about the path to take to reach a specific device when it is not physically connected to it (i.e. is in another network). Data is organized in units of transmission usually called PDUs. These PDUs consists of Header (the information added by this layer to make the PDU reach the remote destination), Payload (the information itself, coming from the upper layer), and Checksum (a group of bytes used to verify the integrity of the received).
- Transport Layer. It provides a mechanism for exchanging data end to end. The transport service ensures that the data is delivered in order and without loss or repetition.
- Session Layer. It provides the mechanisms to control the dialogue between end-to-end devices. This may be necessary given the different type of interactions devices may have. For example, a remote terminal accessing an application may require a half-duplex dialog while, at the same time, another message processing application may require the possibility of interrupting the dialogue, generating new messages and, later, continuing the dialogue from where it was interrupted. These two simultaneous dialogs must be done without mixing up the messages.
- Presentation Layer. The presentation layer defines the format of the data to be exchanged between applications. The presentation layer defines the syntax to be used in the communication. This is the case for services including data encryption and/or compression.
- Application Layer. Finally, the application layer simply provides a means for software programs to access the OSI stack.

Note that, except for the physical layer, there is no direct communication between the peer layers. That is, above the physical layer, each protocol entity passes the data toward the adjacent lower layer, so that it sends it to its peer entity. Thanks to its design, the OSI model does not require any two parties to be directly connected, but can communicate through a telecommunication network.

2.5.3 The TCP/IP Protocol Stack

The TCP/IP protocol architecture is the result of research and development carried out in the ARPANET experimental network, funded by the Defense Advanced Research Projects Agency (DARPA), and is known globally as the TCP/IP family of protocols. It consists of an extensive

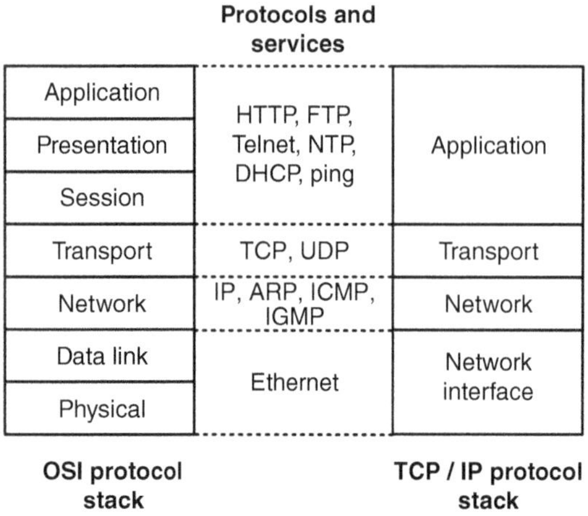

Figure 2.11 Relationship between OSI and TCP/IP layer functionalities.

OSI protocol stack	Protocols and services	TCP / IP protocol stack
Application	HTTP, FTP, Telnet, NTP, DHCP, ping	Application
Presentation		
Session		
Transport	TCP, UDP	Transport
Network	IP, ARP, ICMP, IGMP	Network
Data link	Ethernet	Network interface
Physical		

collection of protocols that have been specified as Internet de-facto standards. The Internet Society (ISOC), the Internet Architecture Board (IAB), the Internet Research Task Force (IRTF), and the Internet Engineering Task Force (IETF) are the core groups behind Internet standards [27].

Although the TCP/IP protocol stack tries to gather the same functionalities that the OSI model does, the former defines a lower number of layers by grouping some of the ones defined by OSI. As shown in Figure 2.11, TCP/IP divides the problem of communications in several different layers:

- Network Interface. It includes the Physical Layer and the DLL. The definition of this layer is very similar to the corresponding ones in OSI model.
- Internet Layer. The Internet layer makes sure that communication occurs between two parties that are not physically connected to the same network. The protocol implemented in this layer provides the routing information to navigate through the interconnected networks to reach the final destination. This protocol is implemented in both the receiving part and in the intermediate systems that are used to route the information. The protocol implemented in this layer is known as Internet Protocol (IP; originally defined in IETF RFC 791).
- Transport Layer. The transport layer transmits data in a reliable and orderly manner. This is, all data must reach the receiving party in the same order that they were transmitted. One common protocol used in this layer is the Transmission Control Protocol (TCP; originally defined in IETF RFC 793). This protocol makes sure that no part of the information is lost in the way thanks to the use of a retransmission mechanism. In contrast, other transport layer protocol is the User Datagram Protocol (UDP; originally defined in IETF RFC 768), which does not guarantee the delivery of the information. This UDP protocol, however, is characterized by a low computational complexity, making it very suitable for low-latency applications.
- Application Layer. Finally, the application layer is the logic needed (software) to support user applications (web browsing, file exchange, etc.).

The useful information or payload from upper layers is increasingly encapsulated into larger data structures that add information in the form of headers (addresses, sequencing, error correction, etc.) to perform its function, as in Figure 2.12.

The OSI model was developed even before protocols were designed and implemented. This means that the model is based on a theoretical exercise, mostly abstract. Therefore, the specifications

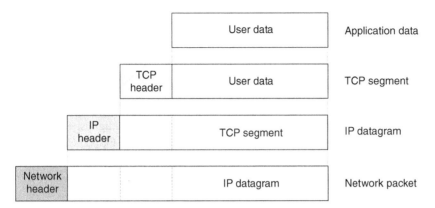

Figure 2.12 Data structures in the TCP/IP layered model.

given in the model were rather general. In contrast with this, in TCP/IP model was a practical description of the already existing protocols and implementations, which were well suited for the scenario they were being applied to.

Nowadays, all the layers defined in the OSI model serve as a good reference to detail the different actions that need to be done to implement a robust and reliable communication between two entities. However, solutions today take the ideas inside the OSI model and apply them into a specific scenario without strict adhesion to "the 7 layers."

2.5.4 User, Control, and Management Planes

The protocol architecture of a telecommunications system is commonly arranged in three separate "planes" to serve different purposes: user plane, control plane, and management plane. Each plane is associated with a different type of traffic and serves a different function.

The user plane (sometimes known as the data plane, forwarding plane, carrier plane, or bearer plane) carries the network user traffic, i.e. the actual information that needs to be sent (e.g. voice, data, etc.). The control plane carries signaling traffic, i.e. data necessary to make user traffic to reach its destination (e.g. control messages to activate the user plane services, control messages to establish the routing paths, etc.). Finally, the management plane carries traffic related to the configuration and monitoring of the network nodes (e.g. performance reporting, alarms events, etc.).

These three types of traffic can be multiplexed over the same transmission links or, less commonly, an additional infrastructure can be deployed given its different transmission requirements. As an example, in optical communication systems (see Chapter 3), a specific wavelength is reserved for the management traffic.

2.6 Transmission Media in Telecommunications for Smart Grids

Any transmission of data between distant locations needs a physical medium support on top of which it can be established. Thus, the foundation of telecommunications lays on the availability of physical media to allow this information transfer to happen.

There is a range of different physical media that can be used in telecommunications. They can be grouped in two broad categories, i.e. wireless and wireline:

- Wireless, radio, or radio communications, use the "air" or "free space" as the "physical" medium.
- Wireline uses optical fiber, coaxial and twisted pairs, and power lines cables.

Depending on the physical media, telecommunication transmission presents different properties:

- Propagation speed. It is the physical velocity derived from the propagation of light speed. In free space (air, i.e. with radio communications), this speed is approximately 300 000 km/s; other media exhibit a lower propagation speed.
- Transmission rate. It is the result of the physical media utilization of the low layers of telecommunication networks. It represents the quantity of information that can be delivered, usually expressed in bits per second.

Physical media are used within the Physical Layer of the OSI model. The main characteristics of the telecommunication transmission, fixed by this layer, are as follows:

- Frequency. It defines the central frequency or frequencies of the transmission.
- Bandwidth. It refers to the range of frequencies around the central frequency that are involved in the transmission. With F as the central frequency and B as the bandwidth, the occupation goes from $F - \frac{B}{2}$ to $F + \frac{B}{2}$. This parameter is directly correlated with the transmission rate.
- Modulation. It is the way in which frequency and bandwidth are used to effectively transmit data.
- Data rate (i.e. transmission rate). It is the result of the application of previous parameters over a certain physical medium, determining in digital telecommunications the speed at which information can be transferred.

Although many media are still used as the support for telecommunications transmission in the utility space, this section focuses on the two main ones, i.e. optical fiber (wired) and radio (wireless). The rest of them will just be mentioned, because, even if they are relevant, they are not so important in terms of presence, uses, and future evolution in Smart Grids. Section 2.7 will cover electricity cables, as the communication supporting mean with PLC technologies.

Metallic cables were the first ones used in wireline transmission (telegraphy). From the old simple data signals, to pervasive telephony applications and modern broadband connectivity, metallic cables are used in short spans and tend to be avoided in applications and premises where safety might be compromised due to the presence of HV electric assets. Their use today is limited to local and access connectivity.

There are two types of metallic means used:

- Multipair copper cables. Simple to install and relatively low-cost. They are normally used for short and medium-length connections (hundreds of meters or even a few kilometers), for both voice and data communication services. They are used in pairs, twisted or untwisted, and these pairs are bundled in cable constructions grouping them in quantities from a few to hundreds of them. When copper pairs are close to each other, the interference ("crosstalk") among them needs to be considered, as it will limit transmit power and will increase noise floor[1] and affect effective reach. Thicker conductors favor propagation, and hence the common reference to the classical American Wire Gauge (AWG) [28].

---

1 Noise floor refers to the addition of all unwanted signals and determines the lowest possible signal that can be detected.</note>

- Coaxial cables. These cables are less popular in utility industry, but common in certain access applications in TSPs. These cables started to be used in long-distance communications when optical fiber cables did not exist and for TV applications in close circuit installations. Their transmission (reach and bandwidth) characteristics are better than those of copper pairs, but their cost is higher. They consist of a central metallic wire, surrounded by a metallic screen fitted over a dielectric material that separates central and shield conductors. Coaxial cables typically follow the Radio Guide (RG) denomination [29] (e.g., RG-11, RG-6, and RG-59).

When discussing local area connectivity applications, metallic cables have been grouped together with optical fiber cables in specific cable standards to allow for the deployment of broadband local networks. Thus, a uniform set of infrastructure elements (cables and ancillary elements), installation instructions, and their corresponding tests have been defined to guarantee an agreed performance. ISO/IEC 11801, EN 50173, and ANSI/TIA/EIA 568-B are widely used in local area cabling (e.g. buildings). These standards define "categories" to achieve a certain system performance expressed in terms of data rate and applications supported. MICE concept, included in these standards, refers to environmental conditions to be supported by the cables once deployed (M = Mechanical rating, I = Ingress rating, C = Climatic/chemical rating, and E = Electromagnetic rating).

2.6.1 Optical Fibers

Optical fiber-based telecommunications, based on the transmission of data as light pulses through plastic or glass cores, have been a breakthrough in bandwidth and range extension of traditional telecommunications. [30] heralded back in 1973 that "the advent of low-loss optical fibers brings new dimensions to optical communication prospects. Fibers may soon be used much as wire pairs of coaxial cable are now used in communication systems." It declared that transmission losses "as low as 2 dB/km" and that "error rates of 10^{-9} at 300 Mbps has been experimentally achieved." It was a breakthrough at the time, humble compared with the achievements today, and far from what the future state of the art will offer.

Optical fibers are made of glass (silica) core surrounded by a layer of glass called cladding, of purer silica. Light (lightwaves) is conducted in the core with a minimum of attenuation (loss of signal) through a mechanism that is commonly explained using the principle of total internal reflection of the light in the core with the wall of the cladding (see Chapter 3). Although there are several types of optical fibers, Single Mode (SM) ITU-T G.652 is the most common and useful one for long-distance transmission.

Fiber optics communications research started in 1975, and its evolution can be divided into six stages [31]:

- The first phase of fiber optics systems operated near the wavelength of 850 nm (called the first window; see bands in Table 2.1). These systems were commercially available by 1980 with MultiMode (MM) fibers (typically used in LANs), and their reach extended up to 10 km with a transmission speed of around 10 Mbps. ITU-T Recommendations G.651 and G.956 (now G.955) standardized them.
- The second phase of fiber optics systems operated near 1300 nm (called the second window). These systems were commercially available by the early 1980s, initially limited to bit rates below 100 Mbps due to the dispersion in MM fibers. With the development of SM fibers (extensively found now in long-range, tens of kilometers, applications), these systems reached distances of around 50 km and 1.7 Gbps data rates by 1988. Some relevant ITU-T Recommendations are G.652, G. 957, and G.955.
- The third phase of fiber optics systems operated near the 1550 nm wavelength (called the third window), where attenuation can get minimal values (0.2 dB/km). These systems were commercially

Table 2.1 ITU-T optical transmission windows.

Designation	Name	Wavelength (nm)
850 Band	850 Band	810–890
O-Band	Original band	1260–1360
E-Band	Extended band	1360–1460
S-Band	Short-wavelength band	1460–1530
C-Band	Conventional band	1530–1565
L-Band	Long-wavelength band	1565–1625
U-Band	Ultralong-wavelength band	1625–1675

Source: Horak [32].

available by 1992 with dispersion-shifted fibers. Distances of around 80 km and 10 Gbps data rates are achieved. Some relevant ITU-T Recommendations are G.653, G. 957, G.955, and G.794.
- The fourth phase of fiber optics systems operated also near the 1550 nm wavelength making use of optical amplifiers to increase distances. Wavelength Division Multiplexing (WDM) techniques multiplied data rates. By 1996, distances of 11 600 km were reached with 5 Gbps data rates for intercontinental applications (obviously with optical amplifiers). Some relevant ITU-T Recommendations are G.655, G.694, and G.695.
- The fifth phase of fiber optics systems included new initiatives to enlarge the transmission capacity of optical fibers with more Dense WDM (DWDM) channels reducing channel spacing; to make the use of some other optical bands possible; to increase the capacity of individual channels up to 100 Gbps; and to reduce the number of optical/electrical/optical conversion needed. Some relevant ITU-T Recommendations are G.656, G.672, and G.680.
- The sixth phase of fiber optics systems includes the transition from the classical digital modulation schemes (an electrical binary bit stream modulates the intensity of an optical carrier inside an optical transmitter – on–off keying, OOK, that is eventually directly detected by the receiver – Intensity Modulation with Direct Detection, IM/DD) to coherent lightwave systems with phase modulation of optical carriers and with homodyne or heterodyne detection. Phase encoding achieves an improved sensitivity of the optical receivers allowing an increase of the length of the regeneration section, a more efficient use of the fiber bandwidth, and enhanced Forward Error Correction (FEC) that with the use of Digital Signal Processors (DSPs) can afford to dispose of Dispersion Compensating Modules (DCM) and the attenuation and nuisance they include. Channels with bit rates up to 400 Gbps and above have also started to be studied.

Optical fiber is the only telecommunication medium that combines high transmission bandwidth, long-distance reach, and a demonstrated guarantee of continuous expansion of capacity and range. Therefore, and in a context where investments have a multi-decade horizon, and where Smart Grids need to adapt to requirements and services needed for the still-to-be-developed grid equipment and utility applications, optical fiber is a secure investment. Many utilities worldwide have managed to develop a business out of their optical fiber network with third-party telecommunication companies [33, 34].

Finally, optical fiber is a nonconductive material. This is highly advantageous when the transmission medium needs to be integrated in HV premises with EMC constraints, which may create disturbances in conductive metallic media. In this context, the telecommunication network can be fully insulated and comply with safety considerations.

2.6.1.1 Optical Fiber Cables for Smart Grids

Cables are the most usual and pervasive asset in utilities. Cables, as a constituent of power lines, can be found suspended on a range of tower and poles (depending on the voltage level within the Transmission or Distribution segments) or walls or underground (most often inside ducts and galleries). The existing rights of way can be used (and are used so by many utilities and TSPs to support optical fiber cables). There are many legislative pieces fostering this use for the expansion of electronic communications, e.g. in Europe [35] and in the United States [36–38]. The intention is to reduce the cost of deploying high-speed electronic communications networks in general and recently to promote broadband deployment in the evolution toward 5G.

There are several types of optical fiber cables, exclusive of utilities, that are found suspended in utility towers and poles. The three first ones are the most common types (OPGW, ADSS, and OPPC, in this order), while the other do have their niche. While overhead cables are unique and different in their composition and construction, there are not so many differences with underground categories:

- OPGW or OPtical Ground Wire.
- OPPC or OPtical Phase Conductor.
- ADSS or All-Dielectric Self-Supporting.
- OPAC or OPtical Attached Cable.
- MASS or Metallic Aerial Self-Supported.

IEEE, IEC, and ITU have standardized the different aspects of this utility specialty cable industry. While electric power cables and optical fiber cables have each their own well-defined markets, utility optical fiber cables are in the intersection of both. Thus designs, construction, and cable-laying need to consider their respective constraints. Apart from the specific communications-related characteristics of the optical fiber carried by the cables (general to all optics industry, and that are not special for utilities), all these standardization bodies have developed specifications to define cable composition and construction, testing procedures, laying techniques and precautions, and operation and maintenance procedures.

OPGW is defined in IEC 60794-1-1 as a "metallic optical cable that has the dual performance functions of a conventional ground wire with telecommunication capabilities." It is also defined in IEC 62263 as a "stranded metallic cable incorporating optical fibres which has the dual performance function of a conventional earthwire with telecommunication capabilities." OPGW (Figure 2.13) is one of the most representative optical fiber cable types for utilities and a very common solution in the HV domain and may be found as the "de facto" cable to be installed in HV

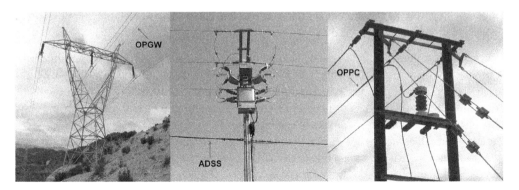

Figure 2.13 Optical fiber cables.

power lines. IEC 60794-4-10, IEEE 1138, and IEEE 1591.1-2012 are complementary specifications for this cable type in full cable systems. For example, IEEE 1138 includes functional requirements, such as electrical, mechanical, optical fiber, environmental and packaging, and test requirements related to design, installation, in-service, and maintenance, including routine tests. IEEE 1591.1-2012 is the hardware test standard; hardware for OPGW is identified as hardware in direct contact with the OPGW cable.

ADSS is defined in IEC 60794-4-20 as a full nonmetallic "dielectric cable that is capable of enduring aerial installation and providing long term service, without any external tensile support." ADSS (Figure 2.13) consists of several protective dielectric fiber optic units surrounded by or attached to suitable dielectric strength members and sheaths. IEC 60794-4-20 covers the construction, mechanical, electrical, and optical performance, installation guidelines, acceptance criteria, test requirements, environmental considerations, and accessories compatibility for an ADSS cable. IEEE 1222 has a similar scope. Additional requirements related to ADSS cable hardware and cable/hardware compatibility are addressed in IEEE 1591.2. ADSS cable is designed to be fastened to power line towers underneath the power conductors, and it is quite a versatile solution when OPGW or OPPC cables cannot be used. ADSS may seem similar to other nonutility dielectric cables, but it is not so due to the effects that are found in HV power lines (e.g. corona discharge effects) and the span or distances to be covered by the cable. The combination of these aspects with the design rules applicable to power lines (regarding the performance of the towers and cables – electric or not – as a durable and resilient system) makes them different from those overhead dielectric cables in telecommunications industry.

OPPC (Figure 2.13) is defined in IEC 62263 as a "stranded metallic cable incorporating optical fibers which has the dual performance function of a phase conductor with telecommunication capabilities." OPPC cables allow for the like-for-like replacement of existing cables in towers, without increasing the requirements to be supported by the system or changing the established easement. However, the fitting needed to access the optical fibers and thermal considerations of the cable are common drawbacks of the solution.

OPAC is defined in IEC 60794-4-20 as a "dielectric, not self-supported, optical attached cable." There are different OPAC type cables, depending on the attachment method:

- Wrapped. It is an all-dielectric lightweight flexible cable, wrapped helically around a conductor (earth wire or the phase conductor) using special machinery. Another name for this cable is Wrapped Optical Cable (WOC) [39].
- Lashed. It is a nonmetallic cable installed longitudinally alongside any wire supported on the towers (the earth wire, the phase conductor, or a separate support cable) that is held in position with a binder or adhesive cord.
- Spiral attached. It is an attachment similar to the lashed cables, where the attachment involves the use of special preformed spiral clips.

MASS cables are defined in IEC 60794-4- as "metallic aerial self-supported cables," and in ITU-T L.26 as "metal armour selfsupporting." A MASS cable is a compact, and generally lightweight solution with no electrical function, designed to provide a telecommunications solution without interfering with the existing power lines or electric infrastructure and minimizing loading. It is usually installed beneath the live phases to ease operation and maintenance.

Underground cables may be laid in shared ducts with power cables. When they are installed near a HV power line, physical separation must exist and, according to ITU L.101/L.43, "a special sheath material should be considered to avoid tracking effects." Typical solutions include the use of a semiconductive over-jacket or a track-resistant jacket compound.

2.6.1.2 Optical Fiber Cables Specifications

Aerial optical fiber cables along electrical power lines specification are well covered in IEC 60794-4 (ITU-T L.26, with the same purpose, does not cover OPGW or MASS). This specification covers cable construction, test methods, and performance requirements for aerial optical fiber cables and their component. Performance requirements include all aspects affecting functionality of the cables, i.e. optical, mechanical, environmental, and electrical. Optical fiber cables carry and protect the optical fibers inside and must be designed and manufactured for a predicted operating lifetime of at least 20 years in all its parts.

2.6.1.2.1 Optical Core

The specification defines "cable elements" as the individual constituents of the different cable designs. These cable elements are adapted in each cable type to its application, operating environment, and manufacturing processes, as well as to the need to protect the fiber during handling and cabling.

The cable "optical core" is the unit where optical elements are found. Optical elements are the cable elements containing optical fibers. Optical elements and each fiber within a cable element shall be uniquely identified (colors, positions, markings, tapes, threads, etc.) so that individual access to fibers is eased.

There are different types of optical elements:

- Ribbon. Optical fiber ribbons are assembled optical fibers. Optical fibers are assembled side by side to be arranged into a strip.
- Slotted core. It is the result of the extrusion of a suitable material (for example, polyethylene or polypropylene) with a defined number of slots, providing helical or SZ configuration along the core (alternating twisting directions, or reverse oscillating stranding as explained in ITU-T L.26). Optical fibers or optical elements are located in each slot. The slotted core usually contains a metallic or nonmetallic central element.
- Polymeric tube. Optical fibers or optical elements are packaged in a tube construction. The plastic tube may need to be reinforced.
- Metallic tube. The tube may be applied over the optical core or can hold optical fibers directly.

The optical elements and the cable core shall contain a filling water-blocking compound, such as grease-like and/or dry-block materials.

To produce the cable optical core with the needed number of optical fiber pairs, optical elements may be laid up as follows:

- Several optical elements without a stranding lay (e.g. a single optical unit in the cable center).
- Several homogeneous optical elements using helical or SZ stranding configurations; in the case of ribbon elements, they may be laid up by stacking several elements.
- Several different configurations in a slotted core (ribbons or plastic tubes).
- Several conductors laid up with the optical elements.

2.6.1.2.2 Cable Strength Members and Sheaths

The core must be accompanied by the elements (strength members) that can provide the needed mechanical and thermal requirements of the overhead lines.

OPGW, OPPC, and MASS cables use stranded wires for armoring purposes and may be round or show other cross-sectional shapes. They can be of different materials including aluminum alloy, galvanized steel, aluminum, and aluminum-clad steel. Corrosion aspects of these metals need to be considered.

ADSS and OPAC strength member consist of aramid yarns, glass-reinforced materials, or equivalent dielectric strength members.

ADSS and OPAC cables need a sheath, and it is common to find an inner and an outer sheath. The inner sheath is usually made of polyethylene. One of its objectives is not to expose the optical core to the mechanical tensions of the cable. The outer sheath needs to be seamless and is usually made of ultraviolet-stabilized weather-resistant polyethylene. It is common to use a tracking-resistant sheath to, jointly with a proper engineering of the position of the conductors, minimize corona effects.

2.6.1.2.3 Cable Specifications and Tests

Each utility specifies its needs depending on many factors particular to its grid, and as a result, even if there are few cable types, there are multiple individual cable specifications and designs that are not equal among utilities. Factory acceptance tests are common to verify that the finished product meets each utility design specification.

In order to have a sound definition of an optical fiber cable, some minimum number of aspects must be defined, including: number and type of fibers; description of the cable design (cable elements, strength members, etc.); overall diameter; calculated mass (i.e. weight per unit of length); modulus of elasticity; coefficient of thermal expansion; maximum allowable tension; allowable temperature range for storage, installation and operation; and strain margin.

OPGW, OPPC, and MASS will need details on the calculated cross-sectional area concerning calculation of Rated Tensile Strength (RTS); the RTS itself; the DC resistance (not for MASS); the fault-current capacity I2t (not for MASS); and the lay direction of outer layer.

ADSS and OPAC may need to include the specification of the tracking-resistant sheath characteristics.

Tests on optical fiber cables are very important. Optical fibers are a very delicate material, and their performance once the cable is laid needs to be granted. Thus, from factory to final location on towers and poles, tests must ensure the integrity of the whole system for its expected lifetime.

Tests are to be performed on cable elements in a finished cable in order to detect any problem in the manufacturing process. Routine tests frequency and sampling vary across utilities, but they are intended to monitor that the process is consistent, as factories produce many different types of cables with the same production lines. The effects of the tests on the optical fiber can be measured as deviation from expected values, within measurement uncertainties (0.05 dB for attenuation is a reference value).

As with the specifications, not all tests are applicable to all cable types. The following list is a non-exhaustive list of the most relevant ones: tensile performance; sheave test; crush; impact; short-circuit; lightning; fitting compatibility; tracking and erosion; repeated bending; torsion; bend; shotgun resistance; aeolian vibration; low-frequency vibration (galloping); creep; stress–strain test on metallic cables; temperature cycling; water penetration; sheath abrasion resistance; aging; etc.

2.6.1.2.4 Cable-laying Operation

The process of cable laying starts with the proper preparation of the drums of optical fiber cable to be transported to the place where they are needed.

Cable should be tightly and uniformly wound onto the reel in layers, with the appropriate length (standard or specified length). Reels are often wooden or steel, nonreturnable or returnable types and should be prepared so that damage will not occur to the cable. Each reel should be clearly identified and tagged with weather-resistant material conveying essential information such as cable specification, cable length, number of fibers, etc. The inner end of the cable should be securely fastened and protected but accessible, to allow the connection of optical measuring equipment. A seal should be applied to each end of the cable to protect it against moisture into the optical fibers or the dripping of the filling compound.

IEEE, IEC, and ITU provide indications for the laying process. All of them acknowledge the special characteristics of some of the cables used by utilities and concentrate on two of them, OPGW and ADSS.

IEEE 524, "Guide for the Installation of Overhead Transmission Line Conductors" provides indications for OPGW and ADSS cables under the Special Conductors section. IEC Technical Report TR 62263, "Live working – Guidelines for the installation and maintenance of optical fiber cables on overhead power lines" expands the scope to other optical fiber cables (adding OPPC and OPAC). And IEC 60794-4-20 has an informative annex for ADSS on this topic. ITU-T L.151/L.34, "Installation of Optical Fiber Ground Wire (OPGW) Cable" provides indications for the laying process of OPGW.

2.6.2 Radio Spectrum

Radio (wireless) is an invisible supporting medium that makes the transmission of data in the form of electromagnetic waves over the air, possible. This "physical" communication medium is apparently present everywhere and does not need a tangible element to connect both transmitter and receiver. However, transmitters and receivers use antennas to interface with this intangible support.

Thus, even though the supporting medium cannot be seen, the rest of the concepts that are applicable to wireline telecommunications apply to radio as well. However, the fact that electromagnetic signals are not confined in a physical element mandates certain precautions around the coordination in the use of the radio spectrum.

Radio transmissions take place in the radio spectrum, "the range of frequencies of oscillations or electromagnetic waves which can be used for the transmission of information" [2]. They may occur at any frequency (see Table 2.2) and with any bandwidth. Different propagation mechanisms come into play in each spectrum zone, and thus reach and usability of the spectrum zone become affected. Spectrum along history, based on the state of the art in each period, has been used from

Table 2.2 Radio frequency ranges as per ITU-R V.431-8.

Symbol	Frequency range	Metric subdivision
VLF	3–30 kHz	Myriametric waves
LF	30–300 kHz	Kilometric waves
MF	300–3000 kHz	Hectometric waves
HF	3–30 MHz	Decametric waves
VHF	30–300 MHz	Metric waves
UHF	300–3000 MHz	Decimetric waves
SHF	3–30 GHz	Centimetric waves
EHF	30–300 GHz	Millimetric waves
	300–3000 GHz	Decimillimetric waves

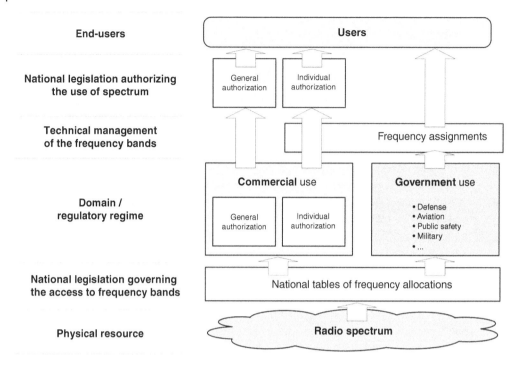

Figure 2.14 Structure to coordinate radio spectrum use. *Source:* Modified from [48].

the lower frequencies to the upper frequencies. Many legacy services and their possible interferences with emerging potential new spectrum uses, limit the capability to use the lower bands and affect the adoption of new technology improvements in spectrum domain.

The regulation of the use of electromagnetic spectrum for radio communications is a complex matter (see Figure 2.14). Radio spectrum is considered in most countries an exclusive property of the state [40]. Although National Regulatory Authorities (NRAs) are commonly established within countries as the competent legal regulatory bodies for spectrum management and regulation, spectrum regulation is implemented at different levels, from the ITU-R as the global reference for worldwide radio frequency coordination, down to regional and country level.

The work of the ITU-R is backed by the Radio Regulations (RR) treaty [41], a global agreement on how different spectrum uses are defined, allocated, and used without harmful interference among them, to ensure coexistence of different services and economies of scale. ITU-R functions are carried out through World Radiocommunication Conferences (WRCs), regularly held every four to five years. Its decisions are incorporated in the ITU RR. The ITU RR (Article 5) contains the so-called Table of Frequency Allocations (TFA) that collates frequency bands and their use. The TFA is particularized for each ITU's world geographical Regions (1, 2 and 3).

Regional spectrum management organizations represent the interests of their member countries and their NRAs along with telecommunications providers and the regional industry:

- Inter-American Telecommunication Commission (CITEL) [42].
- European Conference of Postal and Telecommunications Administrations (CEPT) [43].
- Asia Pacific Telecommunity (APT) [44].
- African Telecommunications Union (ATU) [45].

- Arab Spectrum Management Group (ASMG) [46].
- Regional Commonwealth in the field of Communications (RCC) [47].

In Europe, the Electronic Communications Committee (ECC), within the CEPT, develops common policies and regulations for Europe and is the focal point for information on spectrum use.

NRAs must balance many factors to address each country needs and concerns (e.g. government use, private use, etc.). NRAs start with a National Table of Frequency Allocations (NTFA), being the main instrument of the national legislation to govern the access to frequency bands. A NFTA specifies the radio services authorized by an individual administration along the frequency bands ("allocations") and the entities that might have access to these allocations. From then on, specific national frequency assignments authorize users to transmit on specific frequencies and are awarded licenses ("license" is a broad term meaning that the use of spectrum must be explicitly permitted depending on the frequency planning needed).

2.6.2.1 Radio Spectrum for Utility Telecommunications

Radio-based technologies have been widely used by utilities. Radio was first made available at the end of the nineteenth century and has been present in utility industry since the first remote connection needs. But the fact that utilities have a pervasive network of power lines and access rights on poles and ducts has forced a major trend to favor wireline solutions since optical fiber cables started to be available, more intensely in the last decades. Despite their high data rates, wireline technologies have cost and reach limitations that wireless (radio) solutions can improve. Moreover, radio communications are a perfect match in mobility communications or unaffordable to cover with wireline solutions applications (e.g. widespread rural areas) and have never ceased to be used by utilities.

On the specific utilization of radio communication resources, two major distinct combined uses of radio in utilities must be mentioned: narrowband and broadband. Both are present in voice and data applications.

Narrowband uses of radio have been applied mainly for voice communications in mobility (crews). They have also seen the development of narrowband data communications for command and control type and telemetry applications. In some instances, they have also been used for metering applications, in one-way and bidirectional uses.

Several frequency bands in this domain have been historically reserved on an exclusive basis or allocated to utilities through normal frequency grant processes. Utilities themselves have managed this spectrum, either themselves or through third parties created for that purpose (see the UTC in the United States for the role as "authorized certified frequency coordinator for the Private Land Mobile Radio Services below 512 MHz and 800–900 MHz frequencies," or the JRC in the United Kingdom [49]). Typically, VHF (allocations in 100 and 200 MHz bands) and UHF (in the 400 MHz band) narrowband channels (12.5 and 25 kHz channels) have been used for these purposes.

Broadband radio has been mainly used in point to point applications, to deploy telecommunication solutions where wireline (optical fiber) was not economically feasible. However, the evolution of optical fiber deployments along with the emergence of commercial cellular systems operations has been the major source of a 25-year gap of utility approach to spectrum.

This gap may have grown the wrong idea that utilities do not need radio technologies. The advent of new commercial cellular systems providing unprecedented public coverage of the territory and a set of functionalities that were absent in the mobile private radio systems made utilities focus on extending their optical fiber footprint and start to deprecate part of their Private Mobile Radio

(PMR) systems in areas where public coverage existed. The frequencies were in many cases reused for telecontrol and telemetry purposes, where narrowband was applicable since not many end points needed to be connected. However, commercial cellular systems have been demonstrated not to be prepared for mission-critical applications. There are studies that identify the shortcomings of these systems [50], and there are numerous evidences that commercial systems are not prepared for them (including black starts [51]). This circumstance has its origin in the fact that telecommunication operators "often see the utilities sector as one among many of their end-user industry verticals, and one that sometimes has complex requirements to address, particularly because it presents only a small market opportunity [and] the costs of meeting these requirements could be exorbitant" [52].

Thus, radio communications, and the spectrum needed for it to happen, is recognized as critical for utilities [53], among others, and a proper access and control of it are being pursued by increasingly more utilities and Energy Regulators, counting on Telecommunication Regulators as the ones holding the management of this scarce and valuable (in both monetary and socioeconomic terms [54]) resource.

The spectrum need is becoming more and more urgent and important, due to the evolution of the Smart Grid needs. As it was recorded in [55], there is a strong demand for spectrum for Smart Grid applications, which started with Smart Metering. Although the early implementation of these Smart Metering initiatives somehow made use of commercial wireless solutions, together with other special telecommunication media (i.e. PLC), its expansion into the Smart Grid will need spectrum for utilities. The increasing intelligence being added, enabling automatic responses in routine operations, and reconfiguration needs during emergencies, and the highly distributed nature of utility operations and assets, obliges to a flexible and ubiquitous access that can naturally be offered by radio communications [56], in a context that due to the secure and resilient services needed cannot be found in normal general-purpose commercial systems [57].

2.6.2.2 Radio Spectrum Use

Propagation characteristics are one of the inputs that guide decisions on the frequency bands to be used. However, they are really combined with historical aspects, economic considerations (including existing investments, costs of network rollouts, etc.) and regulation.

Lower-frequency bands favor propagation, do not require strict Line-of-sight (LOS), but offer low-bandwidth channels. Higher frequencies have poorer propagation characteristics, but provide higher bandwidth and higher capacity, given that LOS conditions are achieved (microwaves). Utility use cases are varied and combine different reach and capacity needs, and typically long-reach and moderate capacity needs are the rule. Resilience (even with efficient modulation schemes, possibilities are not taken to the limit [56]) and security in the service offering are a must (licensed spectrum is often associated with this requirement). These general aspects must now be assessed to identify the needs and possibilities for the different Smart Grids implementations, understanding that, even if utilities can accommodate to different options as long as a proper harmonization of spectrum produces a range of available products, as not all spectrum is equally suitable to all Smart Grid applications [57].

The following Smart Grid spectrum use has been suggested [55]:

- VHF spectrum for resilient voice communications and long reach – low-bandwidth data applications. For example, 100–200 MHz frequency range; bandwidth, 1 MHz duplex.
- UHF spectrum for point-to-area and multipoint higher-capacity data applications. For example, 410–470 MHz frequency range; bandwidth, from 3 to 10 MHz duplex.

- Lightly regulated or deregulated spectrum for short-range smart metering connectivity. For example, 870–876 MHz frequency range, as in SRD (Short-Range Devices) regulations [58].
- L-band long-range point-to-point or point-to-multipoint applications. For example, 1.4 or 1.5 GHz (it is a common band for many utilities; up to 4 MHz duplex).
- Microwave bands to access utilities core and complement optical fiber network. For example, 15, 18, 23, 26, 38 GHz backhauling; up to 30 MHz duplex bandwidth.

2.7 Electricity Cables

Power lines are the supporting medium for the utility communications that travel inside the electricity cables, i.e. PLC. Although the electric wires can be used for telecommunication purposes by anyone that has access to them (e.g. in-home appliances connectivity in home networking as the one offered by ITU-T G.hn recommendations over coaxial, telephone wiring and power line electric cables), it is clear that utilities have access to a widespread network of power lines of different voltage ranges.

Power line is an unintentional transmission medium that has been exploited for communication purposes since the beginning of the electricity service. It was quite evident for early electricity and telecommunication pioneers that a pair of metallic cables was sufficient to transmit signals, just as the early telephone service used copper pairs deployed for the purpose. Thus, telecommunications were also started in the power line domain to serve very specific needs.

Power line channels share many characteristics with radio/wireless channels. The main common elements come from the use of an existing transmission medium that was not designed for telecommunications, and that cannot be either changed in its physical design to make it just reach the intended point to communicate or avoid many of its disturbances.

The fact that power lines were not designed for telecommunications, forces us to consider the transmission properties of the cables that can be used to carry telecommunication signals. As it has been shown in Chapter 1, power line cables exhibit different designs in the different voltage ranges. In any power line, several metallic conductors will be found, from two (the minimum to use the cables as if it were the case of a copper pair) to typically four or six. These conductors are bundled in different designs and shapes, with three designs being the most common arrangements: four individual conductors in separated cables; three cables, each with a metallic shield that reminds of a coaxial cable; and four individual conductors in the same cable, tightly packed. Figure 2.15 gives a better idea of the physical appearance of these arrangements in MV and LV.

The variety of available conductors opens the number of options to select two or more of them. Thus, while in LV narrowband PLC applications in the kHz range, it is common to inject PLC between one phase and neutral, in the MV broadband PLC applications in the MHz range, PLC signal is injected in one MV cable, typically between the central conductor and the shield (there are other alternatives, such as using two central conductors of two of the available MV cables). Furthermore, MIMO techniques have been explored and successfully demonstrated in the field to improve performance and availability of PLC systems in the MHz range, using more than two conductors.

The diameters of the available cables depend mainly on the voltage to be carried, and the metal used is either aluminum or copper, with steel used to reinforce the cable when needed. Cable-laying options (main ones being overhead or underground) determine propagation properties as well, and the existence of insulation in the conductor affects the access to the metallic parts.

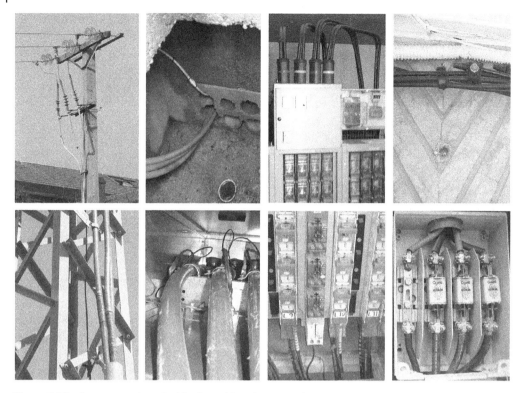

Figure 2.15 Some examples of cable dispositions in power lines (MV to the left; LV to the right).

Thus, the variety of cable types and dispositions suggests a difficulty in understanding the propagation properties, or at least, that the number of different cases to be studied is high. If we consider that the grid voltages in every utility are different, that the interest of registering the detailed aspects of some of the cables in the grid has been often small (e.g. LV grid), and that the propagation properties of the power lines also depend on the grid elements connected to them (transformers, switchgears, etc.), the determination of the cable properties to understand its telecommunication capabilities is difficult and has limited practical interest. Notwithstanding this fact, there are numerous studies on this topic, and its applicability is in the design of PLC systems and the rough estimate of the coverage range in the absence of disturbances.

Power lines as telecommunication media reach every device connected to them, and thus, telecommunication signals in electric wires can potentially reach everywhere. The topology of the grid can be considered a tree, and many segments of it perform as buses. However, there are natural obstacles to the propagation of PLC signals. This is the case of transformers at substations where, as there is no physical connection between primary and secondary windings, not every telecommunication signal will be able to "cross" it. The inherent advantage of such a grid segmentation is that different frequency ranges can be reused within domains where isolation exists. Last but not least, the grid can also be dealt with as a point-to-point connectivity especially in HV, or other lower voltage ranges, if ancillary elements are used to block the propagation of the signals (these elements can also adapt the impedance of the grid to the PLC transmission, so that the power line presents optimal propagation characteristics).

Finally, disturbances in the grid can be hardly avoided. All appliances are connected to it, and while they need to have appropriate compliance with the disturbance limits, their combination in each frequency range may produce disturbance levels that exceed what a more controlled telecommunication medium would present. Aging of components, inadequate disturbance levels, and the big number of elements connected to the grid add up to a noise signal that is detrimental to the performance and/or availability of the PLC channel. However, the noise context of the different grid segments is very different. HV and MV are hardly affected by appliances' disturbances. However, LV grid, especially further down from the SS, usually presents a challenging noise scenario.

PLC channels are, then, difficult to model, harsh, and noisy. The power line channel is frequency-selective, time-varying, and is impaired by colored background noise and impulsive noise. However, PLC addresses these obstacles:

- Theoretical and experimental studies have been developed to assess power line channel transfer functions. Even with the already mentioned difficulties, channel transfer functions exhibit a higher determinism than anticipated.
- Experimental studies have been developed to study the nature of channel noise conditions and the strategies to mitigate it.
- PLC systems are defined, and their network devices designed, to adapt to the signal propagation conditions. Grid impedances have been studied to obtain an optimal injected power transfer to the channel, in different grid locations (e.g. the input impedance is not the same in the SS or in a meter room).
- PLC systems include the state-of-the-art technology features that will make them overcome the noises they will find, with two limits. First, the cost that the PLC application can support (e.g. PLC solutions for Smart Meters are designed to keep costs in the lower ranges); second, a trade-off between low-level strategies (e.g. error correction, coding, etc.) and other strategies that may be detrimental to performance (e.g. segmentation of packets, retransmission of errored packets, etc.).
- PLC spectrum use is coordinated to make the implementation of the different applications in the different grid segments possible. Signal transmit power levels in the different frequency ranges are defined to avoid self-interference and interference of possible nonutility PLC uses (e.g. in-home PLC) and to avoid potential interference with other telecommunication systems that may be affected by unintentional radiation produced by PLC.

2.7.1 PLC Use

The use and applications of PLC can be traced back in history to the nineteenth century, and the different systems implementing it have used the frequencies that were state of the art at the time, and sometimes were already used by other systems using other wireline or wireless media. A simple timeline can be found in Table 2.3.

Early remote meter reading systems and remote load management using used high-power and narrowband PLC signals. Frequencies were in the 125 Hz–3 kHz range, so that very low data rate signals (below 100 bps, so-called Ultra NarrowBand [UNB]) could pass through the distribution transformers and reach end points in the LV grid.

Before PLC was widespread with ripple control (a remote load management system) in the Distribution domain, PLC voice communication over HV became popular in the 1920s. The operation of these systems was in the range from 15 to 500 kHz, and signal bandwidths were of a few kHz. Eventually, some remote control capabilities were added, and protective relaying has been the major application of the evolution of these systems.

The 1980s witnessed a renewed interest in two-way NarrowBand PLC systems (NB PLC) in the Automatic Meter Reading (AMR) domain (see Chapter 5). The automation of elements of the

Table 2.3 PLC timeline.

Period	Application	Standards
1890s	Patents on metering	
1900s	Patents on remote load management	
1920s	Telephony	
1930s/1950s	Remote load management	
1980s	Metering	
	Automation	
1990s	Low data rate Narrowband applications	IEC 61334-5
2000s	Broadband PLC	OPERA[a]
2010	High data rate Narrowband applications	ITU-T G.9903
		ITU-T G.9904
		IEEE 1901.2
	BroadBand PLC	IEEE 1901
		ITU-T G.hn (G.9960–64)

[a] OPERA was a research and development project that produced an industry standard.

Table 2.4 Bands and transmission levels defined in EN 50065-1:2011.

Band[a]	Frequency (kHz)	Max. transmission level (dBuV)	Usage
A	3–9	134	Utility[b]
A	9–95	Narrowband[c]: from 134 till 120 (logarithmic decrease over frequency) Broadband: 134	Utility
B	95–125	122 or 134[d]	Private
C	125–140[e]	122 or 134	Private
D	140–148.5	122 or 134	Private

[a] Although the letters are not used in the norm, they are common use. A band is commonly known as CENELEC A-band.
[b] The norm refers to monitoring and control usage, including metering.
[c] The limit between narrowband and broadband is fixed at 5 kHz.
[d] 122 refers to Class 122 products, for general use. 134 refers to Class 134 products, for industrial use.
[e] CSMA at 132 kHz needs to be used.

distribution grid and, more importantly, industry and building automation were also considered. It was at this time that the European Norm EN 50065 "Signalling on low-voltage electrical installations in the frequency range 3 kHz to 148.5 kHz" was published in 1991, helping these systems to grow. The last published version of EN 50065-1 (a European Harmonized Standard) includes the use shown in Table 2.4. Transmission levels are the most important factor affecting the applicability of these systems, as they limit their range, given channel noise conditions that cannot be controlled inside the system. These systems are known as Low Data Rate (LDR) systems, with speeds of a few kbps.

In the late 1990s, deregulation of the telecommunication and energy markets in Europe stimulated the interest of utilities in other markets and the possibility to offer additional services to consumers.

Broadband PLC (BPL) systems appeared in the market using frequency bands in the range from 1.8 to 30 MHz, providing data rates from several Mbps to hundreds of Mbps. These systems were initially developed and deployed for Internet access and in-home multimedia applications.

In the early 2000s, utilities shifted the focus of PLC into the Smart Grid and stimulated another innovation wave of NB PLC. First innovations came in the Smart Metering domain, with High Data Rate NarrowBand PLC systems (HDR NB PLC) in the 3–500 kHz frequency range. These systems have been extensively deployed in the LV grid and will evolve to cope with LV grid automation.

Thus, starting in the 2010s, and until our days, both BPL and HDR NB PLC systems coexist in the MV and LV grid, combined in different architectures that successfully deliver Smart Grid services. While the trend in HV is to increasingly use optical fiber for communication purpose, some niche HV PLC applications exist as well. The evolution of digital techniques applied to PLC systems, and the better understanding of PLC channels in the different segments of the grid for the different frequency ranges, will push broader band PLC systems closer to the grid edge, ultimately introducing broadband into the last elements of the electricity system, i.e. the Smart Meter.

References

1 *ITU-R/ITU-T: Terms and Definitions*. https://www.itu.int/pub/R-TER-DB (accessed 18 October 2020).
2 IEC 60050 – International electrotechnical vocabulary – welcome. *Electropedia: The World's Online Electrotechnical Vocabulary*. http://www.electropedia.org/ (accessed 18 October 2020).
3 *ITU-T Q.9: Vocabulary of Switching and Signalling Terms*. https://www.itu.int/rec/T-REC-Q.9/en (accessed 18 October 2020).
4 *ITU-T Q.72: Stage 2 Description for Packet Mode Services*. https://www.itu.int/rec/T-REC-Q.72/en (accessed 18 October 2020).
5 *ITU-R V.662-3: Terms and Definitions*. https://www.itu.int/rec/R-REC-V.662/en (accessed 18 October 2020).
6 (2000). IEEE 100: The Authoritative Dictionary of IEEE Standards Terms. Standards Information Network, 7e. New York: IEEE Press.
7 Freeman, R.L. (2004). Telecommunication System Engineering, 4e. Hoboken, NJ: Wiley-Interscience.
8 Sendin, A., Sanchez-Fornie, M.A., Berganza, I. et al. (2016). Telecommunication Networks for the Smart Grid. Artech House.
9 *Telecom Network Planning for evolving Network Architectures. Reference Manual Part 1. Draft version 5.1* (30 January 2008) [Online]. https://www.itu.int/ITU-D/tech/NGN/Manual/Version5/NPM_V05_January2008_PART1.pdf (accessed 28 December 2020).
10 *Telecom Network Planning for evolving Network Architectures. Reference Manual Part 2. Draft version 5.1* (30 January 2008) [Online]. https://www.itu.int/ITU-D/tech/NGN/Manual/Version5/NPM_V05_January2008_PART2.pdf (accessed 28 December 2020).
11 *Mobile Backhaul: An Overview*. https://www.gsma.com/futurenetworks/wiki/mobile-backhaul-an-overview/ (accessed 18 October 2020).
12 *ITU-R F.1399: Vocabulary of Terms for Wireless Access*. https://www.itu.int/rec/R-REC-F.1399/en (accessed 18 October 2020).
13 *ITU-R V.573: Radiocommunication Vocabulary*. https://www.itu.int/rec/R-REC-v.573/en (accessed 18 October 2020).

14 Bidgoli, H. (ed.) (2004). The Internet Encyclopedia. Hoboken, NJ: Wiley.

15 Leiner, B.M., Cerf, V.G., Clark, D.D. et al. (1997). Brief History of the Internet. Internet Society https://www.internetsociety.org/internet/history-internet/brief-history-internet/ [Online] (accessed 18 October 2020).

16 IEEE LMSC (2012). *Overview and Guide to the IEEE 802 LMSC* [Online]. http://www.ieee802.org/IEEE-802-LMSC-OverviewGuide-02SEPT%202012.pdf (accessed 28 December 2020).

17 Hanes, D., Salgueiro, G., Grossetete, P. et al. (2017). IoT Fundamentals: Networking Technologies, Protocols, and Use Cases for the Internet of Things. Indianapolis, IN: Cisco Press.

18 Ho, Q.-D., Gao, Y., Rajalingham, G., and Le-Ngoc, T. (2014). Wireless Communications Networks for the Smart Grid. Springer International Publishing.

19 IEEE Std 2030-2011 (2011). *IEEE Guide for Smart Grid Interoperability of Energy Technology and Information Technology Operation with the Electric Power System (EPS), End-Use Applications, and Loads*, 1–126. https://doi.org/10.1109/IEEESTD.2011.6018239.

20 Freeman, R.L. (2005). Fundamentals of Telecommunications, 2e. Hoboken, NJ: IEEE Press; Wiley-Interscience.

21 Ronayne, J. (1996). Integrated Services Digital Network: From Concept to Application, 2e. London: UCL Press.

22 Pužmanová, R. (2002). Routing and Switching: Time of Convergence? Boston: Addison-Wesley.

23 (2019). Document Name: Study on Edge and Fog Computing in oneM2M systems. Technical Report [Online]. https://onem2m.org/component/rsfiles/download-file/files?path=Draft_TR%255CTR-0052-Study_on_Edge_and_Fog_Computing_in_oneM2M_systems-V0_5_0.docx&Itemid=238 (accessed 18 October 2020).

24 ETSI. Multi-access Edge Computing – Standards for MEC. *ETSI*. https://www.etsi.org/technologies/multi-access-edge-computing (accessed 18 October 2020).

25 Iorga, M., Feldman, L., Barton, R. et al. (2018). Fog Computing Conceptual Model, NIST SP 500-325. Gaithersburg, MD: National Institute of Standards and Technology. https://doi.org/10.6028/NIST.SP.500-325.

26 Stallings, W. (2014). Data and Computer Communications, 10e. Boston: Pearson.

27 Open Internet Standards. *Internet Society*. https://www.internetsociety.org/issues/open-internet-standards/ (accessed 15 November 2020).

28 Stratton, S.W. (1912). Circular of the Bureau of Standards no 31. Copper Wire Tables, 1e. Department of Commerce and Labor. https://ia800600.us.archive.org/21/items/circularofbureau31unse/circularofbureau31unse.pdf (accessed 28 December 2020) [Online].

29 *MIL-HDBK-216 – RF Transmission Lines and Fittings* (4 January 1962) [Online]. https://www.abbottaerospace.com/wpdm-package/mil-hdbk-216-rf-transmission-lines-and-fittings/ (accessed 28 December 2020).

30 Miller, S.E. and Marcatili, E.A.J. (1973). Research toward optical-fiber transmission systems. *Proc. IEEE* 61 (12): 1703–1704. https://doi.org/10.1109/PROC.1973.9360.

31 *ITU-T G Suppl. 42 (10/2018) – Guide on the use of the ITU-T Recommendations related to optical fibres and systems technology*. Guide (October 2018) [online]. https://www.itu.int/ITU-T/recommendations/rec.aspx?id=13824&lang=en (accessed 18 October 2020).

32 Horak, R. (2007). Telecommunications and Data Communications Handbook. Hoboken, NJ: Wiley-Interscience.

33 Blume, S.W. (2017). Electric Power System Basics for the Nonelectrical Professional, *2*e. Wiley-IEEE Press.

34 *Navigant Research Report Shows Adopting a Fiber Optic Backbone Is a Strategic Imperative for Utilities' Long-term Success*. https://guidehouseinsights.com/news-and-views/navigant-research-report-shows-adopting-a-fiber-optic-backbone-is-a-strategic-imperative-for-utiliti (accessed 11 October 2020).

35 P. O. European Union (2014). *2014/61/EU: Directive of the European Parliament and of the Council on Measures to Reduce the Cost of Deploying High-Speed Electronic Communications Networks*.

36 *Communications Act of 1934*.

37 *Federal Pole Attachments Act as of 1978*.

38 *Telecommunications Act of 1996*.

39 Thomas, M.S. and McDonald, J.D. (2015). Power System SCADA and Smart Grids. Boca Raton: CRC Press.

40 Ferrus, R. and Sallent, O. (2015). Mobile Broadband Communications for Public Safety: The Road Ahead Through LTE Technology. Wiley: Hoboken, NJ.

41 *Radio Regulations*. http://www.itu.int/pub/R-REG-RR (accessed 18 October 2020).

42 CITEL. *OAS/CITEL – Inter-American Telecommunication Commission – About CITEL*. https://www.citel.oas.org/en/Pages/About-Citel.aspx (accessed 18 October 2020).

43 *European Conference of Postal and Telecommunications Administrations (CEPT)*. https://www.cept.org/ (accessed 18 October 2020).

44 *APT – Introduction|Asia-Pacific Telecommunity*. https://www.apt.int/APT-Introduction (accessed 18 October 2020).

45 *African Telecommunications Union (ATU)» History*. https://atu-uat.org/history/ (accessed 18 October 2020).

46 *Arab Spectrum Management Group (ASMG)*. http://www.asmg.ae (accessed 28 December 2020).

47 *RCC – Regional Commonwealth in the Field of Communications*. https://en.rcc.org.ru/ (accessed 18 October 2020).

48 *ECC Report 205. Licensed Shared Access (LSA)* (February 2014) [Online]. https://docdb.cept.org/download/baa4087d-e404/ECCREP205.PDF (accessed 18 October 2020).

49 *History of the JRC*. https://www.jrc.co.uk/about-jrc/history-of-the-jrc (accessed 29 December 2020).

50 Forge, S., Horvitz, R., and Blackman, C. (2014). Is Commercial Cellular Suitable for Mission Critical Broadband?: Study on Use of Commercial Mobile Networks and Equipment for "Mission-Critical" High-Speed Broadband Communications in Specific Sectors: Final Report. Luxembourg: Publications Office.

51 *Black Start|National Grid ESO*. https://www.nationalgrideso.com/black-start (accessed 18 October 2020).

52 Adkins, I. (2019). Opinions differ on the need for a dedicated spectrum allocation for the utilities sector. *Analysys Mason* (October 2019) [online]. https://www.analysysmason.com/globalassets/x_migrated-media/media/analysys_mason_utilities_spectrum_quarterly_oct2019.pdf (accessed 18 October 2020).

53 *Spectrum Framework Review. A Consultation on Ofcom's Views as to How Radio Spectrum Should Be Managed*. OFCOM – Office of Communications, UK (23 November 2004) [Online]. https://www.ofcom.org.uk/__data/assets/pdf_file/0014/25403/sfr.pdf.

54 *The Socio-Economic Value of Radio Spectrum Used by Utilities in Support of Their Operations*. European Utilities Telecommunications Council, Release 1 (1 January 2012) [Online]. https://eutc.org/wp-content/uploads/2018/08/Socio-economic-value-of-Spectrum-used-by-utilities-v1.1.pdf (accessed 18 October 2020).

55 (2013). Spectrum Policy Analysis of Technology Trends, Future Needs and Demand for Spectrum in Line with Art.9 of the RSPP: Final Report. Luxembourg: Publications Office.

56 UK Spectrum Policy Forum. *UK Spectrum Usage and Demand*. Consulting 2nd edition (October 2018) [Online]. https://www.techuk.org/component/techuksecurity/security/download/14247?file=Report_on_UK_spectrum_demand_second_edition_October_2018_14247.pdf&Itemid=181&return=aHR0cHM6Ly93d3cudGVjaHVrLm9yZy9pbnNpZ2h0cy9yZXBvcnRzL2l0ZW0vMTQyNDctbmV3LXNwZi1yZXBvcnQtc3BlY3RydW0tdXNhZ2UtYW5kLWRlbWFuZA== (accessed 18 October 2020).

57 Peskoe-Yang, L. (2019). The Invisible Battle for America's Airwaves. *Popular Mechanics* (17 June 2019). https://www.popularmechanics.com/technology/infrastructure/a28039878/battle-radiofrequency-spectrum/ (accessed 18 October 2020).

58 Radio spectrum harmonisation for short-range devices. *Shaping Europe's Digital Future – European Commission* (4 February 2013). https://ec.europa.eu/digital-single-market/en/content/short-range-mass-market (accessed 18 October 2020).

3

Telecommunication Fundamental Concepts

3.1 Introduction

There are a number of elements that are key in the correct understanding of telecommunications systems and services, and specific relationships exist among them. The study of telecommunications involves the analysis of the transmission of information signals (both analog and digital) in an effective and reliable way.

The evolution from analog to digital must be comprehended and specially the implications of the digital world in terms of adaptation of the telecommunication systems. The modulation of signals, together with the access to the medium, affects their capability to share physical media, and it needs to be explained. Some modulations may convey better throughput (data rate) while being more prone to errors as well. Finally, the main concepts and their trade-offs (signal-to-noise ratio, bandwidth, propagation – reach, etc.) of the different telecommunication technologies must be highlighted.

This chapter drills down into the framework provided by Chapter 2 and the previously defined concepts in a detailed manner, while not diving too deep into their mathematical derivation. It also covers the propagation characteristics of specific telecommunication technologies, i.e. optical fiber communications and radiocommunications (wireless).

3.2 Signals

3.2.1 Analog vs. Digital

Chapter 2 has already introduced some general ideas about the different nature of analog and digital signals. This section deep dives into the essence of these two types of information sources, their relationships, and how one may be converted to the other, and vice versa.

3.2.1.1 Continuous vs. Discrete

The very foundations of telecommunications are the transmission of signals from one point to another. A signal is a mathematical representation of the evolution of a physical quantity concerning some parameters (usually time or space). This physical quantity can be voltage, electrical intensity, pressure, light intensity, etc. It may not even have a clear physical meaning, such as the stock price of a specific company. The interesting thing about signals is that they are abstractions of the specific physical magnitude. Thanks to this, they can simply be considered as a mathematical function, $x(\cdot)$, with a given independent variable.

Depending on the nature of the independent variable, signals can be grouped into continuous-time signals or discrete-time signals. The main characteristic of the former is that the independent time can take any real value, typically represented as $x(t)$. In contrast, the independent variable in discrete-time signals can only take real integer numbers, typically represented as $x[n]$. Figure 3.1a shows the graphical representation of the continuous-time signal $x(t) = t$ and the discrete-time signal $x[n] = n$.

From a practical point of view, a continuous signal exists for any time instant, while discrete signals are only defined for specific (discrete) time instants.

A different classification can be performed depending on the granularity of the amplitude values. This classification is another way of looking at continuous and discrete, projecting us into the analog and digital world. Analog meaning that it may take an infinite set of values and digital if it can take a finite set.

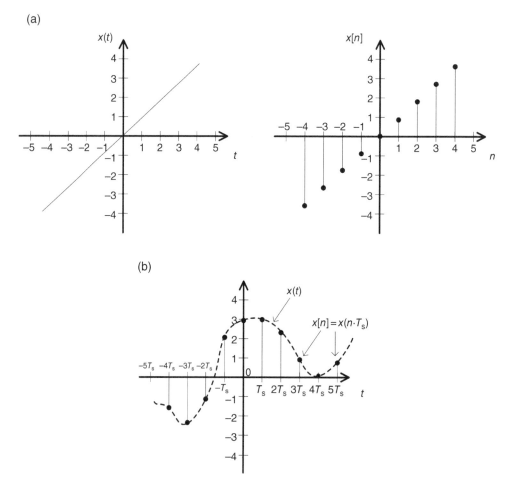

Figure 3.1 (a) Examples of continuous-time signals (left) and discrete-time signals (right). (b) Sampling of a continuous-time signal.

3.2.1.2 Sampling

Despite real world being based on continuous-time signals, in order for computers to process and handle them, they need to convert them into a discrete-time signal so that they can be stored in memory or sent using a digital transmission technique (to be elaborated in Section 3.3 of this chapter). Moreover, this continuous-to-discrete process must be reversible, since the processed signal will return to the real (continuous) world. Consequently, continuous signals are transformed into discrete for processing, storing, or transmitting them and are then typically converted back to continuous domain.

The process of creating a digital signal from an analog one can be divided in two steps: sampling and quantizing.

Sampling refers to the operation of taking the amplitude values of a continuous-time domain signal only at specific and periodical time instants. These uniform sampling times are defined by the sampling period (T_s), i.e. the time increment in two consecutive samples. This idea is shown in Figure 3.1b. The values for T_s must be chosen so that the original analog signal can be restored.

Nyquist–Shannon sampling theorem [1] defines what the maximum time separation of two samples must be for the process to be reversible. The minimum sampling frequency (i.e. the inverse of the sampling period T_s) must be, at least, twice the maximum frequency of the analog signal, as further elaborated in Section 3.2.2.

3.2.1.3 Quantizing and Coding

Quantization refers to the process used to simplify sampled values. Since signals are to be stored in discrete memory or transmitted using digital techniques, the discretization in time domain is not enough and an additional discretization in amplitude values must be carried out. This is known as quantization.

As a result, when amplitude values are quantized, they are assigned to the nearest quantized amplitude level.

Quantized values are the ones that are stored or transmitted. Since this process is carried out by a microcontroller or microprocessor, quantized values are assigned to some fixed-length binary words. The process of assigning a quantized value to a binary word is known as encoding.

This whole process is summarized in Figure 3.2. The top figure represents the continuous signal to be digitalized. On the bottom left-hand side, the sampled version of the signal is shown together with the original continuous time. There, the discretization of the horizontal axis (i.e. sampling) can be observed. To the right (upper), the discretization of the vertical axis (i.e. quantization) is shown. In this example, all the range of the signal is divided into 16 levels and the quantized version (crosses) is chosen according to the level closer to the sampled version (circles). The left-hand side axis of the figure represents the amplitudes of the sampled values, whereas the right-hand side axis represents the corresponding encoding of the quantized levels. In this case, since levels go from 0 to 15, each one of them can be represented by a 4-bit long binary word.

Independent of how many levels or bits per sample, the quantization process is a lossy process due to the impossibility of covering the infinite possible amplitude values of the analog signal. This fact is conceptualized as Quantization Error and can be reduced by increasing the number of quantization levels. Figure 3.2 shows the error signal when comparing the quantized signal (crosses) with the sampled signal (circles).

The device transforming analog signals to digital is named Analog-to-Digital Converter or ADC, while the device doing the reverse operation is called Digital-to-Analog Converter or DAC.

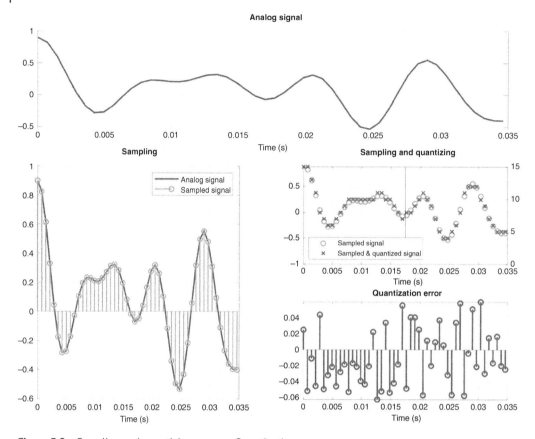

Figure 3.2 Sampling and quantizing process. Quantization error.

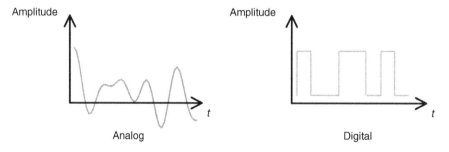

Figure 3.3 Analog (left) and digital (right) signals.

3.2.1.4 Analog and Digital Signals

Once an original signal is converted to sampled and quantized values, it can be accurately represented by its digital version (providing that the Nyquist criteria is met). Consequently, the digital signal represents a serialization of all the bits corresponding to an analog signal or representing any other sort of information (e.g. characters of the alphabet and/or numbers). The digital signal can only take a discrete amount of amplitudes, each one of them representing a bit value (either a logical "0" or a logical "1"). Figure 3.3 shows this idea.

There is a number of benefits in transforming analog signals to digital ones, many of which were mentioned in Chapter 2.

3.2.2 Frequency Representation of Signals

On many occasions, it is not easy (or even possible) to describe the behavior of the signals based on their temporal variation. An alternative and complementary representation of the signals and systems can help us in this work; this is generically referred to as "transformed." Transformed operation exchanges the representation of a signal in one dimension to another.

The Fourier Transform takes a continuous-time signal and creates its equivalent representation in the continuous frequency domain, that is, with infinity frequency precision. In contrast, if we are dealing with digital signals (in the discrete-time domain), it is the Discrete Fourier Transform (DFT) that takes the time signal and translates it into a discrete-frequency domain, that is, with finite frequency resolution.

3.2.2.1 The Continuous-time Fourier Transform

The Continuous-time Fourier Transform (CFT) of a signal can be computed with:

$$X(f) = \int_{-\infty}^{\infty} x(t) \cdot e^{-j2\pi ft} \, dt$$

From its mathematical definition, it is hard to extrapolate some practical ideas. However, there are a few rules of thumb that help to understand how the time and frequency domains are linked to each other:

- The Fourier Transform of a real signal is a complex function. This means that the transformed function will have a real and imaginary part; i.e, a modulus and phase component.
- If the amplitude of the waveform varies slowly, the modulus of its transform will take higher values for frequencies close to 0. An extreme case is the constant amplitude signals whose transform takes a nonzero value only at frequency 0.
- If the amplitude of the waveform varies rapidly, the modulus of its transform will take large values for frequencies far from 0.
- Signals that have an infinite domain in time dimension have a finite domain in the frequency dimension and vice versa.

The following example will illustrate the frequency aspects of the transformed signal depending on the amplitude variation speed. For this purpose, the Fourier Transform of a sinusoid signal of a given frequency f_0 is computed (i.e. $x(t) = \sin(2\pi f_0 t)$). The signal in time domain is assumed infinite.

$$X(f) = \int_{-\infty}^{\infty} \sin(2\pi f_0 t) \cdot e^{-j2\pi ft} \, dt = \int_{-\infty}^{\infty} \left(\frac{e^{j2\pi f_0 t} - e^{-j2\pi f_0 t}}{2j} \right) \cdot e^{-j2\pi ft} \, dt = \frac{j}{2}\delta(f+f_0) - \frac{j}{2}\delta(f-f_0)$$

$\delta(f \pm f_0)$ are Dirac's delta functions placed at $\pm f_0$.

Figure 3.4 visually represents these signals in Time and Frequency domains. Note that, in the frequency representation, only the signal's module is shown. As it can be seen, faster changes create components at higher frequencies (signal at the bottom).

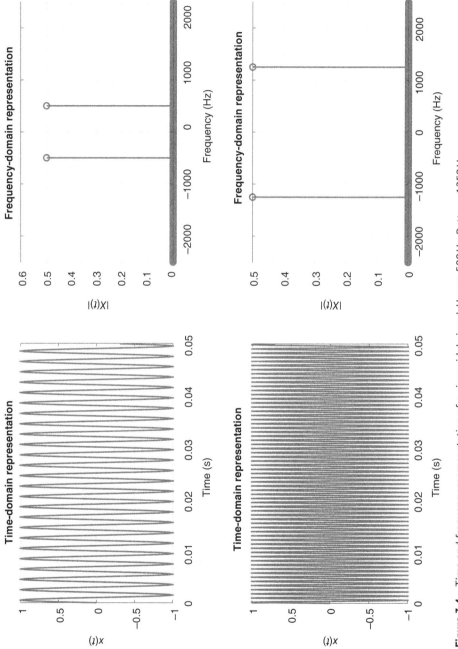

Figure 3.4 Time and frequency representation of a sinusoidal signal. Upper: 500 Hz. Bottom: 1250 Hz.

A different example can be made when signals in time domain are finite, e.g. the same sinusoidal signals above, restricted to times between $t = 0.02$ and $t = 0.03$ seconds. The mathematical derivation is slightly more complex, and it is omitted:

$$X(f) = \int_{-\infty}^{\infty} x(t) \cdot e^{-j2\pi ft} \, dt = \frac{Tj}{2} \cdot \left[\frac{\sin(T\pi(f-f_0))}{T\pi(f-f_0)} - \frac{\sin(T\pi(f-f_0))}{T\pi(f-f_0)} \right]$$

As Figure 3.5 shows, one of the main differences with the previous result is that the frequency representation is not limited to a single value, but it is spread over a wider range of frequencies around the central frequency of the sinusoid signal. Moreover, it can be noted that this spreading is larger with a shorter T.

The main advantage of the Fourier Transform is the opportunity of looking at a signal from the frequency dimension. This aspect is especially useful when representing telecommunication signals that have limited frequency components and are organized in this domain.

Figure 3.6 is an example of this and shows the time-domain and frequency-domain representation of two signals. The first one is a sum of four different sinusoids. While this is not clear looking at the left figure, it becomes evident in the figure to its left, where the frequencies the signal consists of can be seen. This simple example can be generalized in the bottom signal, which corresponds to a human voice signal. It is only with the Fourier Transform that the frequency components can be derived, as represented in the left-hand side figure, where the frequency components are mostly condensed on the spectrum below 800 Hz. This fact is used in telecommunication signal processing to limit the frequency components above 800 Hz to make the signal fit within a "channel," while keeping the content (e.g. the audio of a voice signal) still understandable. This first idea (limiting frequency components) is very useful when more advanced concepts, such as modulation or bandwidth, are explained in the following subsections. The second idea (the signal loosing part of its quality or information) can be handled through the use of digital signals that can transmit the digital information within controlled error (loss) limits.

One final word is for the components in the negative frequency range shown in previous figures, as it may be confusing for readers not familiar with signal processing. Indeed, in real life there is no such thing as a negative frequency. Their representation has to do with a mathematical requirement coming from the Fourier Transform operation: if a signal is real in time domain (i.e. it has no imaginary component), then the positive side of the spectrum must be the complex conjugate of the negative side [1, 2]. Signals that are transmitted over wires or wirelessly are, of course, real, since they take real values of voltage and, as such, must meet this property and count with a negative spectrum.

3.2.2.2 The Discrete-Time Fourier Transform

The previous subsection motivated the need for a tool such as the Fourier Transform in order to understand how a signal is represented in the frequency dimension. However, from a mathematical point of view, the obtained analytical representation of this dimension is still a continuous function (in this case, it is a continuous-frequency function). This is an important limitation if a machine must analyze this information.

Luckily, there is an equivalent expression for the Fourier Transform that deals with discrete sequences (i.e. digital signals) and that provides a discrete representation of the spectrum of a signal. It is known as the Discrete Fourier Transform (DFT) and is defined by:

$$X[k] = \frac{1}{N} \sum_{n=n_1}^{n_2} x[n] \cdot e^{-j\frac{2\pi}{N}nk}; \quad N = n_2 - n_1 + 1$$

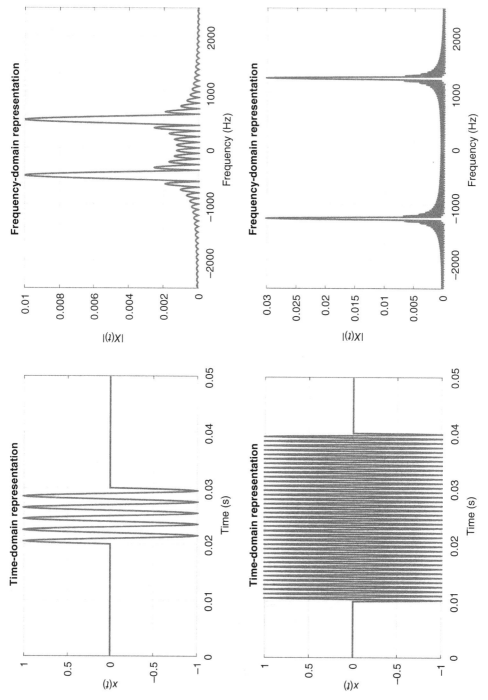

Figure 3.5 Time and frequency representation of a finite-domain sinusoidal signal. Upper: 500 Hz. Bottom: 1250 Hz.

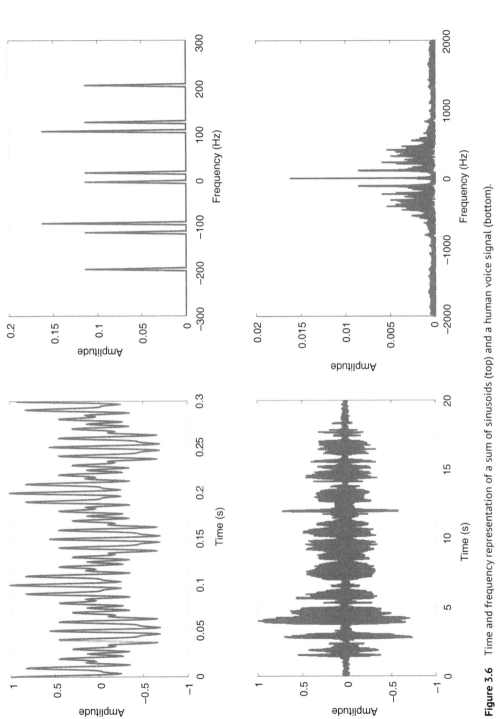

Figure 3.6 Time and frequency representation of a sum of sinusoids (top) and a human voice signal (bottom).

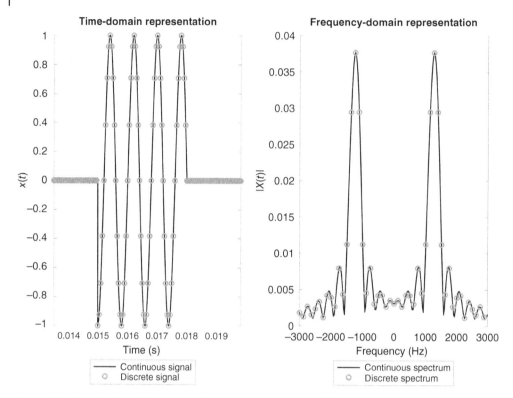

Figure 3.7 Time (left) and frequency (right) representation of a 1250 Hz windowed sinusoid in continuous (line) and discrete form (circles).

Without much detail, the Discrete Fourier provides a sampled version of the Continuous-Time Fourier Transform, which can be interpreted, processed, and analyzed by any kind of Digital Signal Processor (DSP) or microprocessor.

Figure 3.7 shows a representation of a 1250 Hz windowed sinusoidal signal in time (left) and frequency (right) domain and in continuous (line) and discrete (circles) form. As it can be observed, the DFT corresponds to a sampled version of the CFT, suitable for discrete machines such as microprocessors.

The DFT operation is very frequently performed in many devices, given the pervasive presence of digital processing today. However, due to its intensive computational load, a reduced-complexity version of this operation, discovered by James Cooley and John Tukey, is typically implemented. This Cooley–Tukey algorithm is commonly known as Fast Fourier Transform (FFT) [3].

3.2.3 Bandwidth

Bandwidth is defined by ITU-R Recommendation SM.328 as "the width of the band of frequencies occupied by one signal, or a number of multiplexed signals, which is intended to be conveyed by a line or a radio transmission system." However, apart from this formal definition, there is another commonly accepted way of referring to bandwidth when digital information is involved. It considers the capacity of the telecommunication system and expresses bandwidth as the amount of information that a system can manage in bits per second (bps). This data rate is directly dependent on the frequency bandwidth, the modulation, and the coding process.

The term bandwidth is used to define the frequency range covered by one specific signal. This idea is typically referred to as "absolute bandwidth." As a rule of thumb, the more abrupt changes a signal has in time domain, the higher its bandwidth. As the bandwidth is related to the "changes" in the signal in the time domain, it is directly dependent on the data rate. The higher the data rate, the higher the bandwidth needed.

3.3 Transmission and Reception

After this first introduction to the basic concepts and representations of signal, we could dive into specific concepts of telecommunication: how information can be sent efficiently and robustly. Figure 2.1 in Chapter 2 represented a simplified scheme for a communication model. This section zooms in the Transmitter and Receiver blocks of the figure and details the main processes occurring underneath for digital systems. Figure 3.8 shows a more detailed schematic representation of the main blocks present in a generic digital telecommunication process. All blocks will be detailed in the different upcoming subsections, for the time being they will be briefly introduced so that the reader has a vague understanding of their purpose.

Depending on whether the nature of information to be transmitted is analog or digital, it goes through the already mentioned Sampling–Quantizing–Encoding process or not. The channel encoder receives the information bits and includes additional bits to the message. These extra bits will be used at the receiver to both detect errors and, depending on their nature and quantity, correct them. After the channel encoder, the modulator, which is in charge of assigning the different bits or groups of bits to symbols in a geometrical space, can be found. The transmission block creates the final shape of the continuous-time signal to be transmitted. Finally, the last block is the front-end that will adapt the signal to the medium and increase its power so it can propagate as far as the receiver.

The receiver performs the inverse of all the operations carried out by the transmitter. Starting from the channel, the front-end decouples the signal and, in some occasions, amplifies it. Next, the reception block obtains the coordinates of the symbol created by the modulator. Since the received signal will be corrupted by the channel, some corrections will be needed. This is the task of the channel estimator and the equalizer. The demodulator is input with the symbols modified by the equalizer and tries to guess what the most likely transmitted symbol is. The output of the demodulator is the corresponding bits for the decided symbols. These bits are input to the channel decoder, which uses the redundant bits to find possible errors in the data. Finally, the information is ready to be interpreted. In case the source of the information is analog, an extra step is needed to create the corresponding signal based on the bitstream.

3.3.1 Modulation

The concept of modulation can be explained as the process in which certain information signal (analog or digital) is transmitted by changing one or more characteristics of an additional signal, typically referred to as carrier signal.

These modifications in the carrier signal are targeted at adapting the transmission of the message in a given medium taking into account aspects such as its nature, the characteristics of the receiver, the distance, the required transmission rate, its robustness, its quality, etc.

Depending on the nature of the information signal to transmit, modulations can be digital or analog. In the former, variations over the carrier signals can only be discrete, whereas, in the latter,

90 | *3 Telecommunication Fundamental Concepts*

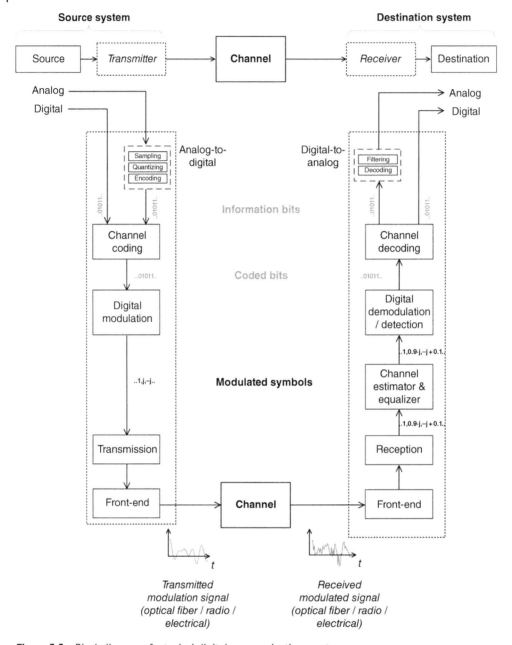

Figure 3.8 Block diagram of a typical digital communications system.

this variation can have continuous values. Although there is a wider variety, a common classification of both modulations can be the following:

- Analog Modulations:
  - Amplitude Modulation (AM). The amplitude of the carrier signal changes with the amplitude of the information signal.
  - Frequency Modulation (FM). The frequency of the carrier signal changes with the amplitude of the information signal.

- Phase Modulation (PM). The phase of the carrier signal changes with the amplitude of the information signal.
- Digital Modulations:
  - Amplitude Shift Keying (ASK). The amplitude of the carrier signal takes discrete values depending on the information message.
  - Frequency Shift Keying (FSK). The frequency of the carrier signal takes discrete values depending on the information message.
  - Phase Shift Keying (PSK). The phase of the carrier signal takes discrete values depending on the information message.

The following subsection elaborates on two examples of simple analog and digital modulations.

3.3.1.1 Example of a Simple Analog Modulation: Double Sideband

In amplitude modulation, the message signal $m(t)$ is coupled into the amplitude of the carrier signal. Mathematically, this is achieved by multiplying both signals. In telecommunications, a commonly used carrier signal is the sinusoid:

$$y(t) = m(t) \cdot \cos(2\pi f_c t)$$

$y(t)$ is the modulated signal and the term $\cos(2\pi f_c t)$ represents the carrier signal (the sinusoid). The parameter f_c is named "carrier frequency" and is typically some orders of magnitude higher than the maximum frequency of the information signal.

Figure 3.9 sheds light over this process showing an audio signal as information to be modulated in Double Sideband (DSB) using a carrier frequency of 10 kHz (i.e. f_c = 10 kHz). Top two figures show the aspect of this information signal in time domain. The two figures in the second row show the time-domain representation when the information signal is multiplied by the carrier signal.

The two figures at the bottom represent the spectrum of the information and modulated signal, respectively. Conclusions are better interpreted in this frequency domain. These figures clearly show the effect of "multiplying" by the carrier signal. The information spectrum is not new, since it corresponds to the same one shown in Figure 3.6. When looking at the modulated spectrum, one can see how it corresponds to two copies of the original information spectrum shifted and centered precisely at the carrier frequency (10 kHz).

Using this modulation method, we are able to place the signal at frequencies where the channel shows better transmission properties (i.e. it may have lower attenuation, lower interferences, etc.) and where it can share the transmission medium with other such similar signals without overlapping (hence the "channel" concept of Chapter 2). A mechanism known as demodulation undoes this process. Further details of these processes can be found in [2].

3.3.1.2 Example of a Simple Digital Modulation: Quadrature-Phase Shift Keying

In contrast with the previous example, Digital Modulations are used to exchange digital signals (i.e. sequences of bits) instead of analog signals.

The fundamentals of the modulation concept are held: the carrier signal is modified depending on the information source. The difference with respect to analog modulation is that the parameter values are discrete. More specifically, in the case of the Quadrature-Phase Shift Keying (QPSK) digital modulation, the phase of the carrier signal changes depending on the bits to be transmitted. Each one of these values corresponds to the pairs of bits in the information sequence: 00, 01, 10, and 11. Thus, bits are taken two at a time to be transmitted at any given instant of time. This is

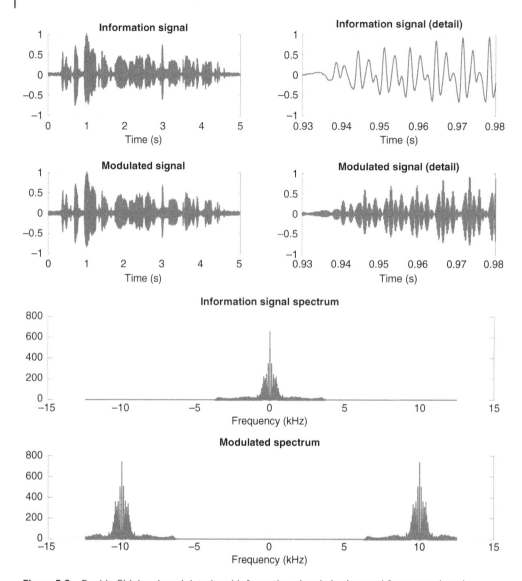

Figure 3.9 Double Sideband modulated and information signals in time and frequency domain.

important, as the transmission speed in channels can be increased by means of increasing the number of bits at a time that can be sent simultaneously at that instant of time.

Mathematically, the modulated signal can be expressed as:

$$y(t) = \begin{cases} \cos(2\pi f_c t), & bits = '01' \\ \cos\left(2\pi f_c t + \dfrac{\pi}{2}\right), & bits = '00' \\ \cos(2\pi f_c t \mp \pi), & bits = '10' \\ \cos\left(2\pi f_c t - \dfrac{\pi}{4}\right), & bits = '11' \end{cases}$$

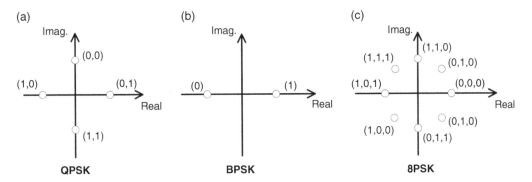

Figure 3.10 Constellation for different Phase Shift Keying digital modulations. (a) Quadrature Phase Shift Keying (QPSK), (b) Binary Phase Shift Keying (BPSK) and (c) 8-Phase Shift Keying (8PSK).

In the case of digital modulation, it is common to visually represent the mapping of bits into the different variables in the carrier signal via what is known as a Constellation Diagram. Figure 3.10a represents the "constellation" of the QPSK modulation in this example. Each point in the constellation (also known as "symbol") represents the corresponding phase for the carrier signal and its designated pair of bits.

Additionally, as Figure 3.10 shows, the higher the number of symbols in the constellation, the more bits can be mapped into the symbols. This has clear implications in the transmission rates, making constellations with a higher number of symbols faster than those that have a lower number of symbols. However, although the noise effect in the transmitted signal will be introduced in Section 3.3.2, it can be intuitive how constellations with a higher number of symbols may be more prone to errors due to noise or channel impairments since symbols are closer together. This trade-off between transmission speed and robustness against errors is common in telecommunications.

3.3.2 Channel Impairments

The communication channel is the transmission medium (wireline or wireless – see Chapter 2) through which the signals propagate from the transmitter to the receiver. No matter what type of communication channel is used, this propagation results in the transmitted signal getting degraded and corrupted as it travels through the channel.

3.3.2.1 Attenuation

The most straightforward form of signal degradations is signal attenuation, which affects the transmitted signal, reducing its amplitude. It normally is a function of the distance and its effects are generally higher at higher frequencies. However, the attenuation strongly depends on the transmission medium. For instance, in the case of optical fiber (see Section 3.4.1) it can go as low as 0.2 dB/km when working at wavelengths of 1550 nm.

Signal attenuation is a central parameter in the link budget computation, addressed in Section 3.4.3.

3.3.2.2 Noise and Interference

The most common type of corruption is caused by additive noise, generated at the receiver's front-end. This noise includes pure noise and interference. The nature of the additive noise is varied, it may be produced due to natural causes, such as the thermal noise or atmospheric interferences in a radio (wireless) transmission; or it can be man-made, as it is the case with the inverters in power lines in the electrical network. In addition to this, out-of-channel interferences from other uses is another form of additive noise that typically affects communications.

Some of the most common types of noise (and interference) are as follows:

- Thermal noise or white noise. Present in the medium and is due to the movement of the electrons in the medium because of temperature. This type of noise is uniformly distributed in the frequency domain, hence its denomination of white. When this noise is limited to a given bandwidth (by filtering), the amplitude of the samples follows a Gaussian distribution.
- Impulsive noise. It is noncontinuous and made up of irregular pulses or peaks of short duration and relatively large amplitude. They are generated by a wide variety of causes that can run from external electromagnetic disturbances produced by atmospheric storms, faults and defects in communication systems, or by starting and stopping machinery processes.
- Intermodulation noise. It may appear when signals of different frequencies share the same transmission medium. This kind of noise affects only frequencies that are the sum or difference of the two original frequencies or multiples of these. Intermodulation noise occurs when there is some nonlinearity in the communication system.
- Crosstalk. It is an effect produced by the energy induced by a line that carries a signal, on another nearby line carrying a different signal.

Noise is added to the transmission signal making the receiver see a different signal to the one sent by the transmitter. To clarify how this affects communications, we will set an example using one of the digital modulations introduced in the previous section. Figure 3.11 shows the same constellation as in Figure 3.10 but including the "decision boundaries" (in dashed lines) used by the receiver of the modulated signal (i.e. the demodulator). It is obvious that the receiver will not receive constellation points at the exact same location set by the transmitter, due to the imperfect propagation of signals. In general, they will be shifted and/or rotated due to the noise and/or attenuation in the "channel." It is the receiver's objective to, upon the reception of a symbol at any location, decide what the most likely transmitted symbol it corresponds to.

The further away the constellation symbols are, the more difficult it will be for a noise or an attenuation to produce an error in the transmission (i.e. move a symbol out of its original area). The top three plots in Figure 3.11 show this idea representing constellations for a different number of symbols (4, 8, and 16). Dots represent the location of the transmitted symbol, whereas the cloud of crosses represent the location of the same symbols after they have been transmitted over a noisy channel. It is important to note that the same amount of noise has been used in the three cases. The figure clearly shows how having a higher number of symbols in the constellation increases the likelihood of having an error in the transmission; it is especially clear in the 16-PSK modulation. In contrast, using a 16-PSK modulation offers twice the transmission speed than a 4-PSK.

This transmission speed vs. error robustness trade-off is a general characteristic in all telecommunication systems.

3.3.2.3 Signal Distortion

Amplitude and phase distortion are responsible for changes in the amplitude and phase of the transmitted signal. The former causes the amplitude of the spectrum to change, whereas the latter creates different phase shifts depending on the frequency. Both have to do with the multipath nature of the transmission media. The multipath effect is produced when there is more than a unique path from a transmitter to a receiver. Consequently, the receiver "sees" a sum of delayed and attenuated versions of the originally transmitted signal, each one corresponding to the

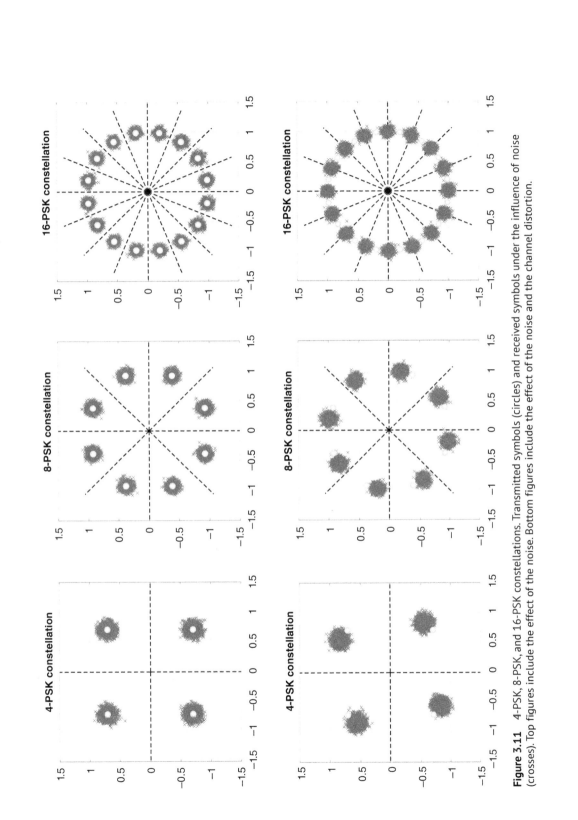

Figure 3.11 4-PSK, 8-PSK, and 16-PSK constellations. Transmitted symbols (circles) and received symbols under the influence of noise (crosses). Top figures include the effect of the noise. Bottom figures include the effect of the noise and the channel distortion.

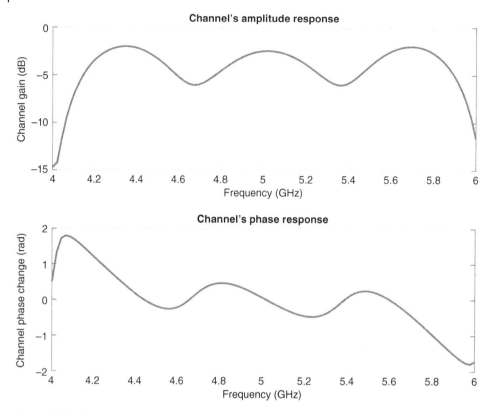

Figure 3.12 Example of a channel's frequency response.

different paths that the signal can take. This effect is common in wireless transmission, where the transmitted electromagnetic wave can bounce in different obstacles within the free space.

One direct consequence of the reception of delayed versions of the transmitted signal is that the duration of pulses is increased (as the receiver sees the beginning of the pulse that traveled the shortest path and the end of the pulse that traveled the longest path). This widening of the pulses may produce some overlapping between two consecutive ones. This effect is referred to as Inter-Symbol Interference (ISI) and may lead to errors in the transmission. ISI's effect is more severe when pulses are shorter (to achieve a high transmission rate) or when the multipath differences are high.

The consequences of multipath are also visible in the frequency domain using what is known as the channel's Frequency Response. This parameter shows how the channel affects the signal as a function of the frequency of the signal components. Figure 3.12 shows an example of Frequency Response of a channel. Looking at the top figure, the negative numbers of the vertical axis show that the amplitude of the spectrum components will be attenuated by a higher or lower factor depending on the frequency. Thus, there will be frequencies more adequate for transmission than others. This was also briefly introduced in Chapter 2 when talking about the different transmission windows on optical communications.

The effects on the phase deserve a special consideration. The effect of the phase can be understood considering the PSK modulation example in Section 3.3.1.2. If we assume that the channel has an amplitude response equal to 0 dB (i.e. 1 in linear or natural units) at all frequencies, and if we were to transmit using a 4.6 GHz signal, the channel would exhibit a channel's phase response of around −0.2 radians (i.e. 11.5°) as shown in the bottom plots of Figure 3.11. Once again, this effect is more damaging with higher-order constellations, such as the case of 16-PSK.

The amplitude and phase distortion are not unique to wireless communications. Multipath effects exist in wireline as well, for example, in interconnected networks with poor impedance adaptation, where electromagnetic waves get reflected in non-adapted junctions producing copies of the signal taking different paths. Additionally, ISI can also occur in the one-to-one direct link, such as in the case of optical communications, as it will be elaborated in Section 3.4.1.2.

In addition to the deformation of the signal's spectrum, another way that the channel may distort the transmitted signal is by reducing its bandwidth (see Section 3.2.3). Real-life channels are band-limited. This means that there is a portion of the spectrum where their transmission capabilities are very limited. As a result, the version of the signal that reaches the receiver does not count with all the frequency components as the original one sent by the transmitter.

3.3.3 Demodulation, Equalization, and Detection

After the signal has traveled through the channel, it arrives at the receiver. There, there are three main tasks to be done: first, shift it to its original frequency location ("demodulation"); then, compensate for the channel distortion ("equalization"); and finally, for digital transmissions, decide the bit or group of bits the signal represents ("detection").

The demodulation process is carried out via a multiplication by a sinusoidal function with the same carrier frequency used in the up-conversion at the transmitter. This multiplication shifts the spectrum of the signal to base band so that it can be analyzed.

The concept of equalization will be further explained in Section 3.3.3.2, since it requires a more detailed elaboration.

The detection process has to do with the decision of what specific digital values the received signal corresponds to. The purpose of the detector is to analyze the region that the received symbol is and decide the most likely transmitted bits. For this decision-making process, the detector considers the decision boundaries shown in Figure 3.11.

3.3.3.1 Signal-to-Noise Ratio and Bit Error Rate

In a telecommunications system, the signal we want to know at the receiver is the message that was transmitted, and the noise is the set of disturbances added to the signal as it is transmitted through the channel between the transmitter and the receiver, which prevents us from knowing exactly the original transmitted signal. Signal-to-Noise Ratio (SNR) is the parameter that needs to be controlled to be able to separate the transmitted signal from the channel disturbances, in the receiver.

Considering the noises mentioned in Section 3.3.2.2, the received signal can be mathematically expressed as the sum of the transmitted signal (affected by the channel propagation) and the total noise signal. One parameter that provides the quality of the received signal is the coefficient between the powers of both of its components: the information signal and the noise signal. This parameter is the SNR and can be computed with:

$$SNR = \sigma_x^2 / \sigma_n^2$$

σ_x^2 and σ_n^2 stand for the powers of both the information and the noise component in the received signal. Indeed, this parameter provides a quality factor since it does not really matter if the noise

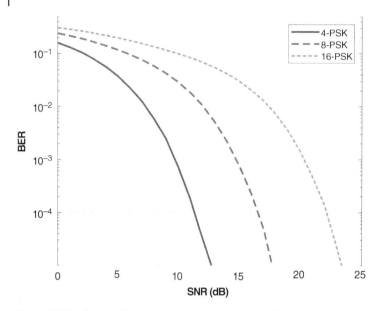

Figure 3.13 Communication performance in terms of BER/SNR for 4-PSK, 8-PSK, and 16-PSK constellation.

affecting the signal is high, as long as the information signal also is of a comparatively high-amplitude value, as determined by the SNR. On the contrary, low noise values can damage the information signal if the latter has very low-amplitude values. This is typically expressed in logarithmic units, i.e. the SNR is in decibels (dB):

$$SNR[dB] = 10\log_{10}\left(\frac{\sigma_x^2}{\sigma_n^2}\right)$$

The SNR in digital systems is tightly coupled to another parameter that also reflects the quality of the communication, the Bit Error Rate (BER). BER represents the number of erroneous bits that arrived at the receiver compared with the total amount of transmitted bits (i.e. it is also a relative parameter). Obviously, dealing with channels with high SNR values, the quality of the received signal is higher and, as a consequence, the amount of errors in the reception must decrease (low BER), and vice versa.

Figure 3.13 shows the relationship between the SNR and the BER for the three different constellations mentioned in Section 3.3.1.2. As explained before, the BER decreases (fewer errors) as the SNR increases (more signal quality). The relationship between BER and SNR evolution is not the same in all constellations (see Section 3.3.1.2 for the trade-off between noise robustness and transmission speed). This can be seen in Figure 3.13: for a fixed SNR value ($SNR = 10\,dB$), the BER value for 4-PSK constellation ($8 \cdot 10^{-4}$) is lower than the value for 8-PSK ($3 \cdot 10^{-2}$), which is also lower than the value for 16-PSK (10^{-1}, 0.1, i.e. 1 error out of 10).

3.3.3.2 Channel Equalization

Noise might be the most difficult effect to deal with, as pure white Gaussian noise is not correlated with itself, and this makes it impossible to predict. Amplifying the received signal does not overcome the effect of the noise because both the signal and the noise get amplified by the same amount, leading to the same SNR ratio and no improvement in the quality of the received signal.

The obvious brute-force approach is to simply increase the transmitted power, as it will make SNR ratio higher. However, there are a number of reasons why this solution cannot always be applied: from regulations about the amount of power to be transmitted or radiated to coordinate and not to affect other channels, to purely technical considerations such as the saturation of some components that can start working outside of its optimal performance point or simply due to energy-efficient issues in battery-powered devices, among others.

With respect to the channel distortion (Section 3.3.2.3), the use of an equalizer at the receiver helps in reducing its effect. The basic idea of the equalizer is that it consists of a block that modifies the received signal in an inverse manner to what the channel did, effectively leaving the signal as if it had suffered no amplitude or phase distortion.

The channel's frequency response is dynamically computed by the equalizer, through a process that makes use of a signal previously known by the transmitter and the receiver. With this procedure, the receiver is always able to compute the channel's effects in a dynamic manner.

3.3.4 Multiplexing

In telecommunications and computer networks, multiplexing is a method by which multiple analog or digital signals are combined into one signal over a shared medium (e.g. in telecommunications, several telephone calls may be carried using one copper pair).

A device that performs the multiplexing is called a multiplexer (or MUX), and a device that performs the reverse process is called a demultiplexer (DEMUX or DMX).

The key concept in multiplexing is that it must be a reversible process, since demultiplexing will always be required. Given this, multiplexing techniques exploit the representation of signals in different domains or dimensions where signals can be easily identified and extracted.

There are different multiplexing techniques, and some of them depend on the type of signal that needs to be transmitted:

- Time Division Multiplexing (TDM). Multiple digital signals (or analog signals carrying digital data) can be transmitted through a single transmission path by temporarily mixing parts of each signal. The mixing process can be at the bit level or in blocks of octets (or higher amounts) and is performed by polling different information sources sequentially. Common systems using TDM are Plesiochronous Digital Hierarchy (PDH – see Chapter 4) or most mobile communications (see Chapter 7).
- Frequency Division Multiplexing (FDM). Several signals are transmitted simultaneously by modulating each of them with a different carrier frequency without overlapping. To avoid interference, channels are separated by guarded or safety bands, which are unused areas of the spectrum. This incurs in an inefficient use of the spectrum, since the guard band area is not used for signal transmission. The use of Orthogonal Frequency Division Multiplexing (OFDM) overcomes this inefficiency, as it will be further discussed in the following.
- Wavelength Division Multiplexing (WDM). It is exclusive of optical transport system. WDM is conceptually equivalent to FDM but using different light carrier signal.

OFDM traditional techniques explained before consider that there are only two ways to increase the transmission speed: transmit using shorter symbols or use constellations with high number of symbols. These two approaches have their limitations. If the symbols become too short, dispersion starts playing a big role since ISI is more significant with respect to the total symbol duration. Moreover, it has also been shown how, when constellations are formed by a high set of symbols, their performance in terms of BER starts to decrease.

So-called multicarrier modulations overcome this limitation. The basic idea in multicarrier modulation is to make use of several carrier frequencies to transmit information simultaneously in time domain and, in parallel, in frequency domain, in a similar way as it is done with the FDM technique, but with information from the same user.

However, if FDM is to be used, there are some restrictions that must be met to avoid power in the side lobes of one carrier leaking to its neighbor. This is connected with the fact that digital signals in frequency domain may have infinite values, thus exceeding their allocated channel.

It became clear in Section 3.2.2.1 that the spectrum used by a non-infinite sinusoidal signal gets spread on both sides of the spectrum. This can be seen in Figure 3.5, where some side lobes appeared around the frequency of the time-domain signal. If we were to use an additional frequency signal to implement FDM, we would have to be very careful not to place it very close to the already existing signal's frequency, or the side lobes of both signals will be mixed up and interfere with each other. This effect is known as Inter-Carrier Interference (ICI).

One solution to avoid ICI is to clear some space around the frequencies that are being used, i.e. Band Guard (a range of frequencies that is not used between adjacent carriers). This approach wastes spectrum and is inefficient.

To overcome this inefficiency in the transmission, Robert W. Chang [4] devised a method to use orthogonal carriers close together and still avoid ICI, using the zero-crossing points of the side lobes to avoid this interference. Using specific frequencies, the effect of one carrier is not present in the rest, so that the carriers are said to be orthogonal and produce no ICI.

This is known as Orthogonal Frequency Division Multiplexing (OFDM), one of the most important multiplexing techniques in modern telecommunication systems.

Orthogonality occurs in frequency domain. Figure 3.14a shows the sum of the spectrum of two non-infinite sinusoidal signals with similar frequencies 3750 and 4843 Hz. Both signals (represented with different tones of gray) have a peak at their corresponding frequencies with an amplitude of 0.5. The black line represents the sum of both signals when transmitted through a medium, which shows the interaction between the two sincs[1]: there is ICI between the two signals due to their side lobes. Choosing frequencies further away (as in Figure 3.14b) does not completely solve the problem.

This situation can be solved with the DFT, which provides a sampled version of the Fourier Transform of a signal. The number and frequency resolution of these samples depend on parameters such as the sampling frequency and the number of samples acquired. Figure 3.14c and d include the values for the sampled version of the spectrum. The realization in these figures uses a sampling frequency of $F_s = 20$ kHz and a number of samples of $N = 32$, which for the purpose of this explanation shows the sampled version of the sum (the signal at the receiver) and does not match the sampled version of the original signal due to the interference.

There are two important points in OFDM. The first one needs the receiver to work with a sampled version of the spectrum, as already explained. The second one is that the position of the zero-amplitude frequencies in the spectrum of each of the transmitted sinusoids needs to be tuned with the sampling frequency (F_s) and number of sampled taken (N). More specifically, the zero-amplitude frequencies are placed at multiple of F_s/N, with the exception of the multiple that corresponds with the signal's frequency; a more detailed explanation of this relationship can be found in [5].

---

1 The Fourier transform of the sinc function is a rectangle, and the Fourier transform of a rectangular pulse is a sinc function.

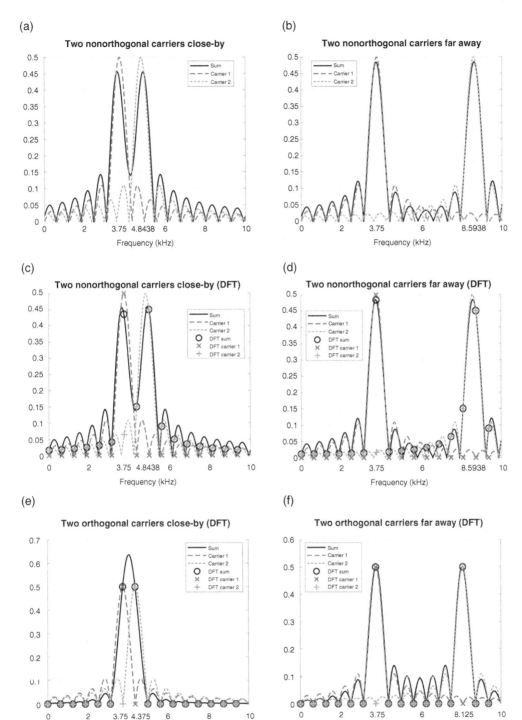

Figure 3.14 Orthogonal frequency representation.

Combining these two effects, the F_s/N factor can be computed and used as transmitting frequencies (multiples of F_s/N factor). By doing so, when the receiver computes the DFT of the summed spectrum, the obtained values would correspond to one of these three cases:

- A frequency where both carrier 1 and carrier 2 have a zero amplitude. Thus, the obtained value would be zero.
- A frequency where carrier 1 is maximum and carrier 2 has a zero amplitude. Thus, the summed value would correspond to the value of carrier 1.
- A frequency where carrier 2 is maximum and carrier 1 has a zero amplitude. Thus, the summed value would correspond to the value of carrier 2.

This effect is shown in Figure 3.14e and f where the frequencies for both carriers have been carefully selected to match different multiples of $F_s/N = 625$ Hz. Carrier 1 is set to 625 Hz · 6 = 3750 Hz and carrier 2 is set to 625 Hz · 7 = 4375 Hz. The amplitude of the sampled summed signal numerically corresponds to the amplitude of each of the corresponding signals, thus eliminating all ICI effects.

This concept can be extended for more than two carriers, for as many carriers as multiples of the F_s/N factor.

To transmit the orthogonal multicarrier OFDM signal, once the multiplex is ready and efficient, each carrier is fed with an independent amplitude value coming from any sort of digital modulators (such as the ones elaborated in Section 3.3.1). Thus, transmission occurs in parallel in frequency and simultaneously in time domain.

Although the spectrum design to obtain the orthogonalization is performed in frequency domain, samples need to be output in time domain for transmission. This is achieved using IDFT[2] (an operation that transforms information from the frequency domain to the time domain). By doing so, a set of time-domain samples are generated, and they can be fed to the communication channel. These samples are transmitted one at a time at $T_s = 1/F_s$ samples per second.

One very important detail when transforming from the frequency to the time domain is that, in general, this transformation can lead to imaginary samples at time domain, which cannot be transformed into voltage levels to be transmitted through the channel. For the time-domain samples to take real values, the spectrum must meet one specific characteristic: the positive side of the spectrum must be the complex conjugate of the negative side of the spectrum, as already introduced in Section 3.2.2.1. From an engineering point of view, this is important since it stops us from using the negative carriers to transmit additional information already present in the positive carriers. In other words, not all N multiples of F_s/N could be used to transmit information but only half of them.

Each of these set of time-domain samples is known as an OFDM symbol, and it lasts for $T_s \cdot N$ samples. Before OFDM symbols are coupled to the channel, a Time Guard window is pre-appended to each one of them. This Time Guard is typically referred to as Cyclic Prefix (CP) and consists of a pre-append of the latest samples of the block. The purpose of the CP is twofold. First, if the CP is longer than the channel's impulse response, it will remove the effect of ISI. Second, this wrapping of samples in the time domain makes the orthogonality of the carriers mathematically easier. Figure 3.15 shows a schematic representation of the operations carried out by an OFDM transmitter. As the figure shows, only N_c out of the possible $N/2$ orthogonal subcarriers is used for data transmission.

---

2 In real-life implementation, this operation would be replaced by its much more efficient version: the Inverse Fast Fourier Transform or IFFT.

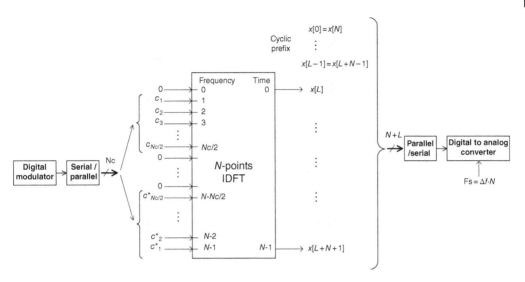

Figure 3.15 Schematic representation of an OFDM transmitter block.

OFDM is highly convenient from that particularity of the channel propagation perspective. If high-bandwidth signals are used, the different frequency components of the bandwidth will be affected by the propagation in a different way, and this affects the quality of the communication and needs to be handled with appropriate and complex techniques. However, in OFDM, the transceiver sees the physical channel as a group of narrowband channels in each of the carriers, each of them much easier to manage. Moreover, propagation constraints affecting some carrier frequencies (e.g. impulsive noise) do not affect the rest, and the communication can be successful.

OFDM also shows some inconveniences. One of the most relevant ones is the Peak-to-Average-Ratio, or PAPR [6–8]. The PAPR refers to the constructed signal after the IFFT, i.e. in time domain. As explained, this signal is created based on the sum of several sinusoids. PAPR produces saturation when there is a constructive interference of many of the data carriers in the spectrum. One of the simplest methods to overcome PAPR problem is to use different phase shifts to the carriers in the same OFDM symbol.

Finally, there are two additional inconvenient effects that damage OFDM performance. One of them is produced when the oscillators (responsible for synthesizing the orthogonal frequencies) at the receiver and the transmitter are not perfectly synchronized. This is referred to as Carrier Frequency Offset (CFO) and produces a slight loss of orthogonality between the carriers, resulting in ICI since the information transmitted in other carriers is sensed as noise.

The other known effect in OFDM transmission is the oscillator's phase noise, which produces that the spectrum of the sinusoidal shape created by oscillator does not look like a delta function but has some small bandwidth. This produces a lack of carrier orthogonality and, consequently, ICI.

Extensive and additional information about OFDM implementation, drawbacks, and advantages can be found in [5].

3.3.5 Channel Coding

Channel Coding is a technique used in digital communications that is focused on the reduction of errors in the receiver side of the systems. As already mentioned, these errors may come from noise sources or interferences in the communication channel.

One straightforward solution for the receiver to fix the errors is to ask the transmitter for a retransmission of the same message. This is commonly known as Automatic Repeat Request (ARQ). It represents a simple solution that is often used and may make sense depending on the scenario. However, it has a clear disadvantage: it incurs in an inefficient usage of the channel, since it is used more than one time to send the same message and needs to wait an agreed time before the lack of reception of a message is confirmed. This kind of techniques are known as Backward Error Correction (BEC) and have several variants such as the Selective Repeat ARQ or the Go-Back-N ARQ.

Forward Error Correction (FEC) techniques are alternative solutions. The basic idea is to extend the original message with some extra bits of redundant information with the intention that the receiver uses them to both detect and, hopefully, correct erroneous bits produced in the transmission. The word "redundant" was not chosen freely in the previous paragraph; indeed, the extra bits need to have some relationship with the original message so that the receiver can exploit them. As a matter of fact, in real-life applications the extra bits are the result of operations over the original message's bits.

As it commonly occurs in engineering, this provides some advantages and disadvantages. One clear advantage is that the message is likely to be sent only once, since the receiver has the tools and the extra bits to correct the errors without asking for a retransmission and waste of time and resources. On the contrary, it has another clear disadvantage that is the extension of the message to transmit. Whether BEC or FEC techniques are more efficient depends on the specific scenario where they are applied to. FEC techniques are widely applied in modern telecommunication systems, and they will be detailed in the following sections.

3.3.5.1 A Simple Example of Coding

An example of a simplified implementation of coding will clarify the technique.

An encoder is a DSP block present at the transmitter. When the encoder is used, two extra copies of each input bit can be produced, with the purpose of adding the "redundant" information. With that, each bit is converted into a "codeword" (i.e. the only possible blocks of bits that can be transmitted) that in this case is 000 (when the input is 0) and 111 (when the input is 1). Indeed, the two extra bits are based on the original message (they are a copy), and they are redundant (instead of 000, a single 0 could be transmitted).

Upon reception, the decoder (the digital signal processing block present at the receiver) takes groups of three bits and decides the original bit each block corresponded to. Given the design of this channel coder, what the decoder would do is to assume that the original message corresponds to the bit that is more repeated in each block. There are several hypothetical scenarios:

- There are no errors in the transmission. In this case, the decoder would have only received sequences of 000 or 111, and it would had decided that a 0 or a 1 had been transmitted, respectively.
- There is one erroneous bit in each codeword. In this case, the possible combinations received at the decoder will depend on where the erroneous bit is located at. To illustrate this, Table 3.1 shows all possible combinations of errors and the corresponding decoder's decision. As it can be seen, using this encoder, all possible errors can be corrected if they affect only 1 bit in each codeword.
- There are two or three erroneous bits in each codeword. Table 3.2 details all combinations and the corresponding decoder's decision. In contrast with the previous case, now, the decoder's decisions are the wrong ones.

Table 3.1 Simulation of receiving one erroneous bit per codeword.

Original message	Tx codeword	Rx erroneous sequence	Decoder decision
1	111	110	1
1	111	101	1
1	111	011	1
0	000	001	0
0	000	010	0
0	000	100	0

Table 3.2 Simulation of receiving two and three erroneous bits per codeword.

Original message	Tx codeword	Rx erroneous sequence	Decoder decision
1	111	100	0
1	111	010	0
1	111	001	0
1	111	000	0
0	000	110	1
0	000	101	1
0	000	011	1
0	000	111	1

The main conclusion from the previous analysis is that the designed code can correct all errors that affected one single bit per codeword. In addition to this, even though we have not considered it so far, is that this code can detect up to two erroneous bits per codeword (although it cannot correct them). The only situation where this code cannot detect or correct the errors is when all bits in the codeword change.

The increment in the message length (to allocate the redundant bits) in FEC techniques is reflected into a parameter know as coding rate (R_c). The coding rate relates the number of bits at the input of the decoder with the number of bits at the output of the decoder. In our specific example, it would be $R_c = 1/3$, which indicates the reduction of the effective transmission rate needed because of using FEC.

Commercial channel codes can be classified in two major classes:

- Block codes. In block codes one of the $M = 2^k$ messages that represent the information message of k bits is mapped to a binary sequence of length n, also referred to as codeword, where $n > k$. One main characteristics of block codes is that they are memoryless. This means that, after a codeword is encoded, when the system receives a new set of k message bits, it encodes them using the same mapping scheme without considering what the result in the previous one was.
- Convolutional codes. These types of codes are described in terms of finite-state machines. At each time instance, i, k message bits enter the encoder, causing n binary symbols generated at the encoder output and changing the state of the encoder from $\sigma^{(i-1)}$ to $\sigma^{(i)}$. The n binary symbols generated at the encoder output and the next state $\sigma^{(i)}$ depend on the k input bits as well as $\sigma^{(i-1)}$.

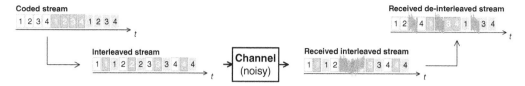

Figure 3.16 Example of how an interleaver scatters burst errors in communications.

The implementation of channel coding is normally asymmetrical in terms of computational complexity, i.e. it requires much less effort in the encoding part than in the decoding part. It is especially asymmetrical in the case of convolutional decoder. In fact, a popular mechanism is used to carry out the decoding in an efficient manner: the Viterbi algorithm [5].

3.3.5.2 Interleaving

The correction performance of normal block or convolutional codes strongly depends on the nature of the errors. Typically, the more scattered the errors are, the better the correction capabilities. For this purpose, the channel coding mechanisms are typically accompanied by an extra component, the Interleaver.

The purpose of the Interleaver is to increase the performance of coding techniques in situations where errors come in bursts (i.e. grouped, not scattered). Figure 3.16 shows how the Interleaver block helps in "scattering" errors that were produced by a noise burst in the transmission. As it can be seen, the decoder is injected with scattered errors even though they occurred in a burst during the transmission.

3.3.5.3 Advanced Coding Techniques

The main limiting factor in obtaining good error correction or noise immunity capabilities is the complexity of the decoder. The major limitation of the block code decoders is their non-optimality, since they have maximum number of correctable bits per codeword, which is directly related to the design of the encoder. In convolutional codes, the complexity of the Viterbi algorithm grows exponentially with the total memory of the convolutional decoder.

To overcome these limitations, there are additional coding techniques that produce better performance with an additional design complexity. Although the literature can offer a more extensive list, for brevity reasons we will briefly elaborate on two of the most common ones: Turbo Codes and Low-Density Parity Check (LDPC) codes.

Turbo codes are built by concatenating two convolutional encoders through a bit Interleaver. Although the variety of turbo codes available is much greater, we will mention exclusively rate 1/3 codes based on a parallel concatenation of rate 1/2 systematic convolutional encoders. Decoding is based on the BCJR (named after its authors L.R. Bahl, J. Cocke, F. Jelinek, and J. Raviv) algorithm [9], a generalization of the Viterbi algorithm that performs a Maximum A Posteriori (MAP) detection of each bit of information (and not of the entire sequence, like the Viterbi algorithm) given the entire received sequence. The idea that underlies turbo codes is to increase the total memory of the code by using the Interleaver, keeping the complexity of the decoder limited by means of the BCJR algorithm.

An alternative to Turbo Codes are the ones known as Low-Density Parity Check (LDPC) codes, introduced by Gallager [10] in 1963 but revisited by the mid-1990s [11, 12]. They exhibit extremely good performance with high correction capabilities. The term "low-density" refers to the *generator matrix* used in the encoding process.

3.3.5.4 Channel Coding in Multicarrier Modulations

In single-carrier systems, channel coding is performed in the time domain, i.e. the encoded bits cover multiple symbols. In multicarrier communication systems, such as OFDM, the frequency domain gives an alternative dimension to which channel coding can be applied to improve the communication performance in the presence of noise or interferences.

One alternative is to perform the encoding over each carrier independently of the others. This is known as time-domain coding and can be applied using any kind of the previously defined coding techniques. By doing so, the coding bits cover a number of different OFDM symbols, giving the code a low probability of showing an error-burst behavior.

This procedure shows two disadvantages. The first one is that both transmitter and receiver need a set of as many encoder/decoders as carriers in the OFDM scheme, since each carrier follows an independent coding flow. The second one has to do with the latency of the received message, since each decoder needs to wait for several OFDM symbols before outputting the corresponding decoded bits.

An alternative approach is the frequency-domain coding. Now coding is performed over all the carriers within the same OFDM symbol. This would reduce the latency. Alternatively, if extra latency can be assumed, the coding may span over several OFDM symbols, thus reducing the chances of facing a burst-error pattern by using an Interleaver. Another clear advantage over time-domain coding is that in this case, only a pair of encoder/decoder elements is needed.

3.3.6 Duplexing

The term duplexing is related to the organization of the transmission and reception flows in a two-way communication system. In general, data exchange can be classified as full-duplex or half-duplex. In half-duplex transmission, only one of the two stations in a link can transmit at a given time. One common example of this kind of systems is push-to-talk devices.

In contrast, in full-duplex transmission the two stations can simultaneously send and receive data. For the exchange of data between computers, this type of transmission is more efficient than half-duplex transmission as communication does not depend on the amount of information coming or going in the opposite direction.

From a physical point of view, full-duplex transmission typically requires two separate paths (for example, two twisted pairs), while half-duplex transmission requires only one. However, this extra communication path does not always need to be an independent communication media (i.e. two separated wires), as we have learnt in Section 3.3.4, there are techniques that allow for reusing the same physical medium:

- Time Division Duplexing (TDD). Where communications in the upward and downward direction take turns to use the channel in corresponding time slots. Consequently, a transmitter device needs to buffer the data until its slot becomes available. The upward and downward slots can be fixed or can change dynamically to adapt and meet the instantaneous communications requirements.
- Frequency Division Duplexing (FDD). Where communications in the upward and downward direction use different frequency carriers to be able to use the channel simultaneously.

TDD and FDD have advantages and disadvantages. Since communications in FDD are simultaneous in time, there is no need for data buffering. However, FDD requires at least double frequency compared with TDD. Dynamic reallocations of TDD slots allow for flexibility in the data transmission; however, it requires for some management mechanism to organize the slot arrangement.

3.3.7 Multiple Access

Multiple access techniques are necessary to manage access to transmission in scenarios where several devices are connected to the same transmission medium (e.g. connected to the same physical cable or using the same radio channel).

A multiple access solution consists of the combination of using a given multiplexing technique to allow for simultaneous orthogonal transmission resources on the same link and a method to fix the rules for using those resources. This is necessary to prevent two or more of these devices from trying to transmit data at the same time over the same resource, resulting in a collision. A collision is the result of adding the energy transmitted by two or more devices simultaneously, giving rise to an unintelligible signal to any of the receivers listening in the medium.

Multiple access methods determine the rules under which different devices can transmit in a channel. These methods are classified, according to the type of access, into:

- Deterministic. This is when access to the bus is predefined, thus minimizing the occurrence of collisions. Some of the most common ones are Time Division Multiple Access (TDMA), Frequency Division Multiple Access (FDMA), and Code Division Multiple Access (CDMA).
- Random. This is when transmitters can access the channel as soon as it is idle. Some of the most common ones are Carrier Sense Multiple Access/Collision Avoidance (CSMA/CA) and Carrier Sense Multiple Access/Collision Detection (CSMA/CD).

All these access methods are not mutually exclusive and are found combined in the different systems.

3.3.7.1 TDMA/FDMA/CDMA/OFDMA

TDMA is based on coordinating the access to the medium in the time domain, i.e. reserving time slots for each of the nodes in the network. The slot structure is predefined, and its schedule is known by all elements in the network. The duration of the slots can be fixed or variable. TDMA does not dynamically adapt to the communication needs of the network elements.

In FDMA, the transmission from different nodes is coordinated in the frequency domain and implemented with different nodes using a different carrier frequency ("channel" in Chapter 2). This particular frequency is preassigned, with modulations that guarantee no interference among channels.

Thus, since each transmitter uses a separated carrier frequency, all transmissions can coexist in time without interference or collision. However, this technique requires a higher bandwidth than TDMA to support all transmission simultaneously.

CDMA method is based in the combined and simultaneous use of the time and frequency domains, with each network device using a unique code to scramble the transmitted signal. The set of codes used in all transmitters in the same network needs to be orthonormal to each other. This way, when a receiver picks up the transmitted signal, it tries to descramble the message with its code and, thanks to the orthonormality, all the information regarding the rest of the transmitter is crossed out, leaving only the information scrambled with the very same code, i.e. the one intended for it.

This technique improves the performance of TDMA and FDMA. However, the amount of orthogonal codes is limited, and its orthogonality is not perfectly achieved in practice, leading to some interference in the transmitted signals.

A final multiple access technique is Orthogonal Frequency Division Multiple Access (OFDMA). As its name suggests (see OFDM), it is a multiple-user version of FDM where each user is assigned with one or several of the orthogonal frequencies of the multiplex. This approach shows very good performance when transmissions are performed over a dispersive channel as the use of narrow-band subchannels reduces the ISI effect.

OFDMA can also be used in systems with multiple antennas per user (also known as Multiple-Input-Multiple-Output [MIMO]). MIMO can take benefit from this technique by multiplexing the transmission of each antenna over a set of orthogonal carriers.

3.3.7.2 Multiple Access Methods
3.3.7.2.1 *Master/Slave*
In this technique, one of the nodes of the network acts as the Master and assumes the control of the exchanges in the medium. The master cyclically polls the rest of the devices, called Slaves, sending a short message to each one of them. If the consulted Slave node has information to return, it will do so by sending a reply message to the previous request. If the slave has nothing to send, it will either send a simple nothing-to-return message or take a previously agreed time-out until the Master considers there is no response and continues to query the next slave node.

The advantages of this type of technique are its simplicity and determinism over the maximum latency time. However, following are some disadvantages:

- Since a Slave node is polled within the cycle, regardless of whether it needs to transmit data to the network, the channel has a minimum unavoidable and, in part, unnecessary traffic load.
- The network depends directly on the operation of the Master. If it fails, the entire system stops working.
- If two Slaves need to communicate between them, the flow of information must pass through the Master.

3.3.7.3 Carrier Sense Multiple Access (Collision Avoidance/Collision Detection)
As mentioned at the beginning of the section, there are alternative media access techniques that allow for random access to a shared medium. One of them is known as CSMA.

This technique is used when several devices share the same transmission medium and there is no specific hierarchy between them (i.e. not a Master/Slave situation). The idea underneath is to "sense" (i.e. listen to) the channel before transmitting in order not to corrupt an already ongoing transmission, due to a collision of data.

Sensing the medium does not completely avoid the risk of having a collision in the channel. To tackle this situation, there exist two variations of this concept: Collision Avoidance (CA) and Collision Detection (CD).

CD technique can be found in Ethernet technology. The working principle is based on sensing of the channel while transmitting. If the information sensed by a transmitting node differs with the message being sent, the transmission of the original frame is aborted since a collision has corrupted the data.

CD technique is difficult to implement in wireless communications as typically the same antenna is used for both transmission and reception (via a duplexer) making it impossible to sense the channel while transmitting. In this situation is where the CA mechanism is used.

CA is based on sensing the channel prior to transmission and, if some communication is detected, the device to wait a random amount of time for the next channel sensing. If no transmission is heard in the channel, the device injects its message. The fact that the waiting period (known as back-off period) is random minimizes the chances of a collision in the medium, although it cannot completely avoid them.

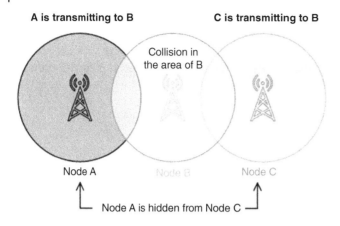

Figure 3.17 Representation of the hidden node problem in a wireless scenario.

This approach does not solve the "hidden node" problem. Figure 3.17 shows this problem in a wireless scenario, where both nodes A and C are transmitting to B and are not aware of the collision since they cannot sense each other's signals. Although this situation may occur in wired and wireless scenarios, it may be easier to understand in the latter. One common solution to this situation is the use of Request to Send (RTS) and Clear to Send (CTS) messages to and from the destination node (Node B in the example of Figure 3.17).

RTS/CTS is a complementary protocol to CSMA/CA. It takes place before data transmission: after verifying that the medium is free, the sender transmits an RTS frame to the recipient, which is also heard by all members of the network within range, indicating their desire to initiate a transmission. After this, the receiver sends a CTS frame as a response to the sender. Only then does the device start sending. It is the receiving node, which is the only one that could observe the possible collision, the one that authorizes the transmission.

3.4 Signal Propagation

3.4.1 Optical Fiber Propagation

Optical communications are based on the transmission of light pulses instead of electrical signals. These light signals may be within the visible spectrum (i.e. having a wavelength between 400 and 750 nm) or, as it is common in high data rate commercial systems, around specific wavelengths (850–1300–1550 nm).

3.4.1.1 Optical Communications Components

Optical communication systems are based on three main components:

- Optical Source. It is the component that creates the beam of light, which is modulated with the information signal. We may differentiate between Light Emitting Diode (LED) sources, which are generally cheaper but allow for a lower modulation speed, and Light Amplification by Stimulated Emission of Radiation (LASER), which are more sophisticated devices and can handle a higher data rate. Optical sources are basically transducers that transform electrical signals to an optical version. This special transformation requires special materials such as some alloy of Indium, Phosphide, Arsenic, and Gallium, where the percentage of them is used to control the specific wavelength of the produced photons.

- Optical Detector. The detector is basically the element that performs the inverse transformation of the optical source based on the "absorption" properties of the same materials used for optical sources but with different concentrations. There are commonly two detectors architectures: PIN photodiode detectors are based on p-type and n-type junction with an intrinsic part in the middle (thus the name, p–i–n) and are able to create one electron per incoming photon; alternatively, the Avalanche Photodiode (AP), which uses a higher supply voltage to accelerate electrons and produces several electrons per incoming photons, thus increasing the performance.
- Optical Fiber. Both sources and detectors could work in a free-space environment. However, to ensure that light is not scattered to the open, an optical fiber is used to capture and control its path.

3.4.1.2 Optical Fiber Propagation Phenomena

As introduced in Chapter 2, an optical fiber consists of two concentrically cylinders of dielectric material with different refractive index. The inner material is referred to as the "core," whereas the surrounding material is called "cladding." There are two main types of optical fibers: the ones made of plastic and the ones made with silicon dioxide (SiO_2). Although plastic fibers are easier to handle due to its higher dimensions (the core is typically 1 mm of diameter), the silicon dioxide optical fibers (core's diameter goes from 5 to 200 µm) allow for much higher transmission rates and lower losses. Thus, they are normally used in telecommunication and Smart Grid applications.

The propagation phenomena in an optical fiber is possible thanks to the Snell's law, which explains the effect occurring when a beam of light reaches an interface between two different media (more specifically, two media with different refractive index). When a beam of light reaches the interface between two media, two effects occur:

- Reflection. Part of the light is reflected on the interface to stay in the same medium. This is shown in Figure 3.18a, where the incoming and reflected angle of the ray is equal. This is:

$$\theta_i = \theta_r$$

- Refraction. Not all the light is reflected to the original medium, as part of it may be propagated to the other medium due to the refraction effect, as shown in Figure 3.18b. In contrast to the reflection, the angle of the refracted ray of light does not generally coincide with the incoming angle, but it depends on the refractive index of both media, following relationship:

$$n_1 \cdot \sin\theta_i = n_2 \cdot \sin\theta_R$$

With these previous formulae the incoming angle that produces a refracted angle equal or higher than 90° can be found. In practice, this means that none of the light is refracted and, thus, all of it gets reflected and stays in the same medium. This is precisely the foundation of the optical propagation using fibers. Thanks to the use of materials with concrete refractive indexes, rays injected within a given range will create incoming angles in the core–cladding interface that will produce total reflection and will travel bouncing from wall to wall. The kind of fibers that have a constant value for the refractive index in the core are known as Step-Index (SI) fibers.

An additional effect to consider in the propagation of photonic waves in the optical fiber is "intermodal dispersion." This effect is produced by the difference in the length traveled and, therefore, the propagation time, between the ray of photons that circulates through the axial axis of the fiber, which will travel the shortest possible distance, and the ray that hits the cladding with the worst possible angle for propagation (also known as "critical angle"), which will travel the greatest possible distance, due to reflection bounces.

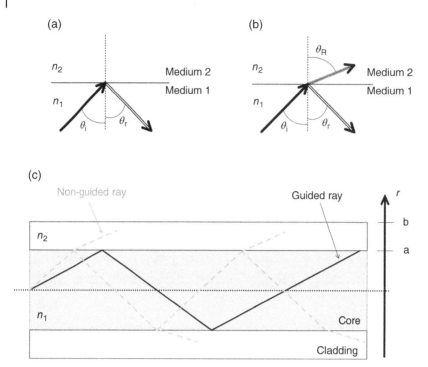

Figure 3.18 Main propagation phenomena in optical fiber.

The effect that is perceived at the fiber output is that the pulses injected to the input of the fiber last longer in time, as part of these pulses follows a longer path. This pulse widening effect causes ISI. One common way to solve this problem is by using Graded-Index (GI) fibers. The core in GI fibers presents a higher refractive index at the axis of the fiber and decreases in the radial direction as we move toward the cladding. This means that rays traveling closer to the axis (i.e. following the shortest path) do it at a slower speed (thus the higher refractive index) than those that travel closer to the cladding (i.e. following a longer path). As a result, the time difference at the fiber end is reduced. In contrast, this kind of fibers are more expensive and present a higher attenuation than SI fibers.

There is an additional effect in optical propagation that limits the maximum transmission speed, the Intramodal Dispersion or Chromatic Dispersion, produced by nonlinearities in the optical fiber. These nonlinearities produce that the phase delay (i.e. the delay introduced by the fiber) depends on the frequency. As a result, some of the frequencies that form an optical pulse of light reach the fiber end sooner than the other. This produces, once again, some pulse widening that, eventually, produces ISI. To tackle this problem, a specially designed fibers is used to compensate this effect by showing a delay complementary to the one in the fiber so that frequencies are grouped back together. Alternatively, there are design methods in the fabrication of the fibers that aim at minimizing the delay difference with the frequency.

Generally speaking, the Chromatic Dispersion is a negligible effect when compared with Intermodal Dispersion.

3.4.2 Radio Propagation

Wireless communications (radio) organize the free-space electromagnetic spectrum in several frequency ranges or bands depending on the propagation characteristics of the electromagnetic waves.

Frequencies ranging from 30 MHz to 1 GHz are suitable for more omnidirectional applications, to cover extended areas. That is, applications where the radiation is performed are a more scattered or not-so-directive way and require not a very strict Line Of Sight (LOS).

The spectrum from 1 to 40 GHz is generally referred to as microwave. In this frequency range, the radiated energy can be steered to aim at a specific direction, which makes it possible to create precise coverage areas and point-to-point wireless links. Its propagation characteristics are not as good as those of the lower bands. However, they can also be used to reach stations not only placed on earth, but also onboard satellites.

Transmission and reception of wireless/radio waves are performed using radiating elements, known as antennas.

3.4.2.1 Antennas

An antenna is defined as an electrical conductor used to radiate or capture electromagnetic energy. To transmit the signal, the electrical energy from the transmitter is converted into electromagnetic energy at the antenna, radiating to the nearby environment. To receive a signal, the electromagnetic energy captured by the antenna is converted into electrical energy and passed to the receiver.

A single antenna is typically used for both transmitting and receiving in two-way communications.

An antenna radiates power in all directions, but this is not always done uniformly in space. The way an antenna radiates depends directly on its physical shape and structure. A common way to characterize the electromagnetic behavior of an antenna is by means of its radiation diagram or pattern, which consists of a graphic representation of the radiation energy of the antenna as a function of the direction.

The simplest radiation pattern corresponds to the ideal case, the isotropic antenna, i.e. one radiating power equally in all directions. In this case, the radiation pattern consists of a sphere centered on the position of the isotropic antenna.

In practical terms, there are omnidirectional antennas, radiating in all directions of a plane; directional antennas, which concentrate the signal beam in a certain direction (as the ones used in VHF, UHF, and microwaves); and sectorial antennas, which cover a sector of space without being as narrow as in the case of directional ones, nor as wide as omnidirectional ones, as it would be the case of cellular antennas.

One popular and important kind of directional antenna is the parabolic-type, which is used in microwave links. The geometrical properties of parabolas are widely used for point-to-point links requiring high directivity. As a rule of thumb, the larger the diameter of the antenna, the more directional the beam will be.

Two parameters define the characteristics of an antenna:

- Gain. The gain expresses how much the antenna is "focusing" the power input from the transmitter. It usually refers to the direction of maximum radiation.
- Beamwidth. It is the solid angle in the three-dimensional radiation pattern at which the gain of the antenna goes below a certain threshold. This threshold is typically set to 3 dB lower than the gain.

As a general rule of thumb, highly directive antennas will have high gain and low beamwidth.

3.4.2.2 Array Antennas and Beamforming

To increase the performance of antennas, further than what can be achieved by "simply" increasing their electrical size, an assembly of radiating elements can be created following specific geometrical configurations. This new group of elements is considered as an array antenna. It is common, although not necessary, that all elements in the array are identical. The key in the radiating performance of the array antenna resides in the geometrical distribution of the individual radiating elements.

As a rule of thumb, there are a number of parameters that can be tuned to control the overall radiation characteristics [13]:

- The geometrical configuration of all elements in the array.
- The relative separation of each element in the array.
- The amplitude of the excitation signal to each element in the array.
- The phase difference of the excitation signal to each element in the array.

Whereas the two first items require some mechanical modification of the array, the two last ones only require a modification in the excitation signal to the elements. In this case, there is a clear advantage in favor of the two last ones since they allow for an easier dynamic reconfiguration of the array and, consequently, the overall radiation pattern. This technique to dynamically reshape the gain and directivity of an antenna is referred to as beamforming.

Having a dynamic control of the beam has numerous advantages. One of them is that antennas can focus their radiation pattern toward the desired users while rejecting unwanted interferences. This allows for the implementation of Space Division Multiple Access (SDMA), i.e. being able to multiplex users simultaneously in time and frequency by using their different locations.

3.4.2.3 Free-space Propagation Phenomena

The propagation phenomena of radio (wireless) signals in open (free) space suffer from specific effects. The main effects can be listed as follows:

- Free-space losses. In any type of wireless communication, the signal gets attenuated with the distance. This type of attenuation can be expressed in terms of the quotient between the radiated power and the power received at the antenna and is calculated as:

$$L = \left(\frac{4\pi d}{\lambda}\right)^2$$

where λ is the central wavelength of the transmitted signal and d is the distance between the two antennas. Or, alternatively, in decibels (d is to be expressed in meters, and f in Hz):

$$L[\text{dB}] = 10\log_{10}\left(\frac{4\pi d}{c_0/f}\right)^2 = 20\log_{10}(f) + 20\log_{10}(d) - 20\log_{10}\left(\frac{4\pi}{c_0}\right)$$
$$= 20\log_{10}(f) + 20\log_{10}(d) - 147.56$$

- Reflection. It occurs when an electromagnetic signal reaches a surface that is relatively large compared with the wavelength of the signal, such as a building. The wave bounces on the obstacle and arrives at the receiver slightly later that the version of the wave that found no object, producing constructive or destructive interferences at the receiver.
- Diffraction. It appears when radio waves reach a corner of an impenetrable body whose size is significantly greater than the wavelength of the radiating signal. When a radio wave finds these corners, they propagate in different directions taking the corner as their new source. This may

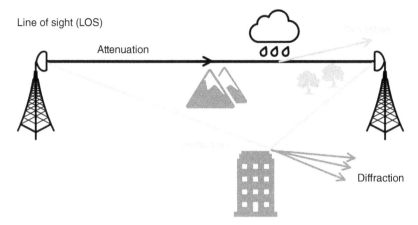

Figure 3.19 Propagation phenomena.

produce situations where signals can be received even when there is no clear LOS from the transmitter.
- Refraction. Radio waves can be refracted (or deflected) when they propagate through a nonhomogeneous medium, such as the atmosphere. Refraction may be due to atmospheric conditions in the troposphere (the lower layer of the atmosphere) that has a layered structure. It results in that only a fraction, or even none, of the transmitted wave following the LOS reaches the receiving antenna.

All these four phenomena are schematically shown in Figure 3.19.

3.4.3 Link Budget

Link budget is a commonly used metric to evaluate the design of any kind of telecommunication system. The basic idea is to account for all the gains and losses that are present in a communication path between a transmitter and a receiver and compare the result with the maximum "budget" that each system can afford. This can be done for both wireless and wireline systems.

Elements that show some gain in the transmitted power are typically amplifiers and antennas. The rest may introduce losses in the system, especially those that are passive (i.e. they do not have a power supply). Some examples are distributors, connectors, splices, splitters, and probably the element that introduces the highest amount of losses: the propagation channel. There are some other power-supplied elements that may introduce no losses or very small losses, such as active distributors or splitters, switches, multiplexers, etc.

When computing the link budget, it is common to add a Security Margin (SM) term to account for fading occurring in the transmission channel or other unanticipated effects. A typical link budget calculation looks like:

$$P_{Rx}\left[\text{dBm}\right] = P_{Tx}\left[\text{dBm}\right] - \sum Losses\left[\text{dB}\right] + \sum Gains\left[\text{dB}\right] - SM\left[\text{dB}\right]$$

Figure 3.20 shows an example of an optical link. There are several components in the path between the transmitter and receiver that both increase or decrease the signal's power. In long communication links (hundreds of km), the signals need to be amplified along the way. In this example in Figure 3.20, the amount of power that reaches the receiver is $-17\,\text{dBm}$: as long as this amount of power is higher than the receiver's sensitivity, the system will work properly.

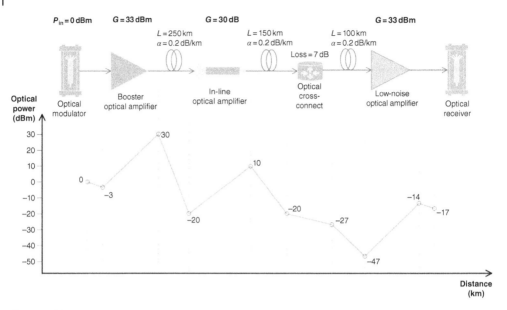

Figure 3.20 Link budget example calculation.

References

1 Oppenheim, A.V., Willsky, A.S., and Nawab, S.H. (1997). Signals & Systems. Prentice-Hall International.
2 Proakis, J.G. and Salehi, M. (2013). Fundamentals of Communication Systems. Pearson Education.
3 Cooley, J.W. and Tukey, J.W. (1965). An algorithm for the machine calculation of complex Fourier series. *Math. Comput.* 19 (90): 297–301.
4 Chang, R.W. (1966). Synthesis of band-limited orthogonal signals for multichannel data transmission. *Bell Syst. Tech. J.* 45 (10): 1775–1796.
5 Haykin, S.S. (2013). Digital Communication Systems. Wiley.
6 Jiang, T. and Wu, Y. (2008). An overview: peak-to-average power ratio reduction techniques for OFDM signals. *IEEE Trans. Broadcast.* 54 (2): 257–268.
7 Han, S.H. and Lee, J.H. (2005). An overview of peak-to-average power ratio reduction techniques for multicarrier transmission. *IEEE Wirel. Commun.* 12 (2): 56–65.
8 Cimini, L.J. and Sollenberger, N.R. (2000). Peak-to-average power ratio reduction of an OFDM signal using partial transmit sequences. *IEEE Commun. Lett.* 4 (3): 86–88.
9 Berrou, C., Glavieux, A., and Thitimajshima, P. (1993). Near Shannon limit error-correcting coding and decoding: turbo-codes. 1. *Proc ICC'93-IEEE Int. Conf. Commun.* 2: 1064–1070.
10 Gallager, R. (1962). Low-density parity-check codes. *IRE Trans. Inf. Theory* 8 (1): 21–28.
11 MacKay, D.J.C. (1999). Good error-correcting codes based on very sparse matrices. *IEEE Trans. Inf. Theory* 45 (2): 399–431.
12 MacKay, D.J.C. and Neal, R.M. (1996). Near Shannon limit performance of low density parity check codes. *Electron. Lett.* 32 (18): 1645–1646.
13 Balanis, C.A. (2016). Antenna Theory: Analysis and Design. Wiley.

4

Transport, Switching, and Routing Technologies

4.1 Introduction

Telecommunication concepts are materialized in technologies. Transport, switching, and routing technologies covered in this chapter have a role in one or several of the core or access domains (and intermediate ones), as in Table 4.1.

Technologies are the result, not only of the conceptual framework but also of the state of the art of the age in which they appeared and evolved. Above all, they are the consequence of the communication problem they intend to solve, and evolve based on its successful adoption and the need to cover new telecommunication needs and use cases.

Technologies overlap but also supersede each other. Technologies complement one another and eventually create hybrid networks that loosely resemble each other, as there is not a unique way to use and combine them. Thus, the assessment of the adequacy of the different technologies to the different scenarios or network domains has to consider the global perspective of any technology within the rest of technologies used, and many more aspects specific to the service needs (bandwidth, latency, symmetry of communications, scalability in terms of users and number of sessions, availability, time-sensitiveness, etc.).

This chapter elaborates on the main transport, switching, and routing smart grid-related technologies. It also includes the main mechanisms that make packet-switching technologies work.

4.2 Transport Networks

Transport function carries telecommunications information in a point-to-point manner, connecting network elements. Transport works from a set of transmitters and receivers communicating with each other to cover the needed distances on top of a physical supporting media. Transport provides the capability to define end-to-end fixed routes (that can be protected so that if one of the paths is down, the alternative path is selected) that will be followed by all data connecting source and destination end-points.

Transport exists at core and access level, covering the distances and data capacities needed in each segment. Vendors differentiate their product portfolio according to this fact, and it is easy to find references to transport solutions in the core, metro or access ranges [1].

The evolution of digital transport technologies has consolidated a group of technologies of different origins that both supersede and complement each other. Along this evolutionary process,

Smart Grid Telecommunications: Fundamentals and Technologies in the 5G Era, First Edition. Alberto Sendin, Javier Matanza, and Ramon Ferrús.
© 2021 John Wiley & Sons, Inc. Published 2021 by John Wiley & Sons, Inc.

Table 4.1 Presence of the different technologies in the different domains of telecommunications.

Technology	Core	Access	Transport	Switching/routing
PDH	X	XXX	X	
SDH	XX	XX	X	
WDM	XXX	X	X	
OTN	XXX	X	X	
Ethernet	X	X		X
IP	X	X		X
MPLS	X	X		X
MPLS-TP	X	X	X	

where these technologies are promoted by different interest groups, much confusion exists [2]. Conceptually, there are two groups of technologies sharing a common background. On one side, we have technologies evolved from the traditional telecommunications (plesiochronous digital hierarchy [PDH], synchronous digital hierarchy [SDH], Dense Wavelength Division Multiplexing [DWDM], optical transport network [OTN]). On the other, other technologies take advantage of the success of packet-based switching technologies and Ethernet-based services (multiprotocol label switching – transport profile [MPLS-TP], Carrier Ethernet).

Transport works over any telecommunications physical supporting media. Historically, copper pairs gave way to coaxial cables, and these were eventually abandoned in favor of optical fiber solutions, prevalent today to cover long distances and manage big amounts of data (see evolution in Figure 4.1). Optical fiber coexists with radio communications and PLC technologies that are also used for the same purpose, but covering shorter distances and providing always comparatively smaller bandwidth.

Transport technologies intend to define standard interfaces that are completely interoperable across vendors. Thus, a transceiver by one vendor can interface another vendor's transceiver of the same technology. The exception to this rule is what happens with radio in terms of point-to-point communication technologies: a microwave point-to-point radio device from one vendor can just be expected to communicate with another one of the same vendor on the other end.

4.2.1 Plesiochronous Digital Hierarchy (PDH)

The first standard transport implementations of digital technologies came using time division multiplexing (TDM – see Chapter 3) in the 1970s, becoming an on-the-field success in the 1980s.

TDM digital data transport made use of the so-called "Plesiochronous Digital Hierarchy" with different implementations worldwide. The "plesiochronous" (meaning nearly synchronous) nature of this transport technology comes from the not-so-strict need of synchronization between the connecting network elements [3]: some reserve "stuffing" bits absorb the slight differences in network element clocks.

E-Carrier, T-Carrier, and J-Carrier [5] were the main TDM-based transport technologies implemented in Europe, the United States, and Japan, respectively (US and Japan's are very similar [6]). These technologies offered interfaces with different standard rate options.

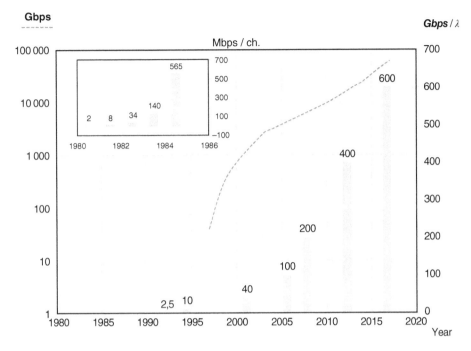

Figure 4.1 Increasing capacity achievement. *Source:* Adapted from[3, 4].

From the lower 64 kbps channels to the higher and uncommon 564 992 kbps maximum data rate, a hierarchy of different capacity channels was created. E-carrier is mainly standardized in ITU-T Recommendations G.704 "Synchronous Frame Structures Used at 1544, 6312, 2048, 8448 and 44736 kbit/s Hierarchical Levels," and G.732 "Characteristics of Primary PCM Multiplex Equipment Operating at 2048 kbit/s." T-carrier is specified as well by several ATIS standards, i.e., ATIS 0600403 "Network and Customer Installation Interfaces – DS1 Electrical Interface" (formerly ANSI T1.403) and ATIS 0600107 "Digital Hierarchy – Formats and Specifications" (formerly ANSI T1.107).

The most pervasively used transport units are probably E1 and T1. Their transport capacity was used for voice communication purposes for decades (and still is). E1 has a total of 32 eight-bit timeslots, 30 of them used for voice and the other 2 for synchronization and control. T1 format has 24 eight-bit time slots and one 1-bit time slot for frame synchronization. Each of the 24 timeslots carries a single voice channel for 5 successive frames; however, in the frame number 6, a 7-bit sample is used, and the eight remaining bit is robbed for signaling purposes.

From these standard structures, two more are derived: fractional E1/T1 and unchannelized framed signal. Fractional E1/T1 allows the use of part of the frame only (a few timeslots; this has commercial meaning when leasing capacity in a link). Unchannelized framed signal, proprietary, allows using the entire frame irrespective of its internal framing boundaries.

The different structures are multiplexed, one inside the others, like in Figure 4.2. Table 4.2 shows the most common data rates offered by International (ITU) and North American (ATIS) standards.

4.2.2 SDH/SONET

SDH and synchronous optical NETwork (SONET) are transport technologies defined as the necessary evolution of PDH technologies to improve their shortcomings, but specifically to address the

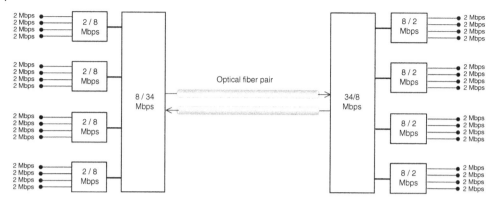

Figure 4.2 PDH link.

Table 4.2 TDM technologies data rates.

ITU-T	ATIS (ANSI)	Data rate (kbps)
	T1 (DS1)	1 544
E1		2 048
	T2 (DS2)	6 312
E2		8 448
E3		34 368
	T3 (DS3)	44 736
E4		139 264

incremental need of higher transmission capacities. These technologies started to replace and complement PDH solutions in the 1990s.

SDH is mainly standardized by ITU-T G.803 "Architecture of Transport Networks Based on the Synchronous Digital Hierarchy (SDH)" and G.783 "Characteristics of Synchronous Digital Hierarchy (SDH) Equipment Functional Blocks." SONET is standardized, among others, in ATIS 0900105 "Synchronous Optical Network (SONET) – Basic Description Including Multiplex Structure, Rates, and Formats" (formerly ANSI T1.105) and ATIS 0600416 "Network to Customer Installation Interfaces – Synchronous Optical NETwork (SONET) Physical Layer Specification: Common Criteria" (formerly ANSI T1.416).

SDH/SONET opened the door to the use of the available optical fiber deployments, providing higher transmission capacities and focusing on optical interfaces. It also simplified the access to the lower level hierarchy transport services (named "*tributaries*"), removing the need of multiplexing blocks with the new simplified add/drop functions. The location of the "tributaries" inside the "aggregate" transport flow is simplified, thanks to a mechanism based on pointers, and bit stuffing is no longer needed. Other capabilities included higher availability, overhead reduction, and traffic encapsulation capabilities.

The hierarchy of interfaces defined by both standards is provided in Table 4.3. Traffic is encapsulated in a hierarchy of transport modules (virtual container [VC] in SDH and virtual tributary in SONET) in such a way that, e.g., PDH can be transported.

Table 4.3 SDH/SONET technologies data rates (STM optical and electrical; OC for optical and STS for electrical).

SDH	SONET	Bit rate (Mbps)
STM-0	STS-1/OC-1	51.84
STM-1	STS-3/OC-3	155.52
STM-4	STS-12/OC-12	622.08
STM-16	STS-48/OC-48	2 488.32
STM-64	STS-192/OC-192	9 953.28
STM-256	STS-768/OC-768	39 813.12

Specifically for packet-based interfaces transport (Ethernet/IP), SDH/SONET rely on a set of new generation functionalities [7] such as generic framing protocol (GFP), link capacity adjustment (LCAS), and virtual concatenation (VCAT) defined mainly in ITU-T G.7041, G.7042, G.707, and G.783 to connect pure packet-based services over existing SDH/SONET networks:

- GFP. It is a framing mechanism to map and transport packet-based client signals, such as Ethernet (but also others), over fixed-data-rate optical channels [8].
- VCAT. It is a mechanism to combine VCs and create larger payload channels (VCAT group, or VCG). VCAT is an inverse multiplexing procedure where the contiguous bandwidth is broken into individual smaller structures that can be transported in existing predefined structures [1].
- LCAS. It is a complementary technology to VCAT, where the size of a VCG may dynamically change.

There are four types of devices in the SDH/SONET architecture, namely Regenerator, Terminal Multiplexer, Add–Drop Multiplexer (ADM), and Cross-Connects. The Regenerator regenerates the clock and the amplitude relationship among signals (attenuated and distorted in the channel). The rest of the elements, although with different functions, in practice, differ in the number of possible interconnections.

Transport protection mechanisms are instrumental in SDH/SONET. Availability targets are 99.999%. Protection mechanisms (Figure 4.3) reserve transport capacity in a redundant route, to use it when the working route fails. Protection mechanisms in SDH are defined in ITU-T G.841 and G.842 (interworking) and are:

- MS-SPRING or Multiplex Section – Shared Protection Ring. MS-SPRING is a complete ring protection. The working channels carry the normal traffic signals while the protection channels are reserved for the protection when needed. Shared protection rings can be categorized into two types: two fibers and four fibers, the latter offering more protection capacity. Switch time can be as low as 50 ms.
- SNCP or SubNetwork Connection Protection. SNCP is a more flexible and granular dedicated linear protection mechanism that can be used on any physical structure (i.e. meshed, rings, or mixed) and can be applied to individual VCs.

4.2.3 DWDM

While until 1990 the telecommunication transport technologies' effort was placed on increasing the transmission capacity in the "time" domain, from there on a parallel effort emerged to create multiple parallel instances of the highest possible speed channels, within the same optical fibers.

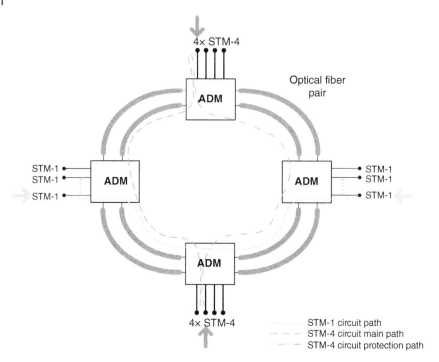

Figure 4.3 SDH/SONET transport network.

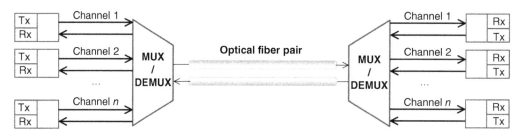

Figure 4.4 WDM link.

WDM multiplies the available bandwidth of existing optical fiber, multiplexing in the "optical" domain hundreds of individual channels, while keeping them separate (see Figure 4.4). Existing SDH/SONET and other less usual data transmission protocols can be transparently encapsulated in WDM.

WDM technology is only applicable to optical fiber, with two flavors: coarse (CWDM) and dense (DWDM). CWDM's capacity, reach, and cost are lower than DWDM's. The standard organization of CWDM or DWDM is defined in the wavelength positions ("grid") defined in ITU-T G.694.1 "Spectral grids for WDM applications: DWDM frequency grid" and G.694.2 "Spectral grids for WDM applications: CWDM wavelength grid." Although CWDM was popular among lower cost applications, DWDM is probably the trend nowadays, with the continuous affordability of optical systems.

Modern systems are year by year improving the data rate – distance product (bps × km), both in real state-of-the-art commercially available systems and laboratory experiments including new multicore optical fibers. Research efforts are focused on the data rate per lambda, the distance to be covered without regeneration (800 Gbps in a single lambda up to 950 km [9, 10]), and new optical fiber types (multicore [11]).

WDM systems count on the optical supervisory channel (OSC) as the way to implement supervision and management needs of transport technologies. WDM network devices integrate several functions:

- Transponder. It is a transmitter–receiver combination that converts an optical signal into another optical signal, by means of a conversion into the electrical domain.
- Multiplexer. It combines different signals from the transponders, each with a wavelength, in the same pair of optical fibers.
- Demultiplexer. It performs the opposite function to the multiplexer, separating wavelengths and sending them to each corresponding transponder.
- Amplifier (optical). It performs the amplification of the input signal in the optical domain.

The different WDM network elements are:

- Terminal network element. It is the origin or destination of signals. It just connects to its adjacent node.
- Repeater network element. It regenerates (reconstructs) the original signal to reach longer distances (3R; re-amplifies and re-shapes optical pulses; and re-times the signal [12]).
- Add–Drop Multiplexer network element. It allows signals to be connected to the adjacent network elements, while they allow for some of them to be extracted or injected in the node. Reconfigurable Optical ADMs (ROADMs) can be dynamically configurable.
- Optical Cross-Connect network element. It manages to optically interconnect more than two network elements.

4.2.4 Optical Transport Network (OTN)

The evolution of WDM systems laid as full optical systems, capable of performing all functions in optical domain, was parallel to the history of SDH/SONET systems. While OTN was being developed to complement simple non-feature-rich WDM solutions, OTN developed similar to SDH/SONET systems.

OTN specifications include framing conventions, non-intrusive performance monitoring and network management, improved error control and correction, rate adaptation, multiplexing mechanisms, ring protection, and network restoration mechanisms operating on a wavelength basis. The main advantages [13] of OTN over SDH/SONET are based on a stronger FEC, transparent transport of high data rate client signals, improved scalability, and tandem connection monitoring.

OTN provides a solution to the pure multiplexing in the optical domain as well, providing a way to multiplex and switch signals in the electrical domain when there is a need to groom them. Hybrid (electrical–optical) switches can do this [1]. The adoption of OTN has been unequal in different work regions.

OTN is often described as a "digital wrapper," allowing different services to be transparently transported over it. OTN is thought to connect over optical fiber links and provide transport, multiplexing, routing, management, supervision, and survivability functions of optical channels carrying client signals. OTN is mainly described in ITU-T G.872 "Architecture for the Optical Transport Network (OTN)," G.709 "Interfaces for the Optical Transport Network," and G.798 "Characteristics of Optical Transport Network Hierarchy Equipment Functional Blocks."

Table 4.4 OTN interfaces.

Service	Bit rate (Gbps)
ODU0	1.25
ODU1	2.5
ODU2	10
ODU3	40
ODU4	100
ODUCn	$n \times 100$

OTN defines a layered architecture with three levels. Client signals are inserted in optical payload units:

- Digital layers:
  - Optical channel data unit (ODU). It provides services for end-to-end transport (see Table 4.4).
  - Optical channel transport unit. It provides transport services for ODU.
- Optical channel layer.
- Media layer.

ITU-T G.798.1 includes examples of OTN equipment illustrating different ways to combine OTN functions:

- OTN network termination unit. It adapts one or more client signals into either a WDM port or a single-channel OTN port.
- Optical Amplifier. An optical amplifier is used to amplify a WDM signal.
- Wavelength ADM (WADM). It consists of two WDM ports with add–drop or pass through capability (in the optical domain or the electrical domain if regeneration is needed) for the individual optical channels within the WDM signals.
- Wavelength Cross-Connect (WXC). It is similar to a WADM, except that it supports more than two WDM ports.
- Sub-Wavelength ADM and Sub-Wavelength Cross-Connect. With several types, its most basic type adds an ODU matrix to their respective equivalent WADM and WXC to provide flexibility in the assignment of clients to wavelengths and multiplexing of multiple clients into a single wavelength.

4.3 Switching and Routing

4.3.1 Switching Principles

Switching occurs at data link layer (i.e. layer 2 of the OSI model; see Section 2.5 in Chapter 2) and solves the interconnection of a number of devices inside a LAN.

As seen in Section 3.3.7 of Chapter 3, shared-medium access techniques are based on splitting the channel resources (time, frequency or code) between the available devices, but this approach is not scalable when the system consists of a high number of devices.

The solution to this situation is the use of a switch device, or simply, a "switch." From the outside, a switch is similar to a hub; however, the difference comes in their functioning: switches only send frames to the output where the destination device is connected to, while hubs repeat all traffic to all ports. Consequently, switches are "smarter" devices than hubs, since they must look at the frame's destination address and forward it to the corresponding port, thus avoiding bottlenecks.

Chances are that two different devices want to transmit to the same destination at the same time. In case of using a hub, since both devices belong to the same collision domain, they must sort the priority themselves using some of the CSMA algorithms described in Section 3.3.7.3 of Chapter 3. However, in case of using a switch, both transmitting devices are isolated from each other and they do not need to contend for the channel since they belong to different contention domains. Still in this situation, the switch would need to send both frames to the same destination device at the same time, which would be impossible. For this purpose, switches make use of some buffering memory where one of the frames can temporarily be stored until the other has been fully transmitted.

Given these advantages when using switches, modern networks make use of them, leaving the use of hubs to legacy system.

There is a third device working at Layer 2 to interconnect different LANs within the same area, the so-called "bridges." Bridges interconnect LANs that for geographical, organizational, logistical, or load-balancing reasons are or need to be kept separated.

Bridges work in a similar manner as switches: they only forward traffic to other LANs if the receiving node is located there. Otherwise, traffic is kept inside the original LAN. This way, each LAN is able to work at its full capacity independently without the bottleneck problem mentioned in case of working with hubs.

Bridges, however, must have a transparent behavior network-wise. From the point of view of connectivity, they should provide it as if both the source and destination device were physically connected to the same LAN. Figure 4.5 shows how a bridge can be placed to interconnect three related LANs.

The following subsections shed some light on how this routing can be performed efficiently and how closed loops can be avoided by organizing the forwarding in a spanning tree shape.

4.3.1.1 Switching Process

Switching mechanism relies on two basic elements. First, each device in the network is univocally defined by its medium access control (MAC) address; all frames sent in the network contain the source and the destination's MAC address. Second, all nodes in the network use a forwarding database (or table) that maps destination MAC addresses with a specific switch port. This database is used in the two main sub-tasks involved in the switching process: frame retransmission and address learning.

Frame retransmission relies on bridges/switches creating and maintaining a forwarding table for each LAN it is connected to or devices inside the LAN. The table indicates the node's MAC address associated with each port. When a frame is received on one of the ports, the switching device must decide whether the frame must be forwarded and on which port.

Address learning relies typically on automatic mechanisms, such as one based on the use of the source MAC address field present in the frames. The switching device learns that the device with the MAC address indicated in the frame must be reachable through the port that the frame entered.

As switching is a dynamic process, a timer is set in each database entry. When the timer expires, the entry is removed assuming that the corresponding device is no longer present in the network. If the destination MAC address in the frame is not in the table, the switching device forwards the frame to all ports except the port where the frame came in. Thanks to this flooding mechanism (broadcast or multicast), it makes sure that the frame reaches its destination.

4.3.1.2 Solving Switching Loops: Spanning Tree Protocol

When a network topology offers alternative paths to connect sources with destinations, loops can be created. This is, e.g., the case in Figure 4.5, with two bridges interconnecting two LANs. In the figure, Host 3 transmits a message destined to Host 1. When Bridge α and Bridge γ receive the message, they forward it to LAN A, where the destination device is. Indeed, the message will be delivered to Host 1, but it will also be delivered to the bridges at the interface located in LAN A. What both bridges will see then, is a message with a source address equal to the Host 3's. With this situation, both will assume that Host 3 has been moved to LAN A and will update their routing table accordingly. Consequently, they will now not be able to forward any frame to Host 3 since they will have misplaced its destination LAN.

A common way to solve these situations uses graph theory so that for each pair of nodes keeps a spanning tree that maintains the connectivity of the graph, but does not contain closed loops.

The spanning tree protocol defined in IEEE Std 802.1D-2004 [14] is used for this purpose. Its algorithm involves exchanging a small number of messages to keep the network loop free and to obtain the least cost spanning tree, while managing to adapt to changes in the topology.

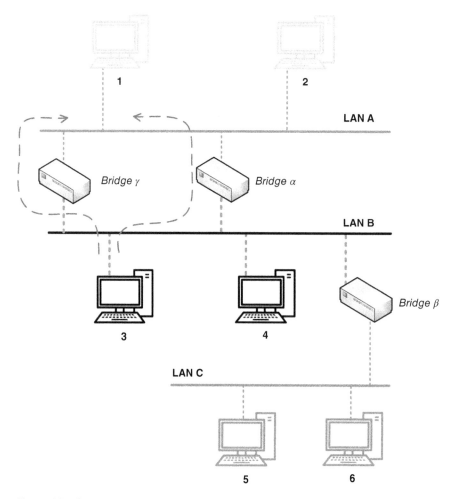

Figure 4.5 Common use of bridges to interconnect three LANs.

4.3.2 Routing Principles

The term routing refers to the process of forwarding a packet from a source node to a destination node when there are multiple routes. Routers are devices with several ports, used as inputs and outputs. Each port connects to different links or paths to other routers. When a packet reaches a router, the devices need to decide which of the ports is most convenient to forward the packet to reach its final destination in an optimal manner.

The routing process consists of two consecutive phases: path determination and packet forwarding, and is implemented with routing algorithms.

The path-determination phase consists in finding the optimal path between the source and the destination as in Figure 4.6 where several host devices act as source and destination.

Routing devices create a routing table where all the information about possible routes to all destinations is stored. Different routing algorithms use different path information (see Section 4.3.2.4) that is typically exchanged between routers, and different criteria (see Section 4.3.2.2) to determine this optimal path.

The packet-forwarding phase is where the packet is actually transmitted. Forwarding is the decision process made by a router to forward an incoming packet to a specific port. This decision is based on both the information inside the packet and the routing table stored at the router. When forwarding, the router needs to change the physical-layer address of a packet to be consistent with the physical address of the device present in the forwarded port. This modification does not change the final destination of the packet since this address is stored in higher-level protocols.

4.3.2.1 Routing Classification

Routing algorithms can be categorized as follows:

- Centralized or decentralized routing algorithms. Centralized routing algorithms compute the optimal route between two nodes making use of the complete global knowledge about the network they are in. The computation of the optimal route can be done in a centralized node or in each individual node; in the latter case global information about the network must have been distributed to all nodes. Algorithms based on the knowledge of all the devices in the network are

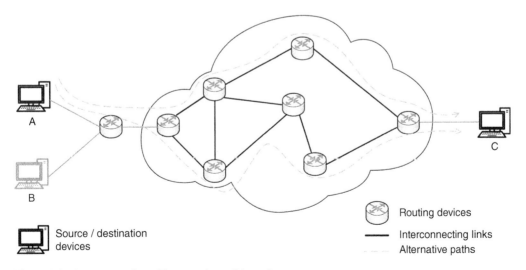

Figure 4.6 Interconnection of hosts and possible paths.

commonly referred to as link-state algorithms and they consist in assigning weights or costs to all network links to find the optimal route.

In contrast, in a decentralized routing algorithm, the calculation of the path is carried out by an iterative and distributed process by all nodes. In this case, nodes do not count with the information of all the links, but the links they are directly connected to. By means of an information exchange process between neighbor nodes, each one of them is able to estimate the corresponding cost of all links in the network without a central entity that provides it.

- Static or dynamic routing algorithms. Route tables in a static algorithm rarely change. In contrast, a dynamic routing algorithm modifies the routing tables based on changes on the topology or traffic in the network. This change can occur periodically or as a response to a change in the network. One direct consequence of dynamic algorithms is that they adapt quickly to new scenarios.
- Load-sensitive or load-insensitive routing algorithms. The main difference between these two types resides in the consideration of the current status of the link or not. Load-insensitive algorithms reflect the current level of congestion of a link into their associate cost, whereas load-insensitive algorithms do not. Load-insensitive algorithms use other link parameters, such as the capacity, to assign the corresponding cost, that could provide a misleading information about the link since a high-capacity link routing a high amount of traffic could be less convenient than a low-capacity one that is dealing with much less congestion.

These categories are not exclusive, and they are used to describe each specific routing algorithm.

4.3.2.2 Routing Metrics

Dynamic routing is key to optimal performance, as the algorithms are constantly assessing if the current path is the optimal one for all possible sources and destinations. This is computed considering a specific set of metrics that are used as objective function in the optimization.

Following are some of the most common metrics used in routing. These criteria are not exclusive and are often combined since, in general, looking at only one of them does not provide optimal performance:

- Path length. The number of devices/routers the path goes through. It is sometimes referred to as the hop count. Routing algorithms try to minimize this metric.
- Bandwidth. The bandwidth of a route refers to its transmission capacity, i.e., its maximum throughput. When the routing path consists of a series of different links, the link with the lowest capacity acts as a bottleneck for the whole path. Routing algorithms try to find the route that uses the highest value of the minimum capacity throughout the route.
- Delay. It represents the time required for a packet to travel from its source to its destination. This metric depends on several other characteristics of the network, such as the bandwidth, the current traffic in the links, or the number of links. Analytical computation of the delay is difficult, and it is often estimated statistically. Routing algorithms try to minimize this metric.
- Congestion. It represents how busy a specific link is. Transmitting over busy link will typically have an impact on the delay, since the available bandwidth is shared over all concurrent transmission using some multiplexing method (see Chapter 3). While bandwidth and path length values do not change over time, congestion can be very time-dependent, since it is affected by the instantaneous transmission needs. Routing algorithms try to minimize this metric.
- Reliability. It refers to the quality of the transmission (typically in terms of BER – see Chapter 3) between the final source and destination. Similar to the bandwidth, the least reliable link within

the overall route may compromise the end-to-end transmission. Routing algorithms try to maximize this metric over the whole route.
- Cost. The cost is not referred to a specific physical phenomenon. In contrast, the cost is a numerical representation of a set of metrics (bandwidth, delay, etc.). Links or interfaces with higher costs provide poorer performance than lower cost ones. Consequently, routing algorithms try to minimize the accumulated cost.

Some of these metrics can easily be derived from physical information like the topology, as is the case of the path length. Others do not have such a direct physical relationship, such as the congestions. Consequently, the former can easily be used by routing algorithms, whereas the latter needs constant monitoring to guarantee that the information used in the path determination is updated.

4.3.2.3 Autonomous Systems
Autonomous Systems (ASs) can be defined as having the following characteristics:

- An AS consists of a set of routers and networks managed by the same administrative control.
- An AS consists of a group of routing devices that exchange information through a common routing protocol. This protocol is referred to as intra-AS routing protocol, in contrast with the inter-AS routing protocol, which is used to interconnect different ASs.

With this consideration of intra and inter-AS, a network may consist of a set of ASs interconnected, as shown in Figure 4.7. Routing protocols inside the AS (intra-AS routing protocol) are decoupled from the routing protocols interconnecting each one of them (inter-AS routing protocols).

4.3.2.4 Routing Algorithms
This section elaborates on the main routing algorithms.

Routing protocols for interconnected networks employ one of three approaches to collect and use routing information: distance-vector routing, link-state routing, and path-vector routing [15].

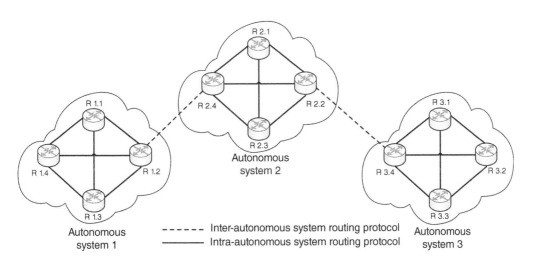

Figure 4.7 Network with inter and intra-autonomous system links.

4.3.2.4.1 Distance-Vector Algorithm

Distance-vector routing requires that each node (router, source, or destination) exchanges information with its neighboring nodes. Two nodes will be considered neighbors if they both are directly connected through a link.

The process is simple: each node maintains a routing table of interfaces (or vectors) from itself to all possible destination in the networks and with the associated metric. With this, all nodes share the information they have for their directly connected nodes. All routers iteratively update the path and cost of forwarding messages to all nodes in the network converging to the optimal routes.

Distance-vector routing requires the transmission of a considerable amount of information by each router since the whole routing table is shared among all neighbors. Additionally, this mechanism causes that, when there is some change in the metric of a specific link, this new status is not known to all nodes in the network instantaneously; on the contrary, it can take a considerable amount of time for nodes to take it into account as updates are sent periodically.

Additional drawbacks are a rather slow convergence to the optimal routes when networks are large and that it does not prevent the creation of routing loops. One of the main advantages of this algorithm is its simplicity.

4.3.2.4.2 Link-State Algorithm

Link-state routing was designed to overcome the shortcomings of distance-vector routing. When a router is initialized, it determines the cost of the link for each of its network interfaces. Then, the router advertises this set of link costs to all other routers in the network topology with a flooding mechanism, i.e., not just neighboring routers. From that moment on, the router monitors the costs of its links. As soon as there is a significant change (the cost of a link increases or decreases substantially, or a new link is created, or an existing link becomes unavailable), the router notifies again its new set of link costs to all routers.

Since each router receives link costs from all routers in the network, each router can build the topology of the entire configuration and then calculate the shortest path to each destination. Once this is done, the router can build its routing table. Unlike distance-vector routing, the router does not use a distributed version of a routing algorithm, since it has a representation of the entire network. With this information, each router can use any routing algorithm to determine the optimal path. In practice, the Dijkstra's algorithm [16] is used in most cases.

In contrast with distance-vector implementations (such as RIP), link-state-based algorithm prevents the creation of routing loops, since each node has a view of the whole topology which makes its detection easier. Additionally, the convergence in the case of the link-state is faster, but it does not make use of a common optimization process. All nodes run their corresponding algorithm to find the shortest path, which can be seen as not very efficient. The flooding mechanism is more efficient in terms of network usage, although the network discovery process at the initial part of the algorithm creates a big spike in traffic load.

4.3.2.4.3 Path-Vector Algorithm

Finally, there is an alternative to previous algorithm known as path-vector routing. It consists of neglecting routing metrics and simply providing information about which networks can be reached by a given router and the ASs that must be gone through, to get there. This approach differs from the distance-vector algorithm in two respects: first, the path-vector strategy does not include a distance or cost estimation; second, each routing information block lists all the ASs visited to reach the destination network via this route. Since a trail vector lists all the ASs a datagram must travel through when following its route, the trail information allows the router to carry out different routing policies. That is, a router can decide a certain path to avoid passing through a specific AS.

4.3.2.5 Routing Protocols
This section elaborates on the most common implementations of the routing algorithms: routing information protocol (RIP) [17], open shortest path first (OSPF) [18], and the border gateway protocol (BGP) [19].

4.3.2.5.1 Routing Information Protocol
RIP [17] is an intra-AS protocol using the distance-vector algorithm to build the routing table and forward messages.

The RIP protocol works as follows. First, each router initializes its routing table with information about its neighbor nodes. Periodically, each router advertises its routing table to the rest of routers in the network that update the information in their corresponding routing tables.

A timer mechanism is implemented in the routing table that removes old routers from which there has been no information for a certain period of time.

The information in the routing table includes a metric value that represents the "distance" to the specific destination.

Further to the periodic advertisement, whenever a router performs a change in its routing table, it is immediately broadcasted to the rest of the nodes. There is also the possibility that newly connected routers send a "request" message asking for all the routing table information instead of waiting for the periodic refreshments.

4.3.2.5.2 Open Shortest Path First
The OSPF protocol [18] is an intra-AS protocol using the link-state algorithm for routing.

OSPF calculates a route through a network interconnection that assumes the lowest cost based on a user-configurable cost metric. The user can configure the cost to be expressed as a function of delay, transmission speed, economic cost, or other factors. Additionally, OSPF is capable of balancing loads between multiple paths of equal cost.

The protocol defines the maintenance of a database that reflects the known topology of the AS it belongs to. This topology is expressed as a directed graph consisting of vertices or nodes (mainly routing devices or other networks) and edges (representing interconnections). The database includes the cost associated with each one of the edges.

Finally, thanks to the information provided in the database which gives a whole view of the network, a routing device calculates the least cost path to all destination networks using Dijkstra's algorithm [16].

4.3.2.5.3 Border Gateway Protocol
The BGP [19] is an inter-AS protocol, used to interconnect different ASs among them. This is typically required in long-distance communications. BGP is based on distance-vector algorithm.

BGP packets do not contain a final destination address but some prefixes identifying AS or subnetworks of ASs. The routing information is stored in the routing information base (RIB).

BGP works in two phases:

- Obtaining neighbor ASs' prefixes. Each AS advertises its existence and its corresponding prefix (or prefixes; ASs may have several) using a TCP connection in port 179.
- Finding the best route to a prefix. Each router runs a local search for a specific prefix using the information in the RIB. The path selection is based on a concept named degree of preference. There are a number of BGP attributes that are considered in the computation of this degree of preference (next hop, administrative weights, local preference, origin of the route, path length, multi-exit discriminator, etc.) [19].

4.3.3 Ethernet

Ethernet is defined in IEEE 802.3 as a packet-based standard for LANs and MANs. Ethernet was designed to be used to access a shared medium (a medium where more than two elements are connected), and nowadays covers a multiplicity of media (coaxial in its early implementations; twisted copper pairs – known as Base-T interfaces – and optical fibers – MM and more commonly SM) and data rates ranging from 10 Mbps to 400 Gbps. The most common data rates are 100 Mbps, 1 Gbps, and 10 Gbps. Ethernet has become the standard for packet-based services and is defined as a switching protocol.

Ethernet offers transmission and reception services, framing and contention resolution services, and a wait function for timing purposes, in two modes of operation, half-duplex and full-duplex. Full-duplex operation uses dedicated channels for transmission and reception, respectively, allowing simultaneous communication between both stations. Half-duplex operation implements a CSMA/CD algorithm (see Chapter 3) to be able to use a shared transmission medium. The algorithm works as follows:

- As there is no central element coordinating access to the medium, any source transmitting station waits for the medium to be silent.
- The station transmits the packet and remains listening to detect possible collisions.
- If the packet collides with another message from other source station, both stations intentionally keep transmitting for an additional predefined period to ensure the propagation of the collision in the medium.
- Both stations wait for a random amount of time (called back-off period) before trying to transmit again the message.

The Ethernet packets transport client data framed in such a way that both destination and source addresses can be easily identified (see Figure 4.8). Preamble and start frame delimiter are binary sequences ("10101010 10101010 10101010 10101010 10101010 10101010 10101010" and "10101011," respectively) that allow the synchronization and identification of those addresses once they are detected. The length/type field helps to delimit and interpret (Ethertype [20]) the content conveyed by the Ethernet packet.

Ethernet frame addresses are known as MAC addresses, are 48 in length, and are usually represented in 6 groups of 2 hexadecimal numbers each (e.g., 14 : AB : C5 : F6 : EF : CD). They identify

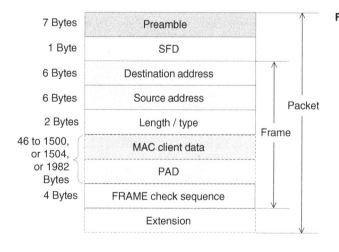

Figure 4.8 Ethernet packet.

individual or group addresses (multicast or broadcast). The IEEE Registration Authority [21] manages the assignment of these addresses, so that each address is uniquely identifying a network interface. When a new device is connected to an Ethernet domain, there is an automatic discovery process involving broadcasting protocols to identify new entrants and in which interface it is available. Each station will thus receive the packet that is intended for it.

4.3.3.1 Carrier Ethernet

The successful and pervasive adoption of packet-based technologies has led to their adaptation to be used in the transport technologies domain.

Carrier Ethernet is one of such technologies. Carrier Ethernet was defined as an alternative to legacy transport technologies (SDH/SONET) when the transport of Ethernet packets was needed within metropolitan distances and eventually longer distances due to the evolution of optic networks.

Carrier Ethernet is defined and maintained by the Metro Ethernet Forum (MEF [22]) as an evolution of the Ethernet protocol providing connection-oriented services. Being the MEF an industry association, it is also focused on providing certification services that intend to guarantee both the interoperability and the availability of the proper ecosystem of solutions and professionals.

MEF 3.0 Carrier Ethernet is the last version of Carrier Ethernet service standards. Carrier Ethernet services are technology-agnostic Layer 2 services, defined for end-user direct consumption (so-called "subscribers") and for operators to use third-party networks.

Subscriber Carrier Ethernet services (MEF 6.3 service definition) are:

- E-Line service. It enables point-to-point connectivity between two end-points.
- E-LAN service. It enables multipoint-to-multipoint connectivity.
- E-Tree service. It is a point-to-multipoint (thus, "rooted") enabling hub-and-spoke multipoint connectivity (see Chapter 7).

Operator Carrier Ethernet services (MEF 51.1 service definition) are point-to-point architecture definitions to enable service providers to deliver Ethernet services across access and transit operators:

- Access E-Line service. It enables point-to-point connectivity to extend a service to subscriber sites outside the service provider network.
- Transit E-Line service. It enables point-to-point connectivity to join multiple provider networks through an intermediate Carrier Ethernet network.

4.3.4 Internet Protocol (IP)

IP is a routing protocol defined in IETF RFC 791 and was originally designed to be used in the Internet. Its simplicity, scalability, and flexibility turned IP into the most successful protocol at the network layer and have eventually been used for non-Internet applications.

IP provides the tools for transmitting blocks of data called datagrams from sources to destinations, identified by fixed length "IP" addresses. IP defines, among other things:

- A packet format with two main parts:
  - Header, including the data needed for the packet to be routed through the network and understood by the destination host (it includes source and destination address, type of service, checksum, etc.).
  - Payload, with the data to be transported.

- An addressing method to uniquely identify a device and to facilitate the routing of IP packets.
- Fragmentation and reassembly mechanism for long datagrams that may need it.

IP addresses (in IPv4) are 32-bit binary numbers representing uniquely a network interface. These 32 bits are commonly represented divided into four groups of 8 bits each and expressed in decimal notation (e.g., as 192.168.2.3). IPv6 protocol (RFC 8200) coexists today with IPv4 and is prepared to solve the scarcity of IP addresses with 128 bits; however, IPv4 is still the most common protocol version.

4.3.5 Multiprotocol Label Switching (MPLS)

MPLS is a "label"-based switching protocol defined in IETF RFC 3031. This protocol simplifies the mechanisms used by routers to forward packets to the next hop from the origin to the destination through the assignment of a label to each packet, that will be used in each network element when selecting the next hop in its way to destination. MPLS packets are assigned the label just once, when the packet enters the network, and MPLS nodes check it to take the "routing" decision to the corresponding next hop. Packet label can also be modified in this process if needed. This process configures a highly efficient and quick packet-switching mechanism that is prevalent in packet networks all over the world.

Every MPLS network device keeps a table of labels, storing the next MPLS node in the path for each particular label. A path defined between two end-points is called the label-switched path (LSP) and a protocol called the label distribution protocol is used to create and manage LSPs. Further sophistication can be introduced with RSVP-TE (RFC 3209, extensions and updates) and eventually segment routing, including traffic engineering capabilities [23].

MPLS has advantages such as its processing simplicity and flexibility, while providing a framework of total packet forwarding control. If specific packets need to follow a strictly defined route, this can be achieved.

MPLS provides several packet-based services that allow Ethernet and IP-based protocols to be efficiently interconnected in the WAN domain with different virtual network schemes:

- Virtual pseudo-wire service (or virtual leased line service), as in RFC 8077. A pseudo-wire is point-to-point connection between two end-points.
- Virtual private LAN service (VPLS), as in RFC 4762 (RFC 4761 is similar but incompatible with RFC 4762). A VPLS is an Ethernet VPN service providing Ethernet connectivity as in an extended LAN domain.
- Virtual private routing network (VPRN) service, as in RFC 4364. This MPLS VPN provides IP connectivity among end-points.

4.3.5.1 Multiprotocol Label Switching – Transport Profile (MPLS-TP)

As a by-product of MPLS, MPLS-TP appears, among other things, to adapt MPLS to a transport function. However, MPLS-TP has not the same market or presence as MPLS.

MPLS-TP is a creation of the packet-based networks, designed to optimize packet transport through tunnels. For this purpose, MPLS had to be enhanced to support static provisioning, QoS support, in-band OAM, fault detection and switchover to backup path within 50 ms, a network management system interface to configure and manage the network, and to separate data and control planes [24, 25]. Thus, MPLS-TP is defined as a subset of MPLS protocols and some others, defined in a joint effort by IETF (RFC 5654) and ITU-T (see ITU-T G.81XX recommendations). MPLS-TP effectively simplifies some MPLS features not needed in a transport technology and defines the new needed features to support traditional transport needs.

References

1 Ellanti, M.N., Gorshe, S.S., Raman, L.G., and Grover, W.D. (2005). Next Generation Transport Networks: Data, Management, and Control Planes. Springer Science & Business Media.
2 Perrin, S. (2008). Packet-Optical Transport Confusion Is on the Rise. *Ligth Reading* https://www.lightreading.com/packet-optical-transport-confusion-ison-the-rise/a/d-id/661198.
3 Valdar, A. (2006). *Understanding telecommunications networks*. IET 52: 92.
4 Winzer, P.J., Neilson, D.T., and Chraplyvy, A.R. (2018). Fiber-optic transmission and networking: the previous 20 and the next 20 years. *Opt. Express* 26 (18): 24190–24239.
5 Horak, R. (2007). Telecommunications and Data Communications Handbook. Wiley.
6 Perros, H.G. (2005). Connection-Oriented Networks: SONET/SDH, ATM, MPLS and Optical Networks. Wiley.
7 Aracil, J. and Callegati, F. (2009). Enabling Optical Internet with Advanced Network Technologies. Springer Science & Business Media.
8 EXFO Electro-Optical Engineering Inc (2005). *Next-Generation SDH/SONET Reference Guide*.
9 Hardy, S. (2020). *Infinera Sends 800G over 950 km in Live Network, Touts Increased Appeal of 800G*. https://www.lightwaveonline.com/network-design/high-speednetworks/article/14169933/infinera-sends-800g-over-950-km-in-livenetwork-touts-increased-appeal-of-800g (accessed 28 December 2020).
10 Mann, T. (2020). *Nokia Cries Foul on Rivals' 800G Claims*. https://www.sdxcentral.com/articles/news/nokia-cries-foul-on-rivals-800g-claims/2020/05.
11 Hecht, J. (2020). *Optical Labs Set Terabit Transmission Records*. https://spectrum.ieee.org/tech-talk/computing/networks/optical-labsset-terabit-transmission-records.
12 Guenther, B.D. and Steel, D. (2018). Encyclopedia of Modern Optics. Academic Press.
13 Walker, T.P. (2005). *Optical Transport Network (OTN) Tutorial*.
14 IEEE (2004). IEEE Standards for Local and Metropolitan Area Networks: Media Access Control (MAC) Bridges. *IEEE Stand. 802.1 D*.
15 Puzmanova, R. (2002). Routing and Switching: Time of Convergence? Addison-Wesley.
16 Dijkstra, E.W. (1959). A note on two problems in connexion with graphs. *Numer. Math.* 1 (1): 269–271.
17 Malkin, G. (1998). *RIP Version 2*.
18 Moy, J. (1998). *OSPF Version 2, IETF RFC 2328*.
19 Rekhter, Y., Li, T., and Hares, S. (1994). A border gateway protocol 4 (BGP-4). ISI, USC Information Sciences Institute.
20 IEEE SA. *IEEE SA – Ethertype*. https://standards.ieee.org/products-services/regauth/ethertype/index.html.
21 IEEE SA. *IEEE SA – Registration Authority*. https://standards.ieee.org/products-services/regauth/index.html.
22 Metro Ethernet Forum. https://www.mef.net/about-mef.
23 Segment Routing. https://www.segment-routing.net/.
24 Joseph, V. and Mulugu, S. (2013). Network Convergence: Ethernet Applications and Next Generation Packet Transport Architectures. Newnes.
25 CISCO Systems Inc (2009). *Understanding MPLS-TP and Its Benefits*. https://www.cisco.com/en/US/technologies/tk436/tk428/white_paper_c11-562013.html.

5

Smart Grid Applications and Services

5.1 Introduction

Utility telecommunication needs stem from its operational activities, spanning a large variety of system functions, components, and processes.

On the one hand, telecommunications are necessary to solve both conventional business needs (as in any other industry), and more importantly, power systems' operational needs. On the other hand, telecommunications are also instrumental to cope with the necessary changes, both evolutionary and disruptive, to enhance operations efficiency, as well as to complement the new grid components' introduction (e.g. DER, EVs, etc.).

This chapter shows the different domains within Smart Grids, where Telecommunications play a key role to turn the Smart Grid vision into a reality, solving current scenarios and enabling new future-proof applications.

5.2 Smart Grid Applications and Their Telecommunication Needs

There is a broad range of operational processes that utilities need to perform, proportional to their different assets (nature, number, dispersion, purpose, etc.), their impact in the different grid segments, and the various parts of the business they run. The realization of these operational processes is carried out through, or assisted by, different operations support systems, which are commonly referred to as applications.[1]

Figure 5.1 presents some of the most relevant applications classified according to the grid segment (Generation, Transmission, Distribution, and Customer[2]) and the operations domain (*Operations Management, Customer Management,* and *Smart Grid*) they assist to. It is to be noted that the assignment of the applications to the Smart Grid domain is somehow subjective and specific to each utility in its evolution into the Smart Grid. In this respect, some of the applications shown in Figure 5.1 under the Smart Grid domain may be novel for some utilities, and consequence of the new needs and services it will find in its way forward. However, many of them are already business-as-usual for advanced utilities.

---

[1] The terms "application" and "system" are often used interchangeably in this context.
[2] While Chapter 1 referred to Consumption Points, Smart Grids should be addressed referring to the Customer connected to the Consumption Point.

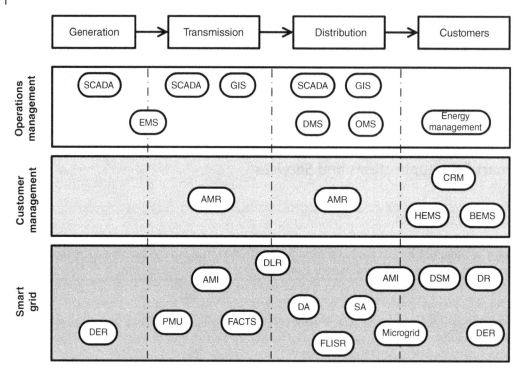

Figure 5.1 Utility applications.

The most representative applications of traditional utility operational needs are those classified under the *Operations Management* domain. They are directly connected with the basic processes of taking electricity to Consumption Points from Generation sites; they are probably the ones with a larger tradition and presence in all utilities. The first applications of this kind were deployed for *Generation* and, progressively, expanded to *Transmission* first (e.g. the Energy Management Systems [EMS]), and eventually to *Distribution* systems (e.g. Distribution Management Systems [DMS]). The connection of these applications with infrastructure field assets is a must, and therefore GISs (Geographical Information Systems) appear in this domain as well. The evolution of these applications under the *Operations Management* domain is also determined by the evolution of the asset control, moving from a local to a wide area perspective. Asset control started locally in Generation plants and transmission substations, and was eventually separated from the on-premise operation, to concentrate part (if not all, depending on the relevance of the asset) of its control functions in remote UCCs. Today both types of local control and remote control (telecontrol) systems coexist [1].

EMSs and DMSs are probably the most relevant systems within the Operations Management domain, since they are a single point of operations' control in the monitoring, supervision, analysis, optimization, simulation, and control of utilities' assets (Transmission and Generation; and Distribution network, respectively). In the core of EMSs and DMSs, Supervisory Control and Data Acquisition (SCADA) systems [2] can be found (Figure 5.2). As a matter of fact, EMSs and DMSs are built from SCADAs [3], as extensions of these SCADAs once advanced computing and applications made it possible to monitor systems in real-time, predict evolution, and trigger automatic actions. SCADAs monitor and operate Remote Terminal Units (RTUs), Programmable Logic Controllers (PLCs), or Intelligent Electronic Devices (IEDs) as they are ultimately referred to nowadays. These elements are deployed on the field, and act as the interface of the telecommunications network with the power system equipment, linking them to the control system. They acquire the data from the field devices and pass on the control commands from the control system to the field devices.

Other specific components within the Operations Management domain are Outage Management Systems (OMSs) that are used to manage the power restoration processes in the grid during service interruptions, in order to reduce the economic impact of power outages. OMSs are full system components that analyze outages (location, extent, etc.) to help dispatchers and crews to solve them. Additionally, they can also predict outages through the analysis of information collected from the Distribution grid.

Another important set of applications are those centered on the customer, the long forgotten central object of the energy system. In this respect, Automatic Meter Reading (AMR), Customer Relationship Management (CRM), and EMSs with a focus on consumption rather than on the network, are the most common baseline system components of the Customer Management domain.

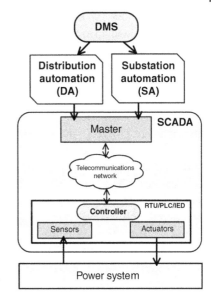

Figure 5.2 The relationship among utility management systems, automation domains and SCADA in the Distribution grid.

They handle the relationship and the contact of utilities with their customers. These systems manage meter information both for grid and customer management. Examples of these systems are CRM, adapted to utility space (mainly for Commercial and Industrial [C&I] customers) and energy management programs.

Finally, the Smart Grid domain brings the attention to the newer and more advanced applications. Some of these applications are evolutions of existing ones (e.g. AMI, as evolution of AMR); others come to focus mainly on the customers, although they may have impact on the grid (e.g. DSM, DR); while some others are the consequence of the need to integrate new energy system elements (such as DER) and pursue a real automation of operations (such as DA). These components will increasingly be consolidated in existing applications, while others will produce full new standalone applications.

ICTs are central for the realization of all these applications. A more detailed description of the elements in Figure 5.1 follows, with a focus on the telecommunications aspects associated with these applications. Some components connected to these applications (e.g. protection and teleprotection) will also be covered, as they are instrumental for a comprehensive understanding of each application and system domain.

5.3 Supervisory Control and Data Acquisition

SCADA systems, at the core of EMS and DMS systems, are the basis of the remote intervention capabilities of utilities over their assets, and the progressive evolution into the automation of remotely executed operations.

SCADA systems' origins were stimulated by the quest for automation of operations, and supported by the evolution of ICT technologies in general, and telecommunications in particular.

Up until the late 1940s [3], many utilities had personnel stationed 24 by 7 at substations to maintain manned operation of substations. The introduction of automation was an opportunity to improve overall efficiency.

Automation has existed in different forms, starting with analog means and at a local level. However, its application flourished when computing and telecommunications started to be available at affordable costs. First, systems were based on electromechanical and narrowband communication technologies, limited to basic remote control and/or supervision (monitoring) applications over distant assets. Progressively, microprocessors, digital solutions, and broader-band communications allowed modern SCADA systems to be the basic mean to operate (monitor and command) utility grids, overcoming prior limitations and expanding the scope of the first SCADA systems.

In essence, the basic function of SCADA systems is to remotely control equipment in each substation from distributed or centralized UCCs. Any critical substation equipment is measured, monitored, and controlled remotely. Information coming from the substations is displayed and stored. This information is used to generate alarms if anything abnormal happens, and remote intervention is made possible to initiate changes to regain substation normal operation. Substation information is normally updated every few seconds, either through a process of sequential scan, or event-triggered reporting when relevant changes happen. Information update frequency and delay are normally tolerable within certain time limits (delays of hundreds of milliseconds, and information refresh in the order of several seconds). However, during disturbances or emergency situations, frequency, and delay needs to be improved (information retrieved more often, and delay down to a minimum), and this imposes constraints to telecommunication and information systems alike.

5.3.1 Components

SCADA systems (Figure 5.2) follow the simple idea of a "brain" controlling remote "hands," and getting information from remote "eyes" and "ears," through a "nervous system":

- Central infrastructure with master station(s) ("brain"). This is a collection of computers, peripherals, and appropriate input and output (I/O) systems [2] that acts as the core of the system enabling UCC operators to monitor the state of the power system processes, and control them. This infrastructure can be monolithic, distributed, or networked, according to the evolution of the state-of-the-art and utility circumstances or preferences. It may have a single master or a multiplicity of them; these masters may be located in the same site, or they can be found distributed in different sites. It will be accessible to operators (people) through a Human-Machine Interface (HMI), for the needed operation interaction.
- Remote units at substations ("eyes," "ears," "hands"). Remote Terminal Units (RTUs), as they are commonly known, are the basic element providing the control at the substations. RTUs acquire all relevant field data from field devices, process the data and transmit what it is considered relevant to the central infrastructure. Likewise, but in the other direction (from central to remote), they distribute the control signals received from the master station to the field devices. RTUs, Programmable Logic Controllers (PLCs[3]), and IEDs are considered evolutions of RTUs; PLCs are generally considered an intermediate step in the evolution of RTUs toward IEDs. PLC acronym is not so common when discussing power system SCADAs.

---

3 PLC more often means Power Line Communications in utility literature.

- Telecommunications infrastructure ("nervous system"). This refers to the set of telecommunication means (e.g. network equipment, transmission media; most often, integrated in a telecommunications network, and not used exclusively for SCADA purposes) enabling the exchange of data messages (i.e. SCADA protocols) between the central infrastructure and remote units as well as between the components of the central infrastructure that may be distributed in different locations. In this respect, one or more *communication channels* have to be arranged between the master station and each individual or group of RTUs. The service characteristics and achievable performance of these communication channels (e.g. bandwidth, latency) may constraint SCADA overall performance and ultimate application.

The number of central infrastructure hardware elements, their location, and their complexity depend on the size of the utility. Software includes data acquisition and control modules, databases (storing data), and reporting and accounting modules. Relevant hardware elements are:

- Computing infrastructure, typically in the form of physical and/or virtualized servers used to host the different specialized functions.
- Communication front-ends. They are communication cards with typically serial or Ethernet interfaces. They interface remote RTUs connecting to the telecommunication network that provide the "channels" in which RTUs communicate with the different SCADA protocols. The servers in the central infrastructure have physical interfaces to connect to these channels (low-speed serial interfaces or regular TCP/IP communications).
- Intercontrol Center Communications. They support data transmission between the master and a higher hierarchy master, or masters in different sites among them, for data exchange purposes. They use Inter-Control Center Protocols (ICCP, see Section 5.3.2.2) for this purpose.

Within substations, RTUs have evolved from electromechanical, to electronics and analog-to-digital converters. Eventually, microprocessors opened a new field of evolution with the incorporation of logic into the RTUs. From then on, since the 1980s, we refer to them as IEDs. IEDs consolidate a myriad of functions on a single device, through a complex software, that can communicate locally with similar IEDs through a Local Area Network (LAN). A generic RTU/IED typically consists of:

- A communication subsystem. Its main function is to manage SCADA protocol messages with the central infrastructure or master station(s). Legacy RTUs used to act as "slaves" of master stations; nowadays, RTUs can send messages to master stations without being polled.
- A logic subsystem. It handles all local processes and actions, such as time keeping, sensing, analog-to-digital conversion, etc.
- A termination subsystem. It is the interface between the RTU and external equipment (substation devices), through analog and digital terminations.
- A Test/HMI subsystem. It allows local control and access to the RTU, for testing, maintenance or operation purposes.

5.3.2 Protocols

The SCADA protocols define the format and sequences of the data messages exchanges between a master station and the RTUs as well as among master stations. The evolution from all analog SCADAs to digital created a myriad of protocols (over one hundred) that was not sustainable in a context of cost rationalization, best practices adoption, and system interoperability. The creation of standards has followed different paths including the adoption of industry standards, and collaborative efforts inside standardization bodies or industrial consortia.

5.3.2.1 Central Infrastructure to Field Protocols

There are two consolidated SCADA protocols that stand out from the rest: Distributed Network Protocol (DNP3) and IEC 60870-5. DNP3 is popular in the Americas and Australia, whereas IEC 60870-5 is popular in Europe. The rest of the world (Asia and Africa) does not show a particular trend.

These protocols do not use the full OSI layer model. In particular, in situations where short reaction time is needed, and bandwidth available might be constrained, IEC developed the Enhanced Performance Architecture known as the EPA model, using just three layers (physical, data link, and application).

IEC 60870-5 series is a set of protocols published by IEC TC 57 WG 3 "Telecontrol Protocols," chartered to develop protocol standards not only for telecontrol but also for teleprotection, and associated telecommunications for electric utility systems. IEC 60870-5-101 and IEC 60870-5-104 are the most popular standards for serial and Ethernet/IP-based communications, respectively (TCP/IP). The working group also developed other relevant standards, such as IEC 60870-5-102 for metering and IEC 60870-5-103 for protection. All of them are developed as "companion standards" based on IEC 60870-5-1 to IEC 60870-5-5 standards.

The Distributed Network Protocol (DNP3) corresponds to IEEE 1815. It is specifically designed for data acquisition and control applications in the area of electric utilities. The DNP3 protocol is built on the framework specified by the IEC 60870-5 documents, and it is quite similar to IEC 60870-5-101. DNP3 has also evolved to support communications over networks using the TCP/IP protocol suite.

At application layer, both protocols use the concept of the Application Service Data Unit (ASDU), but in a different way. In IEC 60870-5-104 (Figure 5.3), we find the Application Protocol Data Unit (APDU) with a maximum length of 255 Bytes as the container of the ASDU, together with the so-called Application Protocol Control Information (APCI). The scheme includes the message delimiting elements, length of the ASDU, and the control fields. ADSU in DNP3 is the application layer "fragment," that can be larger than the APDU in IEC 60870-5-104, and includes the control part of the protocol in the so-called application header.

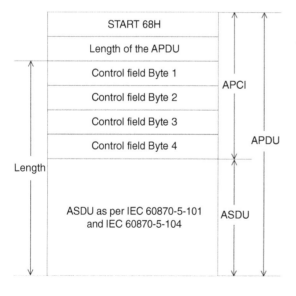

Figure 5.3 Application Protocol Data Unit (APDU) in IEC 60870-5-104 protocol. It is transported over TCP/IP.

5.3.2.2 Central Infrastructure Protocols

Inter-Control Center Communications (ICCP) is a widely adopted protocol in the electric power industry today, for communications within a distributed central infrastructure of SCADA systems, or with third-party SCADAs. Its origins and evolution date back to the 1990s to overcome the multiple proprietary protocols common in industry at that time [4].

ICCP is covered in IEC 60870-6, which was developed by the IEC TC 57 WG 3. TASE (Telecontrol Application and Service Element) was the name given to it in IEC, and first TASE.1 in 1992, and eventually TASE.2 in 2002 (widely used today), were issued. TASE.2 is known as ICCP.

ICCP is a client-server protocol, where UCCs can act both as clients or servers to exchange data. UCCs can communicate among them or with power plants, and even in some cases with transmission substations, using this protocol. The protocol can exchange real time and historical data, measured values, scheduling data, energy accounting data, and operator messages; it also includes a mechanism for exchanging time-critical data between sites.

TASE.2 uses Layer 7 (application layer) of the OSI model and MMS (Manufacturing Message Specifications; ISO 9506-1 and 9506-2) for messages. Together, they describe UCCs with respect to their external visible data and behavior, with an object-oriented format. Supported data types include control messages, status, analogs, text, files, etc. for data exchange and optional functions such as remote control (telecontrol).

Following are the relevant sections of the protocols:

- IEC 60870-6-2. Use of basic standards (OSI layers 1–4).
- IEC 60870-6-501. TASE.1 Service definitions.
- IEC 60870-6-502. TASE.1 Protocol definitions.
- IEC 60870-6-503. TASE.2 Services and protocol.
- IEC 60870-6-504. TASE.1 User conventions.
- IEC 60870-6-601. Functional profile for providing the connection-oriented transport service in an end-system connected via permanent access to a packet switched data network.
- IEC 60870-6-602. TASE transport profiles.
- IEC 60870-6-701. Functional profile for providing the TASE.1 application service in end-systems.
- IEC 60870-6-702. Functional profile for providing the TASE.2 application service in end-systems.
- IEC 60870-6-802. TASE.2 Object models.

5.4 Protection

A fundamental priority of any power system is the safety of system components and people. Thus, power systems need to be designed and maintained to limit the number of faults that might occur, and their derived power outages and voltage fluctuations. A fault can make whole sections of the grid unavailable with a cascading effect. This can affect a relevant area of the grid, if not properly isolated.

A fault is a flow of current that is large enough to cause damage. The typical fault is a short circuit that can happen between a phase and ground or between two phases of a power line. Possible causes could be the contact of one phase with ground through an object or a too short distance between two phases. Faults are unavoidable, as lightning discharges, tree branches, animals contacting circuits, vandalism, unintended incorrect operation, or the deterioration of insulation in power lines can cause them [5]. Faults fall into two broad categories, i.e. short-circuit faults and open-circuit faults, being short-circuits the most severe kind [6]. Most faults are of a transient nature and this allows power supply recovery with short interruption periods.

The role of the protection mechanism is to quickly detect faults and disconnect ("clear") the faulted segment minimizing the impact to the grid. Protection systems consist of five components, namely, current and voltage transformers, protective relays, circuit breakers, telecommunication channels, and batteries. Current and voltage transformers step down the values to be fed to protection relays as input data; protection relays (electromechanical in the past; now solid-state microprocessor-based, allowing complex protection arrangements) sense the fault and initiate a trip or disconnection local signal to command circuit breakers to close or open (connect or disconnect, respectively), the affected section of the grid. Telecommunications are used when the analysis of current and voltage needs to happen at remote terminals of a power line, and/or to allow remote tripping of equipment. Protection relays need to have battery power available to operate on power disconnection situations.

Protection relays' main functional characteristics are:

- Speed. Operation must be executed in the shortest possible time, ranges being in the order of 30–100 ms depending upon the voltage level of the section involved.
- Selectivity. Relays must clearly discriminate between normal and abnormal system conditions, to prevent unnecessary activation.
- Sensitivity. Relays must accurately detect signals crossing predefined configured limits.
- Reliability. Protective equipment must be ready to actuate when needed.

Automatic operation is instrumental to isolate faults as quickly as possible, to minimize damage and the costs of nondistributed energy. Protective devices are embedded in Smart Grid for this purpose, and telecommunications systems assisting them in the protection function will enable them to communicate in real-time to utilize information from other power system sections for self-healing purposes. Fault information from the protective devices is used by utilities to identify the location, time, and number of faults in the grid and resume normal operation after the faults have been cleared. Literature often refers to FDIR (Fault Detection Isolation and Restoration) and FLISR (Fault Location, Isolation, and Service Restoration). FDIR is a more traditional and decentralized approach, while FLISR is often centralized [7–9] (see Section 5.5.3).

Protection in power systems may work to protect individual elements of the system, locally without the intervention of telecommunications in the process (e.g. the protection of a transformer). In its simplest operation, protection relays sense any change in the signal they are receiving (from a current and/or voltage source), assess the magnitude of the incoming signal to check if it is outside a pre-set value, and carry out a specific operation (e.g. tripping a circuit breaker as in Figure 5.4). Thus, this simple protective system consists of the combination of protection relays (detecting the fault) and circuit breakers (isolating the fault).

The other field where protection is important is when substations are connected by means of power lines. In these cases, protection systems are used to isolate faulted sections of the network, in order to maintain power supply in the rest of the system. As distant elements of the connected substations are involved, telecommunications are needed in the protection process to send and receive commands, and to share information that needs to be analyzed at the distant end. Substation protective elements must be connected remotely to be able to open the circuit breakers, considering protection zones, in a coordinated way. The protection domain is in fact where the most demanding requirements for telecommunications exist. There are two protection concepts to be considered.

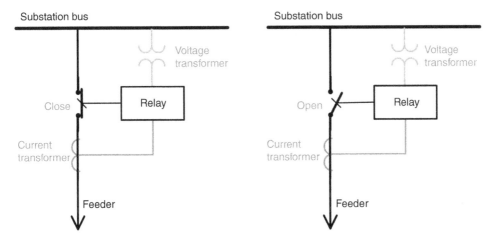

Figure 5.4 Basic protection representation of a feeder.

The first concept is the distance protection. Distance protections (distance relays are device number 21 as per IEEE C37.2, "IEEE Standard for Electrical Power System Device Function Numbers, Acronyms, and Contact Designations") intend to detect the location of the fault in a power line, based on the impedance measured in the line. The impedance of a power line is proportional to the line length, and voltage and current measurements are used to calculate the apparent impedance. The distance protection relays are placed on the two ends of the power line at the substations. They face each other and can typically measure three so-called zones of protection that cover the whole power line length (e.g. the first protection zone measures 80–90% of the protected line length).

The second concept is the differential protection. Differential protection schemes (differential protective relays are device number 87 as per IEEE C37.2; specifically, 87P – phase comparison – and 87L – differential line) are based on Kirchhoff's law, meaning that the algebraic sum of the currents entering and leaving a node is zero, and any deviation from this is the evidence of an abnormal current path. In practice, this means that measurements from the distant end must be compared almost instantaneously.

Distance protection is associated with teleprotection devices capable of sending commands to the other end with very low latency; and differential protection is associated with being able to send information that can be compared to the other end, with a very low latency as well. IEC 61850-90-1, "Use of IEC 61850 for the Communication Between Substations", summarizes the use cases for protection in the communications between substations. Among them, distance line protection with permissive overreach teleprotection scheme, distance line protection with blocking teleprotection scheme, directional comparison protection, transfer/direct tripping, interlocking, multiphase auto-reclosing application for parallel line systems, current differential line protection, and phase comparison protection can be found.

Teleprotection equipment (IEC 60834-1) is a very special telecommunication element associated with distance protection (see Figure 5.5). It is specially designed to be used along with a protection system, and it works connected to a telecommunication link between the ends of the protected circuit. The teleprotection equipment transforms the information given by the protection equipment into a format that can be transmitted to send signals according to the different protection schemes. The data information transfer delay must not typically exceed 8–10 ms covering distances between adjacent substations, which may range from a few kilometres to close to 100 km. The use

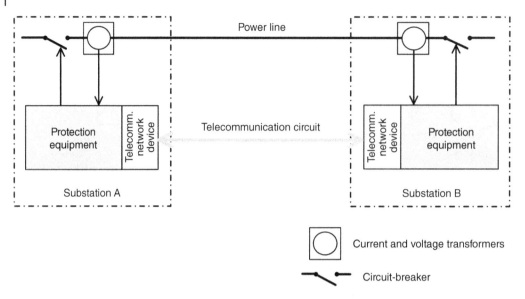

Figure 5.5 Teleprotection concept.

cases covered in IEC 61850-90-1 typically refer to "less than 5 ms," although the value depends on the voltage level (the higher the voltage, the lower the delay). The time constraint is connected to the value of the semi-period of the nominal grid frequency (50 or 60 Hz). IEC 60834-1 fixes the performance requirements for teleprotection equipment and test methods for the different telecommunication media to be used. The total time needed for the protective system to act is critical, and considers the contribution of its different components, not only the telecommunication channel's latency. The teleprotection communication interfaces are described in IEEE C37.94, as the standard interface for $n \times 64$ kbps optical connectivity (single-mode and multi-mode) that the teleprotection device offers to the telecommunication network. This interface is compliant with ITU-T G.704, partially with G.775, but not with G.706. There are still some other interfaces in use, such as analog 4W E&M (4 Wires Ear and Mouth), RS-232, and ITU-T G.703 codirectional 64 kbps.

Differential protection transmission delay aspects are equally important, and of similar magnitude to the ones in Distance protection. Due to the fact that reference signals at both ends need to be compared, a synchronization accuracy of the data is needed as well.

The reliability needs of protection relays permeate into the reliability of the teleprotection equipment. Reliability involves the two important classic requirements of dependability and security. Formal definitions are provided in IEC 60834-1, based on probability calculations that have to be addressed for the different communication channels to be used. Dependability evaluates the certainty that a protection system will operate when it is supposed to, while security evaluates the certainty that a protection system will not operate accidentally when it is not supposed to (e.g. if no command was sent). IEC 61850-90-1 also adds the data integrity consideration, expressed in BER in digital telecommunication channels (see Chapter 3), and assigns a stricter requirement for differential protections than for distance protection.

Telecommunications for protection functions are in a process of change. With the evolution of telecommunications into packet networks, dedicated[4] TDM-based circuits or even optical fiber are

---

4 *Dedicated* in this context means that the telecommunication channels are not shared with any other communication service.

less and less available to provide protection services, for cost or technology obsolescence reasons. When migrating into nondedicated packet-based telecommunication services, differential protection services are more sensitive than distance protection ones. Special precaution needs to be taken to the communication latency difference in both communication directions (that typically needs to be within a certain difference range), and the likely delay change if back-up routes exist [10].

5.5 Distribution Automation

DA covers all aspects of the automation of the entire distribution system operations. DA remotely monitors, coordinates, and operates distribution grid components in real-time, to eventually solve any distribution grid component malfunctioning (grid-"healing"). DA is not a single application or technology component. It is a canopy for different technology elements that are combined to get more efficient operations (time and cost) in the Distribution segment of the grid, and for the customers to get an objectively assessed better service. DA needs to count on the physical elements (e.g. transformers, power lines, switchgears, etc.) along with the Distribution grid, SCADA systems and other applications to fulfil its purpose. DA often includes Substation Automation (SA in Section 5.6), but generally focuses more on what happens down the feeders, i.e. to the customers. Occasionally, DA includes Customer operations (e.g. metering); it is, however, more common to find them within Metering systems and Customer Management domains.

Apart from the substations themselves, there is a number of Distribution grid elements (e.g. circuit breakers, switches, etc.) that, when operated in a smart fashion, can accomplish the increased automation objective in the Distribution grid [11]. This operation can happen autonomously, and often has fall-back mechanisms available for the periods when communications may not be available. However, the interconnection of these elements with central systems for command and information exchange, boosts their capabilities beyond the simple local data-based decisions. These locally taken decisions may trigger actions which, when observed from a complete grid perspective, may not be as accurate as if they would have relied on the wider picture.

At the same time, together with the grid "healing"-related functions, Volt/VAR control activities are important features [12]. "Volt" refers to the voltage in a power line; "VAR" refer to the Volts-Ampere Reactive. In some references, Watts are also included in the concept (Volt, VAR and Watt control), as the purpose of this control is to ensure that the various electric quantities at different points in the utility grid remain within acceptable operational ranges. Thus, voltage drop along the power line is to be controlled (since voltage decays with distance), and reactive power control minimizes losses and maximizes power delivery to the loads.

The grid elements that need be mentioned in DA are:

- Switching devices. These elements connect, disconnect, and/or reconnect segments of circuits when needed, to protect feeders and/or redirect electricity flow. This need may derive from faults in the grid or from load shedding conditions. They can be designed to operate (i.e. break) under load or no-load conditions. They can be classified as:
  - Circuit switches, connecting and disconnecting line segments (load break or non-load break type).
  - Circuit breakers, disconnecting load on pre-set operating conditions (load break).
  - Circuit reclosers and switchers, protecting and reconfiguring circuits. They are installed along line segments (load break and non-load break, respectively).

The circuit recloser or simply recloser is a special device that monitors the status of a feeder to break it when a fault occurs and includes automatic reconnection capabilities, to check if the fault was a transitory condition. The device will attempt to reconnect for a number of configurable times, before it decides that the faulty condition is permanent and "locks out" [13].

- Sensors. Pure sensing devices that monitor the state of different points of the Distribution grid, detecting and recording variables associated to its state. They include voltage and current transformers.
- Voltage regulation devices. These elements control voltage levels along feeders for power quality assurance. Voltage variations may occur due to the continuous changes of the network load, and voltage values need to remain between certain thresholds. Transformers in substations may also have mechanisms available to help in this purpose; this is the case of On-Load Tap-Changers (OLTCs or simply LTCs).
- Volt-amperes reactive (VAR) support devices. These elements, mainly shunt capacitor banks, provide power factor correction to control reactive power. Reactive power is power in the system that is moving back and forth (i.e. it does not do useful work) and it should be minimized.
- Distribution transformers. Although in some power systems transformers are almost always associated with the substation concept (with all different typical functions of a substation), there are other systems where very small capacity transformers are used. In these cases, monitoring parameters such as voltage, current, and temperature is important to balance loads in the grid and to control the lifetime of the transformers themselves.

Finally, there is a trend worth mentioning related to the application of FACTS-like technologies (FACTS are Flexible AC Transmission Systems, see Section 5.10.1) for Distribution grids. Although the cost is still not affordable and the market not yet mature, there is an opportunity for these devices to be used in MV and even LV, due to the wide-spread emergence of DER. D-FACTS [14] and RACDS (Resilient AC Distribution Systems) [15] acronyms are used to refer to them.

5.5.1 Distributed Energy Resources Integration

DER (or DG, used interchangeably, see Chapter 1) needs to be considered within the scope of DA, as components that are connected to the Distribution grid.

DER complements and may offer significant benefits over the conventional power systems, not only due to its more environment friendly origin, but for the avoided costs associated with the Transmission and Distribution of power over long distances [16], the decreased technical losses from this avoided transport, and the help provided to local utilities (especially in remote areas) in times of generation shortage [11]. The integration of DER offers certain benefits to grid planners as well, as it can reduce the peak demand, can be implanted in a quick way due to their lower gestation periods compared to conventional generation, and can help to improve the reliability of the grid.

As already mentioned, the integration of distributed sources of energy is a fundamental aspect of the Smart Grid paradigm. But it needs to be achieved without major impact to the grid operation and maintenance, its fundamental magnitudes (power, voltage, and frequency), and the stability of the electricity supply. And here is where most of the challenge comes from.

Conventional generation integration is radically different to DER integration. The integration of DER requires appropriate protection mechanisms and control by the utility:

- From a volumetric perspective, conventional generation numbers are low, while DER can exceed conventional generation numbers by several orders of magnitude.

- From a connectivity perspective, conventional generation is connected to HV while DER connects to MV and LV.
- From the nature of DER, much of it is not directly connected as synchronous or asynchronous machines but coupled to the network via inverters (static or electronic DC to AC converters), and this fact affects how the interconnection needs to be managed. In this case, the short-circuit capacity of grids dominated by inverter short-circuit current sources is significantly lower than that of grids with rotating machines of the same rating [10].
- From an operational perspective, bidirectional fault currents will be found in a grid where DER is present [10].
- From a visibility perspective, some DERs (e.g. those connected in LV) may not be declared and known to the utility, and therefore no control will exist over them.

From a management perspective [11], DERs can be handled as either managed devices or intelligent autonomous devices depending only on the intelligence built into the DER components. DER will be integrated either through a power electronic interface or a number of DER units may be connected through a Microgrid [13] (see Section 5.9.3). From a control perspective, SCADA will integrate the DER with the DMS to get information and deliver remote commands. The system that manages DER is sometimes referred to as DERMS (DER Management System) [17].

From a protection perspective, DER forces more complex and pervasive protection systems on the Distribution network than conventional generation, due to the current flows in the system.

The deployment of DER is both an opportunity to deploy a telecommunications network associated with the required control of the DER devices, and a must due to the need to have this control over some devices that can cause havoc [12] if not properly integrated in the grid.

To minimize the negative effects and maximize the positive impact of DERs, IEEE 1547 covers the technical specifications for, and testing of, the interconnection and interoperability between utility electric power systems and DER, providing the requirements relevant to the performance, operation, testing, safety considerations, and maintenance of their interconnection. In addition to this, IEEE 1547 includes several documents in the series (IEEE 1547.1 to IEEE 1547.4, and IEEE 1547.6 to IEEE 1547.7) that increasingly cover the evolution of DER-related needs. For instance, IEEE 1547.2 is an application guide that among other aspects, includes the areas that need to be considered when effects of DER connection are to be analysed (from grounding and current flows, to grid fundamental magnitudes, all though the operation procedures, etc.). It is a valuable resource to read together with the IEEE 1547. IEEE 1547 also addresses control aspects, and mentions protocols that can be used (IEEE 1815 [DNP3], IEEE 2030.5 [SEP2] and SunSpec Modbus). Interestingly, IEEE 1547 is not very strict when referring to their performance characteristics, mentioning "availability" (needs to be available while the DER is operational) and "information read response time" under 30 seconds. Typical parameters that are used to remotely control DERs are connection status, real and reactive power output, and voltage.

Of late, attention is being placed over the use of DER for "distributed restart", or "black start", i.e. the support that DER can offer when restoring power in a grid after a total or partial shutdown [18]. The operational procedures that exist today in TSOs and DSOs are based on the use of the bulk generation that has a solid resilience (availability, based on the use of optic fiber-based, highly protected-redundant routes, and properly power backed-up). If DERs are to be used, or incorporated in these procedures, their connectivity should possibly be closer to these highly available and flexible assets, than to a normal IED that is not to be involved in a grid restart.

5.5.2 Electric Vehicles Integration

Electric vehicles (EV) can be classified in three main categories: plug-in hybrid EVs (PHEV), battery EV (BEV) and hybrid EVs (HEVs). A PHEV is a type of EV that has an internal combustion engine (ICE), an electric motor and a high-capacity battery pack that can be recharged by plugging-in the car to the electric power grid. A BEV is a type of EV that uses rechargeable battery packs to store electrical energy and an electric motor for propulsion. Both PHEV and BEV are the types of EVs considered in this section, which will be referred to as PEV (plug-in electric vehicles). HEV are excluded because they use a combination of a conventional ICE and an electric motor for propulsion, but do not need to connect to the grid.

The need to charge the PEVs comes with an important impact on the grid, given that PEVs [12, 19]:

- Could cause a major change in the load profile of residential customers when charging an electric vehicle at home.
- Could increase the peak demand and maximum network loading of areas where they connect to the grid, due to the high correlation of energy usage (PEV charging periods).
- Could cause a major increase in the electrification needs of distribution transformers, that may force utilities to upsize them.
- Could make demand more unpredictable, as although charging points are static, PEV can connect to the grid at different locations in mobility.

Thus, PEV are a source of opportunity for electricity industry, as they will not only stimulate the demand and growth of the system but also an element that, when integrated in the grid, needs to be analyzed. For PEVs as loads, grid planning needs to consider their moving nature that will make consumption highly unpredictable. As a consequence, significant investments in Distribution network equipment will be needed to increase grid capacity [19]. For PEVs as DS elements (batteries on wheels; Battery Energy Storage Systems are generically known as BESS), PEVs can be used as distributed energy storage to provide support of grid operations [19] through the aggregation of a large number of individual PEVs as small batteries configuring a grid-scale BESS (for energy storage, frequency fluctuation control [20], etc.). There are a number of opportunities of the PEV as a DER that have been analyzed by the industry to provide more incentives to transport electrification and use PEV for grid improvement [12].

From the standardization stand point, disperse standards-producing sources configure a convoluted set of different options (one example of this is the diversity of EV charge connectors). The most relevant sources for standards in the EV and charging domains are Society of Automotive Engineers (SAE) and IEC, with SAE J2836, J2847, J2931, and J2953 for communication purposes [16]; and IEC 15118 and IEC 61851 series, among the relevant ones.

The interaction of the PEV with the grid is commonly covered under the V2G (Vehicle-to-Grid) concept. It addresses the control of the PEV as an element of the grid, both as a load and as a battery to transfer energy and supply it to homes, loads, or the grid. This is still in its infancy for grid operation use [21]. If we take IEC 15118 series as an example, IEC 15118-1 defines the requirements and use cases for conductive and wireless (depending on how the PEV charging occurs) "High Level Communication" (HLC; this is referred to as "digital communication" in SAE) between a EV Communication Controller (EVCC) inside the EV and a Supply Equipment Communication Controller (SECC) in the charging station. IEC 15118-2 specifies the communication between PEV and the Electric Vehicle Supply Equipment (EVSE) up to the application layer message level. IEC15118-3 specifies the requirements of the physical and data link layer of a HLC based on a wired communication technology and the common fixed electrical charging installation.

IEC 15118 specifically mentions that the SECC may communicate with the so-called "secondary actors" including utilities under the DSO concept. For this purpose, it includes some utility-specific requirements, such as "power limiting for grid control or local energy control." The intention is to minimize overloading situations and local perturbations that may happen in LV and MV grids due to a large number of EVs charging at the same time so that energy transfer schedule and maximum power profile can be controlled.

5.5.3 Fault Location, Isolation, and Service Restoration

FLISR is a good example of how DA works. It comprises the different elements to identify a faulty condition and take the necessary steps to get it solved minimizing harm to the grid, time-to-restore and customers' service interruption.

Although there are many strategies to achieve this with purely electromechanical means, the use of telecommunications achieves an optimal solution, alone or combined with them.

Utilities take decisions on how much automation they need in their operations, and how many automated devices they need to deploy based in their fault track records, their regulatory targets and incentives.

When a fault occurs in a feeder (see Figure 5.6), there is a sequence of events that need to be triggered to isolate the faulted part of the feeder, to restore power. Figure 5.6 shows a ring configuration, where there is a "Normally Open Point (NOP)" that makes one side of the ring to be fed from one transformer, and the other side of the ring from the other. There are also Circuit Breakers and Reclosers at every SS connected to the MV circuit (represented with the *Connect* and *Disconnect* elements):

- A fault occurs, and it is detected by several elements that, while trying to isolate it, report the measurements and status to a central logic ("application").
- Using the connectivity model of the utility's infrastructure (i.e. how PSs and SSs are connected), the application locates the fault.
- Leveraging *Connect* and *Disconnect* elements, the application isolates the damaged portion of the feeder sending commands through the SCADA to these remotely operated devices.
- The NOP is operated to re-energize the feeder via the other side of the ring.

5.5.4 Indices for Operations Performance

In order to track and improve distribution grid operations, utilities have defined a common set of indices that, despite the nature of their grid and operation differences are used to benchmark their operations and measure their progress improving the service, especially as a result of DA programs.

IEEE 1366-2012 provides a standard definition of these indices. However, there are regions, countries, and utilities that use slightly modified versions of these indices for historical reasons.

Indices are based in a common set of definitions:

- Outage. It is the "loss of ability of a component to deliver power." Outages may be forced or planned:
  - Forced (outage). It is the "state of a component when it is not available to perform its intended function due to an unplanned event directly associated with that component."
  - Planned (outage). It is the "intentional disabling of a component's capability to deliver power, done at a preselected time, usually for the purposes of construction, preventative maintenance, or repair."

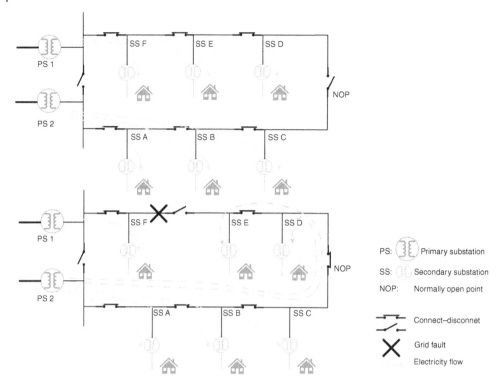

Figure 5.6 Distribution automation example in a ring protection.

- Interruption. It is the consequence of an outage, and it is defined as the "total loss of electric power on one or more normally energized conductors to one or more customers connected to the distribution portion of the system." It does not include any of the power quality issues (sags, swells, impulses, harmonics, etc.). Interruptions can be momentary, sustained, planned, and unplanned. Their "duration" is the "time period from the initiation of an interruption until service has been restored to the affected customers":
  - Momentary (interruption). It is the "brief loss of power delivery to one or more customers caused by the opening and closing operation of an interrupting device" (considered to be less than five minutes). The "event" concept is defined to count the number of these occurrences, not the minutes.
  - Sustained (interruption). It is "any interruption that lasts more than five minutes."
  - Planned (interruption). The loss of electric power to one or more customers that results from a planned outage.
  - Unplanned (interruption). The loss of electric power to one or more customers that does not result from a planned outage.
- Major Event Day (MED). A day in which the daily SAIDI exceeds a MED threshold value.

The metrics (indices in Table 5.1) are classified as "Sustained Interruption," "Load Based" (instead of customers, it considers load), and "Momentary Interruption." The most commonly used indices are:

- SAIFI. It indicates how often the average customer experiences a sustained interruption over a predefined period of time.

Table 5.1 Electric Power Distribution Reliability Indices (IEEE 1366-2012).

Classification	Index acronym	Index
Sustained interruption	SAIFI	System Average Interruption Frequency Index
	SAIDI	System Average Interruption Duration Index
	CAIDI	Customer Average Interruption Duration Index
	CTAIDI	Customer Total Average Interruption Duration Index
	CAIFI	Customer Average Interruption Frequency Index
	ASAI	Average Service Availability Index
	CEMIn	Customers Experiencing Multiple Interruptions (n)
	CELID	Customers Experiencing Long Interruption Durations
Load based	ASIFI	Average System Interruption Frequency Index
	ASIDI	Average System Interruption Duration Index
Momentary Interruption	MAIFI	Momentary Average Interruption Frequency Index
	MAIFIE	Momentary Average Interruption Event (E) Frequency Index
	CEMSMIn	Customers Experiencing Multiple Sustained Interruption (n) and Momentary Interruption Events

- SAIDI. It indicates the total duration of interruption for the average customer during a predefined period of time. It is commonly measured in minutes or hours of interruption.
- CAIDI. It represents the average time required to restore service.
- CAIFI. It gives the average frequency of sustained interruptions for those customers experiencing sustained interruptions.
- ASAI. It calculates the fraction of time (often in percentage) that a customer has received power during the defined reporting period.
- MAIFI. It indicates the average frequency of momentary interruptions.

5.6 Substation Automation

SA intends to achieve a more efficient operation and maintenance of substations and the infrastructure that connects to it. SA can be part of the DA, as described in Section 5.5. However, SA is focused on the substation itself, as a complex environment with a major influence in grid performance, and a set of elements and internal connections that can leverage ICT evolution within the substation itself. The concept relies on the substation as a heavily automated environment, where there should be minimum or no human intervention (locally and remotely). For this purpose, it needs to combine intelligence, sensors, and actuators deployed at the substation, with connection to UCCs.

IEC 61850 series is probably the most successful standard for complete SA, in a framework of interoperability within all the elements in the substation (Figure 5.7). It specifies the functional characteristics, the structure of data in devices, the naming conventions for the data, the communications requirements, how applications interact and control the devices, and how conformity to the standard is to be tested.

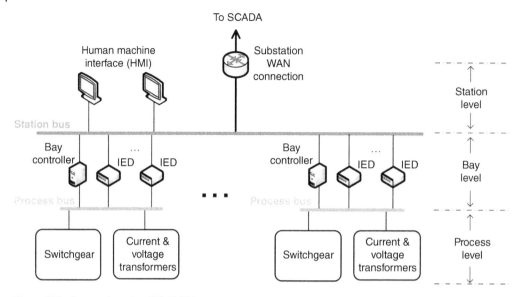

Figure 5.7 Intra-substation IEC 61850 components.

SA Systems (SASs) are used for the monitoring, protection, control, measurement, and alarming of substation functions using modern IEDs [6]. In the context of IEC 61850, we refer to a digital substations, when IEDs operate using a communications network.

IEC 61850 overcomes and improves the traditional protection and control operations (including metering) in substations, where prior to IEC 61850, there was a multiplicity of dedicated single-function relays, electronics, and push buttons requiring complex interwiring between the equipment at the substation yard and the equipment in the substation local control house. Modern computing and progress in communications have produced multifunctional programmable devices; these devices connected using serial interfaces (RS-232 interface) and proprietary protocols and have eventually evolved toward interoperable standard protocols over modern LANs. In the protection domain, IEC 60870-5-103 was developed as the informative interface for protective devices. In the control domain, Modbus protocol was especially successful at substation level, being part of the "fieldbus" domain. Modbus started as a proprietary protocol in 1979, and is an application layer messaging protocol, positioned at Layer 7 of the OSI model, to provide client/server communications between devices connected on different types of buses or networks. The specifications can be found at [22]. There are other fieldbuses available, both in the proprietary and standard domain. A reference to those, together with a short historical evolution, can be found in [23].

Apart from the obvious benefits of standardization (interoperability, interchangeability, reduction of costs, etc.), there are some explicit benefits around IEC 61850:

- Standardized object-oriented data modeling and a standardized configuration language, supported over the Logical Nodes (LNs) and Logical Devices (LDs) concepts.
- Automatic configuration of devices and applications, using Substation Configuration Language (SCL) files.
- Data exchange across devices using Generic Object-Oriented Substation Events (GOOSE) messages over the common substation LAN for fast response. Communications in substations occur at the process level, the bay level, and the station level; the LANs interconnecting them are the process bus and the station bus, respectively.

Table 5.2 IEC 61850 parts

Standard part	Description
IEC 61850-1	Introduction, overview and guidelines
IEC 61850-2	Glossary
IEC 61850-3	General requirements
IEC 61850-4	System and project management
IEC 61850-5	Communication requirements for functions and device models
IEC 61850-6	Configuration description language for communication in power utility automation systems related to IEDs
IEC 61850-7	Details the basic communication structure for substation and feeder equipment. It includes subparts to define the principles and models, common data classes, compatible logical node classes and data object classes, etc.
IEC 61850-8	Provides mapping to MMS, ISO/IEC 8802-3 (Ethernet), and Extensible Messaging Presence Protocol (XMPP)
IEC 61850-9	Specific communication service mapping (SCSM) and precision time protocol profile for power utility automation
IEC 61850-10	Conformance testing of compliant devices

- Support for both client/server and peer-to-peer communication (in peer-to-peer communications both ends have equal pre-eminence).

IEC 61850 standard is organized into 10 parts as in Table 5.2. Within the parts already mentioned, there are some important documents that extend the scope of IEC 61850 to cover inter-substation applications (IEC 61850-90-1), substation-to-control center applications (IEC 61850-90-2), communication of IEEE C37.118 synchrophasor measurements (IEC TR 61850-90-5), communication systems for hydroelectric power plants (IEC 61850-7-410) and communication systems for the integration of DER (IEC 61850-7-420; [24] discusses the integration of DER with IEEE 2030.5 as well).

SAS functions can be broken down in subfunctions, defined as LNs. At the same time, LDs are compositions of LNs and additional services (e.g. GOOSE, Sampled Value exchange, setting groups) that reside in physical devices. The grouping of LNs in LDs is based on common features of these LNs (e.g. the modes of all these nodes are normally switched on and off at the same time). Additionally, LDs provide information about the physical devices they use as host, or about external devices controlled by the LD. Table 5.3 shows the list of LNs groups where substation elements and functions can be easily identified.

One of the most prominent features of the IEC 61850 standard is the representation of the IEDs and the SA functions using SCL files. SCL files are based on eXtensible Markup Language (XML) and are used in SA engineering processes. IED configuration consists of the automation/protection configuration, and the IEC 61850 communication configuration (network settings). Each device must provide an SCL file that describes its own configuration. SCL files contain four subsections:

- The substation section, describing the single line diagram and its LNs.
- The communication section, linking the IEDs.
- The IED section, describing the capabilities of the IEDs.

Table 5.3 List of Logical Node (LN) groups (IEC 61850-7-4).

Group indicator	Logical Node (LN) groups
A	Automatic control
C	Supervisory control
D	Distributed energy resources
F	Functional blocks
G	Generic function references
H	Hydro power
I	Interfacing and archiving
K	Mechanical and nonelectrical primary equipment
L	System logical nodes
M	Metering and measurement
P	Protection functions
Q	Power quality events detection related
R	Protection related functions
S	Supervision and monitoring
T	Instrument transformer and sensors
W	Wind power
X	Switchgear
Y	Power transformer and related functions
Z	Further (power system) equipment
B, E, J, N, O, U, V	*Reserved*

- The binding to LNs belonging to other IEDs, and the LN type section, defining the data objects within the LNs.

Message transmission is defined in terms of timing. The so-called "transfer time" measures the time it takes to complete a transmission, including the processing at both ends (i.e. it is not just dependent on the telecommunication infrastructure). Messages are classified in transfer time classes, as in Table 5.4.

Table 5.4 Transfer time classes.

Transfer time class	Transfer time (ms)	Transfer of (application examples)
TT0	>1000	Files, events, log contents
TT1	1000	Events, alarms
TT2	500	Operator commands
TT3	100	Slow automatic interactions
TT4	20	Fast automatic interactions
TT5	10	Releases, status changes
TT6	3	Trips, blockings

The message types and performance classes are specified in IEC 61850-5. Performance classes within each type are associated to a certain transfer time class:

- Type 1. Fast messages ("protection"). It includes Type 1A and 1B.
- Type 2. Medium speed messages ("Automatics").
- Type 3. Low speed messages ("Operator").
- Type 4. Raw data messages ("Samples").
- Type 5. File transfer functions.
- Type 6. Time synchronization messages.

IEC 61850 standard defines several communication service types, namely GOOSE, Generic Substation State Events (GSSE) – now deprecated, Sampled Values (SVs), Time Synchronization, and MMS. The GOOSE, GSSE, and SVs are mapped directly to Ethernet, while the time synchronization and MMS use UDP and TCP over IP, respectively:

- GOOSE is used for Type 1.
- SV is used for Type 4 messages.
- Time Synchronization is used for Type 6 messages.
- MMS is used for Type 2, 3 and 5 messages.

The implementation of telecommunications in IEC 61850 is based on high-bandwidth Ethernet LANs using optical fiber as the telecommunication medium, to be immune to electrical surges, electrostatic discharges, and other EMC substation environmental aspects. Reliability of the networking devices is achieved through the hardening of the devices to perform in substation environment (see also IEEE 1613). Traffic prioritization and isolation capabilities are included as defined in IEEE 802.1Q-2014.

The physical implementation of the Ethernet station and process buses is achieved through conventional Ethernet switching. Station bus is an OSI Layer 2 Ethernet LAN that provides connectivity for IEDs and other systems in the substation, including the IP routing element for SCADA master station external communication. Process buses are implemented with the same communications technology, and in each of them, IEDs connect among them and with the bay controllers.

This LAN connectivity intends to offer redundancy. This is achieved with the solutions specified in IEEE 62439-3, instead of using the classical pure networking communication mechanisms. There are two redundancy protocols, Parallel Redundancy Protocol (PRP) and High-availability Seamless Redundancy (HSR), designed to provide seamless recovery in case of a single failure. Both are based on the physical duplication of the LAN and the transmitted information and can be implemented alone or combined.

PRP protocol is based on the use of two independent parallel and redundant networks (avoiding any single point of failure) in which Ethernet frames are duplicated and sent simultaneously over them. Although PRP allows several bus topologies (such as linear or ring), and star configuration is the typical one. PRP networks consist of four node types depending on the implemented redundancy when attached to the network. A node will have the same MAC address on both ports and only one set of IP addresses assigned. A node that does not offer double attachments can be connected to these topologies with the so-call redundancy boxes, or "RedBox" for short.

In contrast, HSR protocol is based on the dual transmission of message frames over ring-topology networks in both directions. HSR protocol works under the assumption that each device connected in a ring network is a doubly attached node, each having two ring ports interconnected by full-duplex links. HSR nodes support the IEEE 802.1D bridge functionality and forward frames from one port to the other, except if they already sent the same frame in that same direction or if it is its own injected frame. Hierarchical ring topologies are also supported.

5.7 Metering

A meter is an electricity instrument that measures the amount of electric energy consumed over a given period of time. This measurement is expressed in "billing units," as this is the main purpose of these devices. One of the most common billing units is the kilowatt hour (kWh), i.e. the amount of energy consumed by a 1 kW load in 1 hour [12]. Other meters measure different parameters: the watt meter measures power; multifunction meters measure both active and reactive power, etc.; meters can also record consumption and/or power over specific periods (e.g. 15 minutes).

Meter are an essential element of any utility, as the consumption by each customer needs to be billed, and they have been used since the end of nineteenth century. Meters are found at customer premises (residential, but also commercial and industrial premises), and at points of the grid where energy transit needs to be controlled (e.g. border between Transmission or Distribution grids, where they interconnect). Meters need to be designed for the different needs, and can be found as single-phase (most residential customers'), two or three-phase metering types.

Meters started being electromechanical devices. Many ideas and alternative designs [25] were tested until 1890, when patent US423210 by Ottó Bláthy proposed a meter design based on the Ferraris principle. Induction, "Ferraris" type, electromechanical meters have been the most successful meters, reaching our times. The principle of the electromechanical induction meter operation is to count the revolutions of a metallic disk that rotates at a speed proportional to the power going through the meter to the load.

The electric energy noted by the meter needs to be "read" on a regular basis (monthly, bi-monthly, quarterly, etc.), and the comparison of the last meter read with the previous one, fixes the money to be paid by the customer.

The advent of electronics motivated the design of electronic (solid-state) meters. Electronic meters are not only a consequence of the state-of-the-art evolution but also offer some inherent advantages: a wide range of electrical parameters can be recorded; a higher degree of accuracy can be obtained; functionalities can be upgraded; integration with other electronics (e.g. telecommunications) can happen. Electronic meters started to appear in 1970, and their introduction in the market has been slow. This has been due to the factors influencing the cost of these new meters (low consumption, high reliability, and accuracy) subject to the control of local authorities that need to grant that consumers are charged fairly.

Commercial and industrial environments, and the utilities substations, were the first to host electronic meters, while residential customers (the ample majority of the meters) have had to wait until the digitalization wave has impacted utilities, leveraging the opportunities to improve the traditional meter reading services, and the added value of AMR systems for the grid operation and customer engagement. Indeed, electronic meters together with telecommunications, have changed the approach to the metering services, and have provided new tools to control grid operations.

Specifications for meters and metering services exist in ANSI and IEC. Both standard bodies specify accuracy classes and associated tests for electromechanical and solid-state meters. ANSI C12.1-2014 is an equipment performance standard for electricity meters, comparable in scope to the combination of IEC 62052-11 and IEC 62053-11. ANSI C12.20 specifies metering performance for 0.1, 0.2, and 0.5% accuracy meters (class numbers represent the maximum percentage of metering error at normal loads [26]) and is comparable to IEC 62053-22:2003. From a purely mechanical perspective, ANSI meters have a typically round form factor, while IEC meters are square.

Thus, with the evolution and capabilities upgrade of meters came the evolution of the metering operations. AMR systems started in the 1960s in water, gas, and electricity utilities with the purpose of decreasing the costs derived from the need to send crews to manually take meter readings, often in difficult-to-reach and access meter locations (e.g. private property). Any telecommunication technology available was proposed and tested. However, availability of the technology (both in terms of being physically present, and "available" in telecommunication terms), costs factors in scalable scenarios, security and safety concerns, have conditioned the successive AMR systems, from the semi-automatic walk-by or drive-by type systems, to the fully automated and centralized meter reading systems.

AMR has been improved and overtaken by the Advanced Metering Infrastructure (AMI) concept. AMI concept is commonly replaceable with the Smart Metering concept. AMI, as a difference with AMR, is connected with an ICT infrastructure that gets the system to work in an automated fashion, retrieving readings and controlling meters remotely.

Some of the drivers and benefits for AMI can be shortlisted [12]. Many countries and supranational entities are either mandating, favouring or subsidizing AMI roll-outs, and continuously assessing the results [27] against their benefits. However, some of these drivers are not present in all Smart Metering systems, but they can be added to improve Cost-Benefit Analysis (CBA) studies:

- Reduction (direct and indirect) in costs, compared to manual meter reading, and manual connection and disconnection of consumers.
- Outage detection, through integration of metering data with OMSs.
- Granular consumer data information, to evaluate individual and aggregated load profiles for grid operations and commercial offers.
- Power quality information for grid operations.
- Reduction in consumer complaints (accurate and updated consumption information accessible to the customer in real time).
- Customer engagement through DSM (see Section 5.9.1) programs.
- LV feeder and phase identification of meter connections.
- Tamper and theft detection.

AMI systems include functionality that exceeds billing purposes [28]:

- Two-way secure data communications.
- Real-time information (or quick enough to allow energy-saving programs – there is a wide consensus that a minimum update time of 15 minutes is required) delivered to the utility, to the customer, or to a third party designated by the consumer.
- Wide range of parameters measured in the smart meters (instantaneous voltage, current and power, cumulative energy consumption, active and reactive components, etc.), including import/export and reactive metering to allow DER data integration.
- Generation and communication of specific events and alarms of the smart meters.
- Remote connection and disconnection of supply points (new connections, bad debt, change of energy supplier, etc.), and power limitation control.
- Remote update of smart meters.
- Advanced tariff systems support to reduce peaks in energy demand.

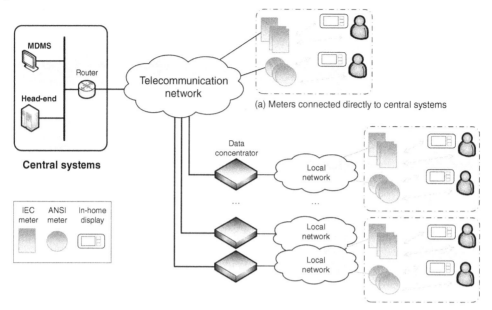

(a) Meters connected directly to central systems

(b) Meters connected to data concentrator, and data concentrators to central systems

Figure 5.8 AMI common architectures.

The architecture of AMI is shown in Figure 5.8. Its more representative IT component is the Meter Data Management System (MDMS) that collects and stores the large quantities of data obtained from the smart meters, through a Head-End system. The Head-End is the assembly of hardware, with the appropriate software, interfacing telecommunications networks, to adapt to their characteristics. The MDMS processes the information (evaluates the quality of the information; generated estimates if information is not present; etc.) and makes it available to other applications that will make use of it (billing systems, customer information systems, OMSs, etc.).

Depending on the historical background, the type of smart meters (residential vs commercial and industrial) and the telecommunication technologies and networks used, the smart meters are directly connected to the Head-End, or use an intermediate element called the Data Concentrator (DC). In case a direct connectivity between Head-End and smart meter can be established, there is no intermediate function or storage. However, if a distributed architecture requiring DCs is used, smart meter readings are typically polled from them, and the data are stored there till the Head-End retrieves them [29]. The direct-access architecture is more common when low quantities of smart meters are to be connected to the system (e.g. commercial and industrial cases), while distributed architectures are used in large quantities environments (e.g. residential) or when AMI is the starting point of an evolutionary process leading to deploy a multiservice Smart Grid, rather than a Smart Metering-only network.

There are numerous proprietary application-level protocols in AMI systems. However, there are two major trends as the more widely accepted standards. ANSI defines the communication between smart meter and the MDMS in ANSI C12.22, while in ANSI C12.19 the data structures to be transported are defined. A similar approach can be found in IEC 62056 series, which basically covers the Device Language Message Specification/COmpanion Specification for Energy Metering (DLMS/COSEM) suite [30].

DLMS/COSEM has a larger implantation than ANSI C12.22 and C12.19. DLMS/COSEM family of standards defines both the data information model and the procedure to exchange this information. These protocols can run over most telecommunication networks, as they are transparent to the underlying telecommunications connectivity among AMI nodes (e.g. meters, DCs, and Head-Ends). DLMS is the application layer protocol that turns the information held by the objects into messages. COSEM defines the object model; OBject Identification System (OBIS) codes are the objects classifying smart meter data in a hierarchical way. These sets of rules and concepts are supported by the DLMS User Association [30], which maintain the blue, green, yellow, and white books [31] with protocols information.

Communication in DLMS/COSEM follows a classical client–server paradigm. The client is placed at the central part of the system (the Head-End or the DC) and communicates with a number of servers implemented in each of the smart meters. Client–server communication is connection-oriented, so prior to the transmission of the data, an "association" needs to be established (authentication, encryption, etc. are included). Both the information exchange and the object definition and implementation are heavy in terms of data exchange, but reliable.

AMI systems have extensions to reach out not only to the smart meters but also to the end-customers as well. A central idea of metering systems is to enable customer interaction and participation. This requires his access to the meter in a quick and efficient way. The are some considerations to be made:

- Smart meters, in the different regulatory regimes, are the responsibility of distribution companies or of the retailers[5] (that do not operate any grid). Moreover, these devices may or may not be the property of the utility, as customers can buy their own meter.
- Smart meters are not usually close to the customer. In many cases, they are in hidden places in building premises, or inside cabinets, as they are not devices designed to face customers.
- Direct smart meter connectivity depends on in-home or in-building telecommunication networks.
- Data from smart meters may come directly from the meter to the customer locally, or can be accessed by customers directly from utility systems. The trend is that no specific devices are deployed in customer premises.
- Traditional meters systems and applications have often neglected the customer segment of AMI. Consequently, not many standards exist to cover this gap.
- Consumption in a house, measured through inaccurate devices, may not be the same as metered consumption. Meters are metrologically controlled devices, subject to a heavy regulation as measurement instruments that must reliably and consistently report accurate consumption. However, many systems directly connecting to appliances or with nonregulated means may offer alternatives for the customer to control his consumption (see Home Energy Management [HEM] systems in Section 5.9.2).

5.8 Synchrophasors

Synchronized phasors (synchrophasors) are the evolution of the phasor concept in Chapter 1 to express and manipulate voltage and current grid values. Synchrophasors are phasors calculated from data samples using a standard time signal as the reference for the measurement, thus having all of them a defined common and unambiguous phase relationship.

---

5 *Retailer* refers to energy suppliers to end customers; in this case, specifically electricity. They may not own or operate a distribution grid.

Thus, synchrophasors measure not only the amplitude but also the phase of sinusoidal signals, with reference to a unique and precise global time reference. Phasor measurement units (PMUs) are the measurement devices that measure synchrophasors [32]. When deployed across the broad geographical area of a power system, PMUs are the source devices of a wide-area synchrophasor network that has many applications, including Wide-Area Measurement (or Monitoring) Systems (WAMSs); Wide-Area Monitoring, Protection, And Control (WAMPAC); and Wide-Area Situational Awareness (WASA).

WAMS is probably the most common application. WAMSs gather real-time synchronized measurement data from PMUs across a broad geographical area, mainly to monitor and control dynamic performance of the power system.

The description of the states of the power system is instrumental to keep stability of the system [2]. SCADA systems, besides its purpose to send commands to RTUs and IEDs, deliver valuable system data that helps to determine the actual state of the power system. WAMS add a new dimension to the measurements, and situational awareness helping to visualize the system better and helping to understand the actual status in the field before actions over it are taken.

PMUs use a global time reference to produce synchrophasors' phase value. This reference is often taken from the well-known precise reference of the Global Positioning System (GPS), although other timing sources can be used [33]. This estimated phase can be used relative to any other such measurement across the complete power system, as the phase-angle is calculated using Coordinated Universal Time (UTC) as a time reference. Thus, two synchrophasors, calculated at different points in a network, can be easily compared because they are related to common instants.

PMUs were introduced in the 1980s and had their origin in the need of synchronized sampling in the design of protection systems for data samples of signals in different substations far apart [33]. Applications of PMUs are quite common in transmission, though they can also be found in distribution, often in connection with DG [34, 35]. PMUs can be realized as stand-alone physical devices or as a part of a multifunction device (e.g. protective relay, digital fault recorder, meter, etc.). PMU measurements are used to support many applications, ranging from visualization of information and alarms for situational awareness to applications that provide sophisticated analytical, control, or protection functionality. Applications, such as dynamics monitoring, use full-resolution real-time data, along with grid models to support both operating and planning functions. Applications display measured frequencies, primary voltages, currents, real and reactive power flows, and other quantities for system operators. Their use improves various grid operations [2]:

- State estimator performance improvement (accuracy and speed) if enough observable PMU data are available. Other EMS functionalities can take advantage of this.
- Improved visualization of the system for UCCs (situational awareness) to make better assessments and be able to take more informed decisions during procedures to avoid potential disturbances.
- Stability margins improvement, as PMUs can monitor small signal oscillations in the system.
- Disturbance location, jointly using PMU data and SCADA information.
- Post-disturbance recovery, in cases of islanding recovery or a black start of the system, due to the improved situational awareness.
- Disturbance analysis for post fault diagnostics, due to the possibility of recreating precise sequence of events.
- Improved protection schemes, as protection relays can have real time comparable PMU measurements.

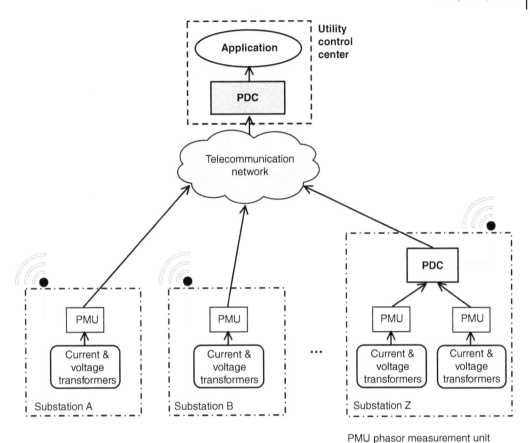

Figure 5.9 Phasor Measurement Units (PMUs) and Phasor Data Concentrators (PDCs) architecture.

Phase measurements in the system are instrumental to improve the availability and quality of the power system. PMUs offer valuable information that can also be used for other important purposes (e.g. quality – harmonic distortion). When PMUs are or have not been available, phase information is estimated at UCCs by means of the state estimation process, a cumbersome and time-consuming process, that cannot be used at local level (i.e. where the PMU could be installed).

In addition to PMUs, the architecture of a synchrophasor network (see Figure 5.9) includes phasor data concentrators (PDCs). PDCs are nodes in a communication network where synchrophasor data from a number of PMUs or PDCs are collated, processed (with its time-stamp), and sent as a single stream to the higher level PDCs and/or applications.

PMUs send data in real time to PDCs. If multiple IEDs in a substation provide synchrophasor measurements, a local PDC may be deployed inside the substation. The data collected by PDCs may be sent to other PDCs/synchrophasor systems and/or directly to central systems.

The first synchrophasor standard was IEEE 1344 and certain limitations forced the creation of IEEE C37.118 [33]. IEEE C37.118.1, IEEE C37.118.1a and, IEEE C37.118.2 (IEEE C37.118.1 focusing on measurements, and IEEE C37.118.2 focused on data transfer) complete the series. IEC TR 61850-90-5 appears as the connection of synchrophasors with the IEC 61850 ecosystem (see Section 5.6). PMUs are installed within substations and may communicate inside or beyond it. Although IEC 61850 was originally designed for SA, SV could be used for sending streaming data, and GOOSE specified under IEC 61850 could be used to send synchrophasors for event-driven applications. However, being typically used to send information within a LAN, SV, and GOOSE protocols are designed to run directly onto Ethernet LANs and cannot be directly used to send synchrophasors over a WAN without the adaptation provided in IEC TR 61850-90-5 [32].

IEEE C37.244 sits in an important position in synchrophasor measurement systems as it is a guide mainly for functional, performance, and conformance needs of PDCs. Among the functional requirements, data communications, communication media, and cybersecurity aspects are described. As per PDC performance, requirements are discussed referring to latency, robustness, availability, reliability, etc.

There are PMUs spread all over the world, specifically integrated in WAMS [33]. In the United States, the pioneer and arguably the more important reference is the North American Synchrophasor Initiative (NASPI) [36, 37], to integrate PMUs across North American power system. NASPInet is the network developed for this purpose, and it has proven to be beneficial to accurately analyze and control the North American power grid performance, including post-disturbance data analysis and early warning systems. The architecture for the NASPI WAMS is based on a data bus connecting the phasor gateways (the primary interface between the utility, or another authorized party, and the data bus [12]) that get connected to the utilities' PDCs. In terms of the data services needed for the PMU data exchange, NASPI defines several data classes (A through E). Class A service the most stringent requirements: e.g. PMU measurements are taken at a very high rate (30, 60, or even 120 measurements per second – 60 Hz-frequency line); it expects a very high reliability (99.9999%) and a network delay of less than 50 ms. In Europe, TSOs report usage of PMUs to record transients, power line thermal monitoring, voltage stability, and power oscillation monitoring, etc.; of late, they have started to interconnect their wide-area measurements, exchanging just a relevant subset of all the data available [38].

5.9 Customers

The customer is not a grid service, but the object and ultimate goal of the power system. However, it is the customer the one that decides the loads that are connected to the grid, and their consumption patterns. Thus, customers' will, decisions, and actions determine when, where, and how much power is needed.

Customers are increasingly aware of the potential of their participation in the power system and how their collaboration in the management of the energy they consume affects the economics and sustainability of the system. This awareness needs to be accompanied with the appropriate systems that can make them easily and effectively participate in it and enjoy part of the benefits they will help to achieve. Some of the benefits are not evident, as they are realized collectively and have an impact in the planning and investment needs in the power system. Indeed, this has an elusive and future-looking component not easily evident to the customer in terms of avoided investments and the controlled evolution of tariffs. However, other benefits should be made evident for customers as immediate consequences of their actions, such as savings in his monthly bills.

There is a number of elements that intend to get the active participation of the customer in the power system. Their real implementation is assisted by different combinations of the applications and services included in this chapter, and the telecommunications needed to make them work:

- Demand-Side Management (DSM). It consists of all the activities by which utilities can influence consumer demand, when in need of matching it with the electricity generation.

- Energy Management. It allows customers to have a much greater control of what they consume in their premises, including producing energy for self-consumption and excess selling.
- Microgrids. As a representative example of small-scale electric systems, that could afford to be independent from large, public electric power systems.

5.9.1 Demand-side Management

The typical yearly and daily aggregated electricity consumption patterns present periodic behaviors that help planners to predict the power to be generated, transported, and distributed to the different geographical areas. Despite this predictable periodicity, special circumstances (e.g. an unexpected heatwave) can make them change, and if this happens within an already planned day, it creates stress in the system.

Aggregated electricity consumption patterns also show peaks and valleys representing periods or times of high demand, followed with the contrary. This requires the power system and all its components to be dimensioned for the highest peaks, so that grid assets and generation resources can cope with all the electricity needed. However, it becomes evident that, from a system dimensioning and resource utilization efficiency perspective, it would be better to reduce the variability between peaks and valleys and so have the same energy demanded in a flatter way (see Figure 5.10).

Multiyear energy needs' control is not the direct target of DSM. DSM focuses on the shorter-term consumption peaks reduction, understanding that habits need time to be transformed (this is why it is referred to as targeting long and medium-term). The idea of DSM is to stimulate customers to change the way in which they consume energy over the day with dynamic energy prices. If the consumption of energy has a lower cost at a certain period, the customer may see value in changing his/her consumption habits [39]. One example is "delay-tolerant" consumption such as washing machine and other specific appliances that could be scheduled to work at different times of the day [40]).

DSM activities are heterogeneous and, apart from the energy efficiency initiatives, Demand Response (DR) appears as an outstanding one. DR is often considered interchangeable with DSM, although it just refers to a particular area of DSM. While DSM includes all the activities designed

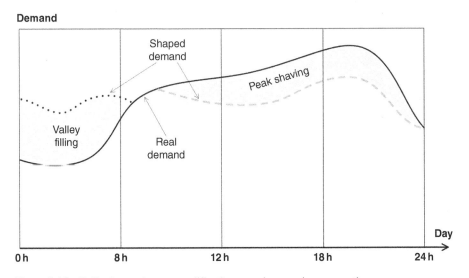

Figure 5.10 Daily demand curve modification to reduce peak consumption.

to influence customer use of electricity to produce the desired changes in the load shape (pattern and magnitude), DR is focused on the short-term needs.

DSM (and DR) diverse initiatives can be implemented without the intervention of the utility or with the utility taking control of part of the customers' loads. This classification has to do with the "price-based" or "incentive-based" type of programs that could be developed [16].

Price-based programs give customers the opportunity of taking advantages of better prices depending on when they consume energy. There could be exchange of information with the utility, but it is mostly not mandatory, and price-based programs can be implemented easily. With nonflat or dynamic pricing tariffs, customers become active participants in the grid [41]. From the consumer behaviour perspective [42], possible price-based schemes include the following (the first three pivot on the price; the last two focus on refunds):

- Time-Of-Use (TOU) pricing. Different prices are fixed for different blocks of hours.
- Real-Time Pricing (RTP). Different pricing rates are fixed per hour.
- Variable Peak Pricing (VPP). It is a mix of TOU and RTP for on-peak and off-peak periods.
- Critical Peak Pricing (CPP). The price for electricity rises during specified periods of time when utilities observe or anticipate high-energy prices or power system stress conditions.
- Critical Peak Rebates (CPR). For similar situations to CPP, the customer gets refunds for any reduction in consumption relative to what the utility considered the customer was expected to consume.

In contrast, incentive-based programs are usually connected with the control of the load from the utility. This is enabled through the support of two-way telecommunications up to the consumer for the utility to be able to shift loads in a real-time manner. Direct load control is a representative example, with the utility directly managing some consumers' appliances that are relevant from the collective consumption standpoint (e.g. heat pumps, water heaters, etc.) to re-shape consumption patterns in selected grid areas. Consumers may sign a contract with a service provider to let them decide when the appliance is to be connected or disconnected under certain conditions, with the costumer having the right to override the orders from service providers under pre-established conditions. A well-known direct load control implementation is ripple control as described in [42], which is based on a simple technology that uses voltage pulses carrying control messages up to LV loads, in specific audio frequencies over the MV grid. Newer strategies exist with different degrees of success and grid presence [43].

DR involves all types of customers, Depending on the scale (residential, commercial, or industrial), the programs for each group have different telecommunication requirements. Typically, residential environments may be the most challenging due to the massive amount of end-points. This is why these programs for residential customers are typically aligned with other initiatives such as Smart Metering. Other than that, individual customers or elements can be superseded:

- Customers can be addressed collectively either through third-party aggregators (as in [44] and [45]) or because they are part of a smaller grid entity (e.g. a Microgrid).
- Intermediate systems may facilitate the management of individual loads inside customers' premises. This can happen at different levels with Home Energy Management Systems (HEMSs) and Building Energy Management Systems (BEMSs), as addressed in the Section 5.9.2.

5.9.2 Energy Management

In the context of customer-side electricity control, the advent of automated systems capable of controlling electricity-related devices in buildings and homes has provided an impulse to DSM in general, and DR in particular, reducing the need for manual intervention. Giving the customer the opportunity to engage with, be informed of, and easily control his consumption, is one of the drivers of DSM programs adoption, as "perceived usefulness" and "ease of use" are necessary to change habits [46].

The main energy consuming appliances in houses and buildings can be categorized as follows [47]:

- Heating and Cooling. HVAC (Heating, Ventilation, and Cooling) systems used to improve ambient conditions.
- Lighting System. Along with HVAC, lighting systems are responsible for the bulk of electricity consumption.
- Other Appliances. The rest of electrical appliances.

DR cannot be implemented if the user does not have an easy mean to access and control his consumption. Thus, DR is facilitated by specific devices and appliances located beyond the meter (i.e. inside homes and buildings [9]). Both HEMSs for residential customers and BEMSs for commercial customers facilitate operations and, essentially, they ease participation of customers in DR programs.

HEMS and BEMS must be able to manage consumption units (e.g. appliances), monitoring their individual consumption to achieve its reduction. At the same time, they should be user-friendly, easy to deploy, configure and manage, be secure and interoperable, and present data in a usable way [48, 49].

HEMS and BEMS are not new [50]. However, the emergence of a mature private energy management industry offering affordable and easy-to-manage HEMSs and BEMSs (including other services, under the terms "connected home" and "smart building" [9]) have made DR find the vehicle to articulate specific utility proposals.

DR can be automated with real-time communication between the utility systems and HEMS/BEMS over a communications network [34]. UCCs send load control signals to HEMS/BEMS, which in turn control consumption units behind the meter, shutting them down or reducing their consumption as appropriate, to the degree and within the conditions agreed between utility and end-user. With the appropriate information from the utility (the meter and the grid; dynamic energy prices [51]), customer priorities and choices, external information (weather information forecast [52]), and smart algorithms, HEMSs/BEMSs will materialize utility and consumer objectives [9], while presenting the result of the adjustments (e.g. saving) to the customer in real time.

On the protocols' standardization activities, two parts must be differentiated. On the one side, there is a widespread interest in the standardization of both communications and application aspects of the in-home part, and many proprietary and standard options exist for the different components of the HEMS/BEMS. ECHONET [53] is a representative example of the different technical building blocks of the solution. On the other side, the communications with the utility to enable DR are progressing slowly. Today, IEEE 2030.5 and IEC 62746 series are two standard solutions. IEEE 2030.5 Smart Energy Profile (SEP) protocol appears as the most promising alternative. This standard defines a wide scope application layer compatible, with TCP/IP communications, which intends to facilitate the information flow between devices such as smart meters and appliances, PEV, DERs, and EMSs. As a historical note, it evolves from ZigBee Alliance (now the Connectivity Standards Alliance) [54] and HomePlug Powerline Alliance [55] work, updated to work with PV smart inverters. IEC 62746-10-3 is the result of the openADR Alliance [56] work. It is an information exchange model standardizing message formats used DR and DER management. With it, dynamic price and reliability signals can be exchanged among utilities, and energy management and control systems.

The control of the energy consumption is today a trend that exceeds the electricity used in buildings and homes and reaches all energy aspects of cities and industries, for example, energy aspects are integral and fundamental part of Smart City initiatives (they are present at least as one of the four core infrastructure components [57]) and are covered by some ITU recommendations (e.g. ITU-T Y.2070) as well.

5.9.3 Microgrids

A microgrid is as a collection of controllable and physically proximate distributed generation and load resources, where multiple sources of AC power can be found [58]. Microgrid concept is itself evolutionary, as it is today something that, having evolved from those small scale decentralized isolated networks in nineteenth century [59], has grown in interest due to the incorporation of DER resources, without which the concept is not fully materialized today. These DER resources, within a Microgrid, are not presented to the utility as individual elements, but as a higher level entity that may be easier to control and integrate.

Thus, a Microgrid can be better defined as "a group of interconnected loads and Distributed Energy Resources (DERs) with clearly defined electrical boundaries that acts as a single controllable entity with respect to the grid and can connect and disconnect from the grid to enable it to operate in both grid-connected or island modes" [60]. The customer component needs to be added to this definition, to include all aspects that make the concept relevant.

Microgrids offer benefits for customers and utilities, while simultaneously bringing in integration and control challenges that need to be solved [12].

The massive penetration of DG units and the need to integrate them in the electric power system are not a sustainable option when individually taken [36]. However, with a coordinated approach that converts individual units (generators and storage, but also loads) into subsystems (as small-scale electricity networks) than can be connected to the utility network in the so-called Point of Common Coupling (PCC), Microgrids achieve a scale that can be used by the utility to control flexibility while ensuring the system's reliability and power quality requirements. At the same time, customers, many times promoters of the DER elements in the Microgrid, can jointly benefit from market participation.

Microgrids can operate connected to the grid but also islanded. These two operation modes make them useful for remote areas where networked electrification is difficult. It is also useful for disaster preparedness and recovery [58].

IEC has produced some guidelines for Microgrids. One technical specification deals with Microgrid projects planning and specification (IEC TS 62898-1) and the other with Microgrids operations (IEC TS 62898-2). ITU refers to Microgrids in ITU-T Y.2071.

Microgrid Energy Management System (MEMS) will coordinate energy transactions between Microgrids and the utilities' DMS and EMS. There are two IEEE references relevant in this domain. First, the IEEE 1547.4-2011 standard, "Guide for Design, Operation, and Integration of Distributed Resource Island Systems with Electric Power Systems" (the term DR island system, is used here for Microgrids, "intentional islands") that defines the basic Microgrid connection and disconnection process for a stable and secure electrical interconnection of DER to the Distribution grid; it is instrumental for the integration of the Microgrid [61]. Second, IEEE 2030.7, "Standard for the Specification of Microgrid Controllers," is directly related to the Microgrid operation with the MEMS and includes the control functions that define Microgrids as systems in their diverse operating modes. It contains the control approaches required from the DSO and the Microgrid operator.

Microgrids are still the focus of much research, as seen in [62].

5.10 Power Lines

Power lines are a very important asset in utilities for many reasons. Specifically, power lines in the transmission segment of the grid are, in investment terms, a very significant percentage of the total investment. In terms of "cost of opportunity," big transmission projects (including their power lines) take a long time to materialize, and once the infrastructure is laid, it needs to be operated to reach its maximum capacity. Thus, although we refer to power lines generically in this chapter, transmission grid is specifically addressed.

5.10.1 Flexible AC Transmission System

FACTS technology, and specifically FACTS controllers, control the interrelated parameters that govern the operation of transmission systems including series and shunt impedance, current, voltage, and phase angle. FACTS technology was a consequence of the evolution of power electronics. Developed in 1960s, FACTSs were brought to the scene as an integrated philosophy during the 1980s [63] and, by the beginning of the twenty-first century, they have extensively proven their potential [64]. The new paradigm associated with FACTSs of making the power system electronically controllable motivates that power system equipment is designed and built considering the procedures included both in its planning and operation, to enable high-speed control of the path through which energy flows.

Thus, FACTSs are systems based in power electronics and other passive components that provide real-time control over the physical parameters of an AC transmission system's power lines. FACTS controllers [65] are used to improve the power transfer capability of the line, to prevent loop flows, to facilitate fast voltage control and frequency stability, and to balance power between parallel transmission paths (to prevent over or underloading). Traditionally, thyristors have been used for FACTSs, although IGBT (Insulated Gate Bipolar Transistors) in STATCOMs (STATic COMpensators) and other alternatives are increasingly present, as a consequence of the development of HVDC (high-voltage DC) technology in the Transmission grid.

FACTSs can be connected to the grid at several locations (at the interfaces or points of coupling between sources and loads) and in different configurations (shunt, series, or shunt – series), parallel or in series with the power lines.

5.10.2 Dynamic Line Rating

The maximum power flow capacity on a transmission power line is limited by heating considerations to maintain safe and reliable operating conditions. This heating depends on the power line intrinsic conditions (e.g. conductor type, diameter, etc.), but it also depends on the environmental conditions in which the power line operates. These intrinsic and operational conditions determining the heating situation of the power line are commonly referred to as conductor thermal ratings.

Within a complex system such as the power grid, where different pieces are interconnected, the costly investment associated with power lines demands that they are operated at their maximum capacity to maximize investments. At the same time, if that does not happen to be the case, decisions on the power generation sources to be connected to the system may be biased by the congestion situations in the grid, and total system costs may not be optimal. Grid congestion can be managed with new power lines construction, or conductor changes; however, dynamic line rating (DLR) systems are intended to avoid physical network upgrade, due to the cost, time and inefficiency of the brute-force approach.

In this context, DLR is a broad and blanket term for different technologies and methodologies to determine conductor thermal ratings in a dynamic fashion [66]. In contrast to "Static Line Ratings" that use static and conservative worst-case information about the Transmission line operating environment (weather conditions, wind speed and direction, ambient temperature, solar radiation, etc.), DLR uses improved, more granular, or desirably, real-time data. For example, conductors expand when the temperature is high, so that in the case of overhead power lines, they lengthen and the distance to the ground gets reduced. This can result in safety clearance distances not being respected. With DLR, the physical properties of the conductor (electrical resistance, mechanical strength, thermal rating, etc.) and environmental conditions (e.g. ambient air temperature, wind speed, solar radiation) can be estimated and used to calculate thermal limits and consequently maximum supportable current at any point in time.

DLR technologies have traditionally been classified as weather-based and asset-based systems. The former ones collect environmental conditions (wind speed and direction, ambient air temperature, solar radiation, and line current). The latter focus on measurement of the conductor itself, gathering local conductor temperature, position or tension, and line current. Both types use these parameters from sensors and IEDs deployed locally, to calculate the maximum allowable conductor current, so that the ratings can be incorporated into a control system, such as a SCADA system or EMS.

DLR technology has not yet been consolidated [67, 68] as there is no clear best combination of performance, cost, and ease of installation for sensing and monitoring in the huge spans of overhead transmission lines. Direct conductor monitoring tends to offer precision, with the burden of installation and maintenance costs. Environmental parameter monitoring depends on the proximity of weather stations and will not be as accurate, although probably more cost effective. Any of the methods, and specially asset-based ones, need communications connectivity and will require appropriate data transfer readiness and acceptable latency.

5.11 Premises and People

5.11.1 Business Connectivity

HV substations, and non-HV but strategic substations, have been the permanent center of work for many utility employees till recent times. Although it is true that there is no need for many of these premises to have people supporting their operations locally, it was not so in the past. However, even if in recent times, these substations are prevalently unmanned, their location is such that the rest of the Distribution grid around them is the territory where utility crews operate. Consequently, as a matter of convenience, utility crews that may start their work day are close to the grid that many require their intervention on the field if needed. Thus, many substations are nowadays used as permanent utility offices.

Many utilities, especially those more focused on Transmission assets, or those that have developed private networks for their telecommunication needs, possess telecommunication assets that have been developed, and deployed over the years. There are different drivers for these investments, including the operational needs and the lack of cost-effective public infrastructures before the 1990s. Others are a consequence of the opportunity of the global process of telecommunications liberalization that allowed utilities all over the world to build telecommunication infrastructures (in particular, optical fiber cables) over existing utility poles and other rights of way that could be leveraged for commercial use with telecommunication operators needing optical fiber for their networks. And, finally, there are other Smart Grid-related requirements that recommend the use of a private telecommunications network.

Telecommunications infrastructure and network development have evolved in many different ways in utilities all over the world. Utilities' private telecommunications network is combined with commercially provided telecommunications services, where developing a private network is not feasible (e.g. lack of access to licensed spectrum), not cost effective, or not a strategic priority. At the same time, there is a representative number of utilities that developed carriers' carrier models. This opportunity started in the 1990s and is still a possibility and a trend for the Smart Grid readiness [69]. The evolution of these models created a powerful optical fiber cable network reaching many of the substations in the utility.

Business connectivity in a substation combines "corporate" and operational needs. This is, people need to access typical office-type systems as regular office employees (e.g. e-mail access, intranet access, ERP software access, etc.; these applications do not have real-time or critical operational characteristics), while they also need to access an electric operations environment (e.g. GIS, operational databases, UCC applications, etc.). At the same time, the substation is the center of many connectivity needs related to the different services and applications described in this chapter. The three environments need to be separated but combined through the appropriate telecommunications network infrastructure reaching the substation.

Telecommunications connectivity reaching the substation need to be compliant with the appropriate cybersecurity, availability, performance, and resiliency requirements. When combining different corporate, operational, and Smart Grid services, the requirements for the operational context are stricter, while the bandwidth needs are usually higher on the corporate end.

Finally, substations are electromagnetically sensitive environments, and because of the environmental considerations already explained in Chapter 1, tend to be implemented with either optical fiber or radio (microwave) systems.

5.11.2 Workforce Mobility

The efficient organization of work depends on the appropriate use of technology to have the workforce accessible wherever their activity is taking place. The accessibility of the workforce has several facets:

- The location of the workforce needs to be managed in a coordinated way.
- The instantaneous bidirectional communication with the workforce must be granted.
- The access of the workforce to the information they need to perform their job must be granted.

The access to the location of the workforce is used for the proper organization of the different activities. This is no different to any other dispatching activity. When the location (e.g. GPS) and availability for activity are known, different applications can optimize the assignment of planned tasks and incident responses.

The accessibility to voice communication may seem straightforward in this hyperconnected world. The voice communication between crew members and of these members with the UCCs is part of the common operational procedures, to an extent where if no possibility to communicate orally, the operation on a substation or in a power line may need to be aborted (often, the communication needs to be recorded as dictated by either local regulation or utility safety procedures). However, there is a group of challenges for utilities that need to be considered when granting the access to voice communications. The first one has to do with coverage. The activity of the utility may require a job in a place where no commercial mobile network coverage exists. The other challenge has to do with the proper availability guarantee of the service that needs to be operational when difficult power system situations exist. This is, in the case of a local blackout, where electricity is not available to make commercial telecommunication systems work, utility operations need to happen in order to restore the power service. This situation is so critical that there are national and international regulations that rule how the voice service access needs to be for blackout recovery situations.

Utilities have a long tradition of running private mobile voice communication systems, parallel to public commercial network services. These systems started when cellular systems were not the rule (before the 1980s) and have been developed ever since (with differences in different utilities) to offer a service with characteristics (specific coverage, availability, and features) that cannot be complied with commercially available systems. Thus, the early adoption and advances on private mobile radio (PMR) systems, a consequence of public safety needs' adaptation (police, firefighters, blue-light services in general), had utilities as another heavy user. Traditional features of PMR systems have been different to the ones of public commercial mobile telecommunications ([70] which includes a list of requirements typical of PMR systems – push-to-talk [PTT] operation mode for fast channel access, closed group of users, etc.). Although PMR systems have in many places been displaced by public mobile (cellular) systems, they still cannot offer what it is being required for mission critical applications, the most important being coverage and availability. To fill this gap, apart from PMR systems, utilities also revert to complement land-mobile systems with satellite-based voice communications [46], which despite their shortcomings, are used when terrestrial systems may be facing operational difficulties or are not available at all.

Lately, data mobile communications are attracting the attention of the utility sector. With a world of possibilities opening up with the access to data when away from the office and doing a job in a substation or in power lines, utilities (as some other industries) are developing applications to help their crews' activity on the field. Beyond the fact of being able to access the data, they need as if they were in the office, augmented reality, remote-hands, and such new applications for the digital worker come both as an opportunity and as a greater burden for utilities' telecommunications' needs, as they many need to happen where there is no commercial service available, or the one that exists, is either narrowband and/or not cost-effective.

5.11.3 Surveillance

The physical protection of substations and other strategic utility premises such as UCCs from threats (internal or external; physical or cyber) is instrumental in the correct operation of the system and fundamental for a reliable electricity delivery.

Among these assets, substations in general and PSs in particular, deserve a special treatment in all power systems. The impact of a PS malfunction is huge in the overall system. Substations are usually unmanned and the first defence needed is the physical one. There are many substations that go unnoticed, and thus they attract less attention from curious people. However, there are a minimum number of indications that need to be followed in order to prevent, control and minimize the effects of unauthorized access to the substation (see IEEE 1402). Finally, most of all newly built substations follow construction criteria to gain community acceptance and improve environmental impact (IEEE 1127). These guidelines refer to selection of the site, design and construction phase, and include routine inspection of physical site security elements mentioned in IEEE 1402.

Until cybersecurity concerns attracted the attention of security professionals, security in substations had solely been associated to the prevention of human intrusion and the intrusion of animals to a lesser extent (IEEE 1264). Intruders include inadvertent people, thieves, vandals, disgruntled employees, or terrorists.

The physical elements typically included to control physical security, are

- Physical access prevention elements (building design and construction materials, fences, gates – locks, barriers to energized equipment, etc.).
- Hiding of premises and assets.
- Signs and lighting of the property.
- Security patrols or onsite physical security presence.

The surveillance methods above do not need telecommunications to work and may be just considered physical. However, they are accompanied and complemented by technological elements (sensors and alarms) that may need some sort of local or remote telecommunications connectivity.

Substations' security needs must be always assessed considering their criticality for the complete system, to evaluate their vulnerability and analyze risk scenarios that may require specific actions. When the substations are within the context of what can be understood as critical infrastructure for a nation, there are usually local regulations that define how this assessment needs to be performed and which kind of actions must be enacted in the different situations. In order to have an idea of the implications of all this, NERC CIP (North American Electric Reliability Corporation Critical Infrastructure Protection) standards [71] can be mentioned, as they are one of the most relevant worldwide references. NERC CIP 014-2, e.g. refers to Physical Security and develops a physical security reliability standard "to identify and protect Transmission stations and Transmission substations, and their associated primary control centers that if rendered inoperable or damaged as a result of a physical attack could result in instability, uncontrolled separation, or Cascading within an Interconnection."

Intrusion detection and access control systems are commonly found in substations, independently of the voltage level. Intrusion detection systems typically make use of motion detection systems. Physical presence is sensed, using photoelectric cells, lasers, optical fibers or microwaves, to trigger alarms that can act simultaneously or in a sequence, locally (sirens and/or lights) and remotely (including access to security authorities, e.g. police). The remote alert may be accompanied of sound and video recorded and/or streamed and is sent to surveillance control centers or recently referred to as SOCs (Security Operations Centers).

Video cameras are a very common element of physical security under the surveillance concept. Cameras record actions from fixed positions, remotely operated or more often autonomously. While the recording is often stored locally, it can be transferred to the control centers on-demand, or when local alarms are triggered.

The improvements in the audio-visual (AV) technology, the reduction in cost, and its widespread adoption in many fields have made them be adopted by default in physical substation security. However, AV means and the telecommunications needed for them may be considered a disruption element in terms of bandwidth needs, as AV-based surveillance is probably one of the substations services needing the largest bandwidth [34]. Thus, codecs, motion in the images, frame rate, static-images' refresh, and video resolution need to be controlled not to demand too much bandwidth from a connectivity that may not offer the expected reserved bandwidth. Video rates from 128 kbps to 2 Mbps are often affordable in relevant substations.

References

1 Smith, H.L. (2010). A brief history of electric utility automation systems. Electric Energy T&D Magazine 14 (3): 39–46.
2 Thomas, M.S. and McDonald, J.D. (2015). Power System SCADA and Smart Grids. Boca Raton: CRC Press.
3 Blume, S.W. (2017). Electric Power System Basics for the Nonelectrical Professional, 2e.
4 Becker, D. (2001). *Inter-Control Center Communications Protocol (ICCP, TASE.2): Threats to Data Security and Potential Solutions*. Technical Review 1001977 (October 2001) [Online]. Electric Power Research Institute (EPRI). https://www.epri.com/research/products/1001977 (accessed 10 October 2020).
5 Gers, J.M. and Holmes, E.J. (2011). Protection of Electricity Distribution Networks, 3e. Herts, United Kingdom: Institution of Electrical Engineers.
6 Bansal, R. (2019). Power System Protection in Smart Grid Environments. Boca Raton: Taylor and Francis, a CRC title, part of the Taylor and Francis imprint, a member of the Taylor and Francis Group, the academic division of T&F Informa, plc.
7 Hartmann, W. (2019). FLISR concepts. Presented at the 43rd Annual System Protection and Substation Conference, Nebraska, USA (October 2019) [Online]. https://beckwithelectric.com/wp-content/uploads/docs/events/2019-Iowa-Nebraska/IA-NB-FLISR-Primer-191021.pdf (accessed 5 April 2021).
8 Aurus, A., Oikonomou, T., and Vögele, Y. (2017). *Distribution Automation for Electrical Grids. Fault and Outage Management* (September 2017) [Online]. https://library.e.abb.com/public/bf5991a91709431a99dacdd646616f3f/Fault%20and%20outage%20management%20external%20webinar.pdf (accessed 11 October 2020).
9 Carvallo, A. and Cooper, J. (2015). The Advanced Smart Grid: Edge Power Driving Sustainability, 2e. Boston: Artech House.
10 CIGRE (2001). *Protection Using Telecommunications*. Final Report (August 2001). Paris, France: CIGRE.
11 Godfrey, T. (2014). *Guidebook for Advanced Distribution Automation Communications* Technical Report 3002003021 (November 2014) [Online]. Electric Power Research Institute (EPRI). https://www.epri.com/research/products/000000003002003021 (accessed 8 October 2020).
12 Borlase, S. (ed.) (2017). Smart Grids: Advanced Technologies and Solutions, 2e. Boca Raton: Taylor & Francis, CRC Press.
13 Ekanayake, J., Liyanage, K., Wu, J. et al. (2013). Smart Grid: Technology and Applications. Hoboken, NJ, USA: Wiley.
14 CIEE (2013). *Distribution System Voltage Management and Optimization for Integration of Renewables and Electric Vehicles. Research Gap Analysis*. Final Project Report (December 2013) [Online]. California Institute for Energy and Environment (CIEE). https://uc-ciee.org/ciee-old/downloads/Task%202%202%20VVOC%20Gaps%20White%20Paper%20Final-2014-01-09.pdf (accessed 11 October 2020).
15 Peng, F.Z. (2017). Flexible AC transmission systems (FACTS) and resilient AC distribution systems (RACDS) in smart grid. Proc. IEEE 105 (11): 2099–2115. https://doi.org/10.1109/JPROC.2017.2714022.
16 Kabalci, E. and Kabalci, Y. (eds.) (2019). Smart Grids and Their Communication Systems. Singapore: Springer Singapore.

17 Seal, B., Renjit, A., and Deaver, B. (2018). *Understanding DERMS*. Technical Update 3002013049 (July 2018) [Online]. Electric Power Research Institute (EPRI). https://www.epri.com/research/products/3002013049 (accessed 11 October 2020).

18 National Grid ESO. *Distributed ReStart | National Grid ESO*. https://www.nationalgrideso.com/future-energy/projects/distributed-restart (accessed 10 October 2020).

19 Hatziargyriou, N., Poor, H.V., Carpanini, L., and Sanchez-Fornie, M.A. (2016). Smarter Energy: From Smart Metering to the Smart Grid. London: The Institution of Engineering and Technology.

20 Amini, M.H., Boroojeni, K.G., Iyengar, S.S. et al. (eds.) (2019). Sustainable Interdependent Networks II: From Smart Power Grids to Intelligent Transportation Networks, vol. 186. Cham: Springer International Publishing.

21 SSE Enterprise (2020). *London Bus Garage Becomes World's Largest Vehicle-to-grid Site*. Press releases (11 August 2020). https://www.sseutilitysolutions.co.uk/insights/press-releases/london-bus-garage-becomes-worlds-largest-vehicle-to-grid-site (accessed 10 October 2020).

22 *Modbus FAQ*. https://modbus.org/faq.php (accessed 11 October 2020).

23 Galloway, B. and Hancke, G.P. (2013). Introduction to industrial control networks. IEEE Commun. Surv. Tutor. 15 (2): 860–880. https://doi.org/10.1109/SURV.2012.071812.00124.

24 Simpson, R., Mater, J., and Kang, S. (2019). *IEC 61850 and IEEE 2030.5: A Comparison of 2 Key Standards for DER Integration: An Update*. PacWorld2019, White Paper [Online]. https://cdn2.hubspot.net/hubfs/4533567/IEEE-2030-5-and-IEC-61850-comparison-082319.pdf (accessed 11 October 2020).

25 S. E. International (2006). *The History of the Electricity Meter* (28 June 2006). Smart Energy International. https://www.smart-energy.com/features-analysis/the-history-of-the-electricity-meter (accessed 11 October 2020).

26 Seal, B. and McGranaghan, M. (2010). *Accuracy of Digital Electricity Meters*. White Paper 1020908 (May 2010) [Online]. Electric Power Research Institute (EPRI). https://www.epri.com/research/products/1020908 (accessed 10 October 2020).

27 Tounquet, F. and Alaton, C. (2020). Benchmarking Smart Metering Deployment in the EU-28: Final Report. European Union.

28 Publications Office of the European Union (2012). 2012/148/EU: Commission Recommendation of 9 March 2012 on Preparations for the Roll-out of Smart Metering Systems. European Union.

29 Zhou, J., Qingyang Hu, R., and Qian, Y. (2012). Scalable distributed communication architectures to support advanced metering infrastructure in smart grid. IEEE Trans. Parallel Distrib. Syst. 23 (9): 1632–1642. https://doi.org/10.1109/TPDS.2012.53.

30 DLMS. *What Is DLMS/COSEM*. DLMS User Association. https://www.dlms.com/dlms-cosem/overview (accessed 10 October 2020).

31 DLMS. *Coloured Books*. https://www.dlms.com/resources/coloured-books-extracts (accessed 11 October 2020).

32 Monti, A. (ed.) (2016). Phasor Measurement Units and Wide Area Monitoring Systems: From the Sensors to the System. Amsterdam: Elsevier.

33 Phadke, A.G. and Bi, T. (2018). Phasor measurement units, WAMS, and their applications in protection and control of power systems. J. Mod. Power Syst. Clean Energy 6 (4): 619–629. https://doi.org/10.1007/s40565-018-0423-3.

34 Budka, K.C., Deshpande, J.G., and Thottan, M. (2014). Communication Networks for Smart Grids: Making Smart Grid Real. London: Springer.

35 NASPI (2020). *Synchronized Measurements and Their Applications in Distribution Systems: An Update*. NASPI-2020-TR-016 (June 2020) [Online]. NASPI. https://www.naspi.org/sites/default/

files/reference_documents/naspi_distt_synchro_measure_apps_20200716.pdf (accessed 11 October 2020).
36 Bakken, D. (ed.) (2014). Smart Grids Clouds, Communications, Open Source, and Automation. Boca Raton, FL, USA: Taylor & Francis.
37 *About NASPI | North American SynchroPhasor Initiative*. www.naspi.org (accessed 11 October 2020).
38 ENTSO-E (2015). *Wide Area Monitoring – Current Continental Europe TSOs Applications Overview, Version 5*. Technical Report (September 2015) [Online]. https://docs.entsoe.eu/dataset/wide-area-monitoring-continental-europe/resource/18be44ed-98fe-40ea-a04c-6f12ef071850 (accessed 11 October 2020).
39 Nakano, S. and Washizu, A. (2019). In which time slots can people save power? An analysis using a Japanese survey on time use. Sustainability 11 (16): 4444. https://doi.org/10.3390/su11164444.
40 Mouftah, H.T. and Erol-Kantarci, M. (eds.) (2016). Smart Grid: Networking, Data Management, and Business Models. CRC Press.
41 SmartGrid.gov (2019). *Time Based Rate Programs: Recovery Act* (16 December 2019). https://www.smartgrid.gov/recovery_act/time_based_rate_programs.html (accessed 10 October 2020).
42 Kaempfer, S. and Kopatsch, G. (eds.) (2012). Load management and ripple control. In: Switchgear Manual, ABB, 12e, 769–770. Berlin: Cornelsen Scriptork.
43 Peak Load Management Alliance (PLMA) (2018). *The Future of Utility "Bring Your Own Thermostat" (BYOT) Programs* (March 2018) [Online]. https://www.peakload.org/assets/Groupsdocs/PractitionerPerspectives-UtilityBYOTPrograms-March2018.pdf (accessed 11 October 2020).
44 Emissions-EUETS.com. *Demand Side Response Aggregator (DSR Aggregator)*. https://www.emissions-euets.com/internal-electricity-market-glossary/855-demand-side-response-aggregator-dsr-aggregator (accessed 11 October 2020).
45 OFGEM (2016). *Aggregators – Barriers and External Impacts* (May 2016) [Online]. https://www.ofgem.gov.uk/system/files/docs/2016/07/aggregators_barriers_and_external_impacts_a_report_by_pa_consulting_0.pdf (accessed 11 October 2020).
46 Sendin, A., Sanchez-Fornie, M.A., Berganza, I. et al. (2016). Telecommunication Networks for the Smart Grid. Artech House.
47 Bouhafs, F., Mackay, M., and Merabti, M. (2014). Communication Challenges and Solutions in the Smart Grid. New York: Springer.
48 Karlin, B., Ford, R., Sanguinetti, A. et al. (2015). *Characterization and Potential of Home Energy Management (HEM) Technology* (January 2015) [Online]. Pacific Gas and Electric Company. http://www.cusa.uci.edu/wp-content/uploads/2015/02/PGE-HEMS-Report.pdf (accessed 11 October 2020).
49 Andaloro, A., Antonucci, D., Vigna, I., and Lollini, R. (2018). *Technical Specifications for BEMS and HEMS. Characteristics, Common Functions and Requirements of BEMS & HEMS to Be Installed in the Lighthouse Cities*. STARDUST – Holistic and Integrated Urban Model for Smart Cities, Technical Report D1.6 (September 2018) [Online]. http://stardustproject.eu/wp-content/uploads/2020/06/D.1.6.pdf (accessed 11 October 2020).
50 Lamarre, L. (1991). Building the intelligent home. EPRI J. 16 (4): 60.
51 Siozio, K., Anagnostos, D., Soudris, D., and Kosmatopoulos, E. (eds.) (2018). IoT for Smart Grids. Design Challenges and Paradigms. New York, NY: Springer Berlin Heidelberg.
52 International Energy Agency (2015). *How2Guide for Smart Grids in Distribution Networks. Roadmap Development and Implementation*. Technical Report (May 2015) [Online]. Paris, France: IEA. https://webstore.iea.org/how2guide-for-smart-grids-in-distribution-networks (accessed 11 October 2020).

53 ECHONET. *What Is HEMS?* www.echonet.jp (accessed 11 October 2020).
54 Zigbee Alliance. *Zigbee Alliance*. www.zigbeealliance.org (accessed 11 October 2020).
55 European Innovation Partnership – European Commission (2016). *Homeplug (HomePlug Powerline Alliance)* (26 August 2016). https://ec.europa.eu/eip/ageing/standards/home/domotics-and-home-automation/homeplug-homeplug-powerline-alliance_en (accessed 11 October 2020).
56 *About OpenADR*. https://www.openadr.org/overview (accessed 11 October 2020).
57 Vlahoplus, C. and Litra, G. (2017). *The Smart City Opportunity for Utilities*. ScottMadden (23 May 2017). https://www.scottmadden.com/insight/the-smart-city-opportunity-for-utilities (accessed 10 October, 2020).
58 IEC (2014). Microgrids for Disaster Preparedness and Recovery: With Electricity Continuity Plans and Systems. White paper. Geneva, Switzerland: International Electrotechnical Commission.
59 CEN-CENELEC-ETSI Smart Grid Coordination Group (2012). *Smart Grid Reference Architecture* (November 2012) [Online]. https://ec.europa.eu/energy/sites/ener/files/documents/xpert_group1_reference_architecture.pdf (accessed 11 October 2020).
60 Ton, D.T. and Smith, M.A. (2012). The U.S. Department of Energy's Microgrid Initiative. Electr. J. 25 (8): 84–94. https://doi.org/10.1016/j.tej.2012.09.013.
61 Hatziargyriou, N. (ed.) (2013). Microgrid: Architectures and Control. Chichester, UK: Wiley.
62 Parhizi, S., Lotfi, H., Khodaei, A., and Bahramirad, S. (2015). State of the art in research on microgrids: a review. IEEE Access 3: 890–925. https://doi.org/10.1109/ACCESS.2015.2443119.
63 Acha, E., Fuerte-Esquivel, C.R., Ambriz-Perez, H., and Angeles-Camacho, C. (2004). FACTS: Modelling and Simulation in Power Networks. Chichester; Hoboken, NJ: Wiley.
64 Fairley, P. (2010). Flexible AC transmission: the FACTS machine. Flexible power electronics will make the smart grid smart. *IEEE Spectrum*, Special Report (December 2010) [Online]. https://spectrum.ieee.org/energy/the-smarter-grid/flexible-ac-transmission-the-facts-machine (accessed 11 October 2020).
65 Hingorani, N.G. and Gyugyi, L. (2000). Understanding FACTS: Concepts and Technology of Flexible AC Transmission Systems. New York: IEEE Press.
66 US Department of Energy (2019). *Dynamic Line Rating*. Report to Congress (June 2019) [Online]. https://www.energy.gov/sites/prod/files/2019/08/f66/Congressional_DLR_Report_June2019_final_508_0.pdf (accessed 11 October 2020).
67 Morozovska, K. and Hilber, P. (2017). Study of the monitoring systems for dynamic line rating. Energy Procedia 105: 2557–2562. https://doi.org/10.1016/j.egypro.2017.03.735.
68 Black, C.R. and Chisholm, W.A. (2015). Key considerations for the selection of dynamic thermal line rating systems. IEEE Trans. Power Deliv. 30 (5): 2154–2162. https://doi.org/10.1109/TPWRD.2014.2376275.
69 Guidehouse Insights. *Navigant Research Report Shows Adopting a Fiber Optic Backbone Is a Strategic Imperative for Utilities' Long-term Success*. https://guidehouseinsights.com/news-and-views/navigant-research-report-shows-adopting-a-fiber-optic-backbone-is-a-strategic-imperative-for-utiliti (accessed 11 October 2020).
70 Gray, D. (2003). TETRA the Advocate's Handbook: From Paper to Promise. Looe, Cornwall: TETRA Advocate.
71 NERC. *Critical Infrastructure Protection (CIP)*. United States Mandatory Standards Subject to Enforcement. https://www.nerc.com/pa/stand/Pages/ReliabilityStandardsUnitedStates.aspx?jurisdiction=United%20States (accessed 11 October 2020).

6
Optical Fiber and PLC Access Technologies

6.1 Introduction

Optical fiber-based technologies and Power Line Communication (PLC) are the most relevant access wireline fixed-network solutions for the Smart Grid.

Most of the technologies covered in Chapter 4 make use of optical fiber. However, there are more specific technologies for the access network, such as Passive Optical Network (PON)-based ones. Digital Subscriber Line (xDSL – copper cables) and Hybrid Fiber-Coaxial (HFC) technologies will not be covered in this chapter, as they are less capable and prepared to support Smart Grid services compared with optical fiber alternatives. However, there are Smart Grid services today supported on xDSL and HFC.

PLC will be also covered in this chapter, as the natural option for any access technology reaching devices connected to the power grid. Utilities combine optical fiber and PLC solutions, making optical fiber reach increasingly closer to the grid edge.

This chapter elaborates on PON and PLC technologies, covering the most relevant systems in each of them.

6.2 Optical Fiber Passive Network Technologies

A PON is an optical fiber network with a point-to-multipoint topology used to deliver bidirectional data flows from a single network point in the border of the access network, to multiple user end points [1].

PONs do not typically need power supply in the intermediate points of the network, mainly consisting of optical fiber and splitting/combining components. PONs are intrinsically different from active optical networks, and electrical power is only required at the transmit and receive points, easing the rollout. However, the introduction of active reach extenders (e.g. as defined in ITU-T G.984.6) has created a situation where some infrastructure components of a PON system may not be entirely passive. That is why in some contexts the term Optical Distribution Network (ODN) refers to the mostly passive point-to-multipoint distribution means extending from the core edge to the end users. Thus, active reach extenders and dual-homing concept can be integrated in the ODN definition.

FTTx (Fiber To The x) is the acronym that comprehends PONs, with the PON being the most common implementation of FTTx [2]. The "x" means access to any premise from the network core

---

Smart Grid Telecommunications: Fundamentals and Technologies in the 5G Era, First Edition. Alberto Sendin, Javier Matanza, and Ramon Ferrús.
© 2021 John Wiley & Sons, Inc. Published 2021 by John Wiley & Sons, Inc.

to the end customer (e.g. while the one of the most common use is H, for Home – FTTH, other common meanings are C for Curb or Cabinet, O for Office, B for Building, N for Node, Dp for Distribution point, or P for Premises). FTTx gives also name to different FTTH Councils whose mission is "the acceleration of fibre to the home adoption" worldwide [3].

6.2.1 Mainstream Technologies and Standards

PONs have their origin in the 1980s with the idea of fiber rings dropping service to network users. However, this ring topology idea was abandoned early, to be replaced by optical tree architectures [4]. Today, point-to-multipoint (Figure 6.1) is the most commonly used PON architecture.

A PON network starts at the Access network edge, with the Optical Line Terminal (OLT). If the network is operated by a TSP, the OLT is typically located at a local exchange. From there, the optical fiber cable is driven to a passive splitter. Distribution fibers (one fiber per splitter or user) continue their route and connect to drop terminals to reach end users' Optical Network Terminals (ONTs)/ONUs. Splitters used in series (cascaded splitter architecture) are commonly found. Downstream (from the network to the end users) refers to traffic from the OLT to the ONTs/Optical Network Units (ONUs) and upstream to the contrary.

There are two families of standards in PONs, the ITU and the IEEE families (see Table 6.1). ITU family can be found in the G series of the ITU-T. A group of companies called FSAN (Full Service Access Network) [5] foster the requirements, recommendations, and adoption of these standards since 1995. IEEE family are advocated by the Ethernet Alliance [6], an industry consortium working in Ethernet technologies across markets.

6.2.1.1 PON Technologies Evolution

The earliest PON standard ideas [2] were around APON (ATM [Asynchronous Transfer Mode] over PON) system specified in the 1990s. It offered symmetrical capacity of 155 Mbps (coincidental with SDH). ITU BPON standard (ITU-T G.983.1 to G.983.5) built upon APON, and offered a maximum downstream speed of 622 Mbps and an upstream speed of 155 Mbps.

The real success of ITU's PONs started with ITU-T G.984 GPON standard in 2004. GPON increased the maximum downstream speed to 2.5 Gbps and the upstream to 1.2 Gbps, adding Ethernet and TDM circuit transport (E1/T1), to the ATM transport. Eventually, Ethernet would become the networking standard, and the rest of transport types' support has lost interest. Ethernet frames transport in GPON uses the so-called GPON Encapsulation Method (GEM). GPON began to be deployed by telecommunication operators in substantial volumes in 2008 and 2009.

In parallel, the 802.3 subcommittee of the IEEE (responsible for the Ethernet standard) had been working its own version of FTTH. The Task Force developing the standard was IEEE 802.3ah and gave its name to the EPON standard in 2004. EPON specified the minimum items necessary to be implemented in the PON standard, and aspects such as detailed management protocols and encryption (present in GPON) were left to commercial interests. Thus, EPON standard was easier to implement and spread in different world regions including Asia and America.

GPON has continued its evolution in two domains. One domain referred to technologies that should coexist on the same fiber distribution network with GPON; another domain followed a longer view of technology not necessarily burdened with interoperability constraints. After different adaptations, and driven by market changes, two solutions emerged. The first one is the ITU-T G.987 XG-PON (X for the number 10) designating a 10 Gbps downstream rate and a 2.5 Gbps upstream rate. XG-PON was eventually complemented with the G.9807.1 XGS-PON (the S for

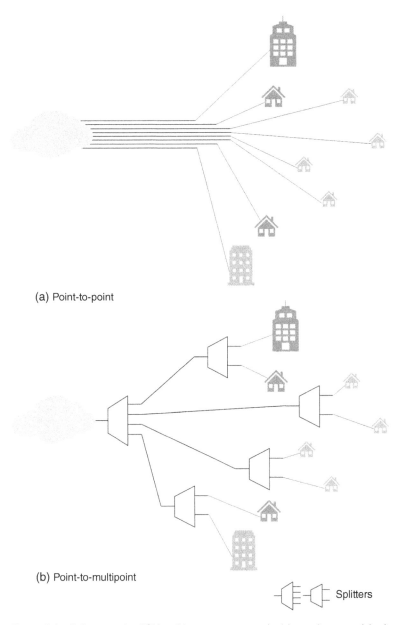

Figure 6.1 Point-to-point PON architecture, compared with a point-to-multipoint architecture.

symmetry) offering a symmetrical 10 Gbps data rate and working in the same frequency space. The second one is the ITU-T G.989 NG-PON2, which uses Wavelength Division Multiplexing (WDM) to configure multiple 10 Gbps wavelengths both upstream and downstream to deliver a symmetrical 40 Gbps service. GPON, XG(S)-PON, and NG-PON2 occupy a different spectrum and can thus coexist on the same PON network.

EPON has continued to evolve as well. Its first realization has been known by several other terms, including 1G-EPON, GE-PON (Gigabit Ethernet PON) and EFM (Ethernet in the First

Table 6.1 Main service characteristics of the different PON standards.

Standard	GPON	XG-PON	XGS-PON	NG-PON2	EPON	10G-EPON
Approved	2003	2010	2016	2015	2004	2009
Family	ITU-T G.984	ITU-T G.987	ITU-T G.9807.1	ITU-T G.989	IEEE 802.3ah	IEEE 802.3av
Standard specification documents	G.984.1/G.984.2/G.984.3/G.984.4/ G.984.5/G.984.6/G.984.7	G.987/G.987.1/G.987.2/ G.987.3/G.987.4	G.9807.1/ G.9807.2	G.989/G.989.1/ G.989.2/G.989.3	802.3ah-2004	802.3av-2009
Transmission speed downstream (Gbps)	2.50	10	10	4×10	1.25	10
Transmission speed upstream (Gbps)	1.25	2.5	10	4×10	1.25	10
Optical power budget (dB) [1]	32	35	35	35	29	29
Split ratio[a] [1]	1 : 64 (128)	1 : 64 (256)	1 : 128 (256)	1 : 128 (256)	1 : 64	1 : 64, 1 : 128

[a] Values in brackets do not imply feasibility of the system. It shows the split ratio supported by the protocols.

Mile). The IEEE 802.3ah defined point-to-point Ethernet to the home (either on optical fiber or twisted pair) as well. After IEEE 802.3ah, a new Working Group was formed under the IEEE 802.3av name to increase the speed to 10 Gbps, eventually producing 10G-EPON standard under the name of IEEE 802.3av. 10G-EPON can coexist on the same PON as EPON, as both occupy a different spectrum. Finally, IEEE P802.3ca 50G-EPON Task Force was created in 2018 to extend the operation of EPONs to multiple channels of 25 Gbps providing both symmetric and asymmetric operations (data rates of 25/10 Gbps, 25/25 Gbps, 50/10 Gbps, 50/25 Gbps, and 50/50 Gbps – downstream/upstream). Alongside these data rate expansion efforts, a SIEPON (Service Interoperability in Ethernet Passive Optical Networks) group was formed to fill in some of the missing pieces of the EPON standard to configure more robust commercial systems, building partially on the work of the Metro Ethernet Forum (MEF) [7].

PON systems usually integrate a radio frequency (RF) television (TV) signal (analogue or digital), that is broadcast over a PON onto a single wavelength (typically 1550 nm). This is known as RF video overlay, and it is used in commercial proposals of cable TV industry and residential Internet access.

6.2.1.2 Supported Services and Applicability Scenarios

PONs provide services that combine the traditional telecommunication TDM (PDH) services inherited from the ITU, with the Ethernet-based ones. While TDM support is classic in ITU family (together with SDH and SONET interfaces as service node interfaces, to connect OLT to the core networks), Ethernet services have become prevalent both as a user network interface and to connect OLT with the core.

Thus, Ethernet (IEEE 802.3) 10BASE-T, 100BASE-TX, and 1000BASE-T interfaces, Integrated Services Digital Network (ISDN; the digital evolution of the PSTN), PDH (E1, E3, T1 and DS3), and legacy old-fashioned ATM interfaces are examples of services supported by GPON and XG(S)-PON. NG-PON2 extends these examples to xDSL, SDH/SONET, OTN, and 1 and 10 Gbps on optical fiber. IEEE family covers the Ethernet user interfaces that can be delivered in IEEE 802.3 specification, within the speeds allowed by EPON and 10G-EPON.

In addition to services, there are a number of benefits to the use of PONs [1]:

- Powerless infrastructure, to lower rollout and maintenance costs.
- Infrastructure that facilitates upgrades, as PONs can offer higher throughputs by simply upgrading the electronic part of its ends.
- Coexistence of services in the PON infrastructure, through technologies such as wavelength multiplexing. Some of the new standards are compatible with existing systems on the same PON.
- Ease of maintenance, both for the good transmission properties of the optical fiber and the non-vulnerability to electromagnetic interference, which favor capacity and reach (especially compared with other copper-based technologies).

However, this passive infrastructure also comes with limitations:

- Distance limitations. PONs achieve ranges between 20 and 40 km, while active optical networks may reach up to 100 km.
- Difficulties to perform maintenance without disrupting the performance of network users. This is inherent to shared infrastructure, although standards have developed solutions to allow for a nonintrusive operational solution (e.g. out-of-band wavelengths).
- Cascading effects of optical fiber feeder line breakdown. A failure in the fiber close to the OLT, or in the OLT itself, will potentially affect users down the feeder.

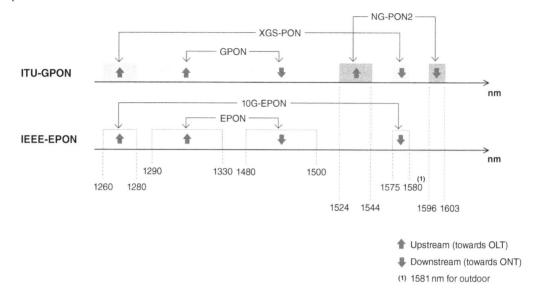

Figure 6.2 Optical fiber spectrum occupation.

6.2.1.3 Spectrum

Upstream and downstream occupied bandwidths share the same optical fiber. Thus, they need to be separated inside it.

Different spectrum occupations differentiate the various generations of ITU and IEEE standards, as shown in Figure 6.2. Coexistence can be achieved, if the bands do not interfere.

The historical selection of the bands is not a simple matter and has become more complex. GPON uses the same spectrum as EPON, and XGS-PON with 10G-EPON.

6.2.1.4 System Architecture

The PON system architecture is described based on three elements [8]. OLT reaches ONTs through the ODN. If the ODN is a tree, the OLT is the root and the ONTs are the leaves:

- ODN. It is the combination of optical fibers and splitters (but also combiners, filters, and possibly other passive optical components) connecting the access network connection point with the end users.
- OLT. It is the PON device generally located at the TSP's local exchange to control the ODN and interconnect with the Core network. Its primary function is to convert, frame, and transmit signals for the PON network users and to coordinate the multiplexing of the ONT shared upstream transmission [4]. The OLT signal may be combined with other OLTs on the same PON, through a Coexistence (CEx) element, i.e. a WDM coupler to combine/isolate spectrum bands to be inserted in the same optical fiber.
- ONT or ONU. It is the powered device to connect end users, acting as their service interface. ONT and ONU terms are slightly different, but they are often used interchangeably. The ONT is both a Customer Premises Equipment (CPE) and a network equipment. Most TSPs give more importance to this last role and develop their operational procedures for the ONT to be fully managed as an extension of the OLT. Even when the ONT is regarded as CPE, some functions must be managed from the TSP, including initialization, software upgrade process, and PON maintenance and diagnostics [4].

Table 6.2 Splitter optical power losses.

Number of ports	Insertion loss (dB)
2	3
4	6
8	9
16	12
32	15
64	18
128	21

The signals on the optical fiber can be split to provide service to a different number of users in each standard. To host more users, either a new parallel infrastructure is deployed or a new coexisting standard is placed on the PON.

6.2.1.4.1 Splitters

Optical splitters divide (split) any optical wavelengths in the downstream direction and combine (couple) them from the end users back to the OLT. Optical splitters are not wavelength selective and work bidirectionally.

Power is lost when an optical signal is split, roughly the same in all ports. This loss depends on the number of ports. Table 6.2 shows typical figures for this so-called insertion loss.

The number of ways the downstream signal is divided (split) before reaching end users is known as the splitter, splitting, or split ratio and is particular to each standard. The split ratio is a key parameter of any PON and determines the number of end users that can be served by an OLT. Thus, if a PON serves 64 end users, the split ratio is 64. This PON can be designed with a different structure of splitters: e.g. 1 : 4 and 1 : 16; or 1 : 2 with 1 : 4 and 1 : 8. In each optical fiber branch, the power is divided by 2, 6 times; and the total power lost for each end user is 18 dB ($64 = 2^6$; $6 \cdot 3\,dB = 18\,dB$).

Splitters require no cooling or any maintenance, leading to a long useful life span.

6.2.1.4.2 PON Range

The combination of the attenuation of the optical fiber and the passive components of a PON determines the distance that each PON can reach, for each end-user number design. The total attenuation must support each PON standard and device category's link budget (see Chapter 3):

Some realistic figures [4] can be used to calculate the distance vs. split ratio trade-offs in PONs:

- Splitter loss: 3.4 dB. Although an ideal splitter should lose 3 dB, real-life ones incur slightly higher losses.
- Attenuation per kilometer: 0.4 dB. Although it depends on the optical fiber and the PON standard, this is a good conservative figure.
- Other losses: 2 dB. There is a multitude of little effects, not foreseeable contingencies, and aging effects, which may prevent the link from operating at the expected level.

From a MAC layer perspective, differential reach is a parameter that needs to be considered to control the multiplexing of upstream frames. It is defined as the difference in optical fiber distance

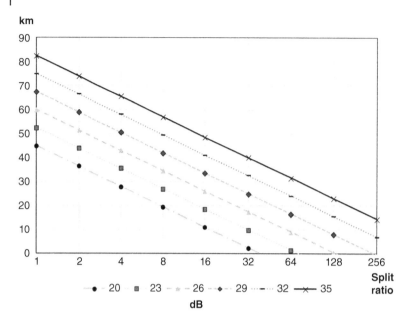

Figure 6.3 Range vs. split ratio in PONs.

between the farthest and the nearest ONT, to the OLT, and is used to calculate the time difference due to the propagation of the signal to adjust the individual transmissions of the ONTs without collision. This parameter can be controlled by a smart location of the splitters.

Figure 6.3 shows reference curves to estimate a realistic coverage of PONs. The optical loss budget sits in the range from 20 to 35 dB for different technologies and classes within each standard [4].

6.2.2 Main Capabilities and Features

6.2.2.1 Time and Wavelength Division Multiplexing

The currently deployed PON systems are TDM PON systems [8], most of them delivering Ethernet traffic to end users. This includes ITU and IEEE families, although both implement it in a very different way. ITU family's architecture relies on framing structures with a very strict timing and synchronization requirements, since their inception they were thought to support TDM traffic; Ethernet traffic is conveyed using GEM. IEEE family focused on preserving the architectural model of Ethernet, and no explicit framing structure exists: Ethernet frames are transmitted in bursts, natively, with standard inter-frame spacing.

In the ITU's family downstream direction, the OLT centralizes the traffic multiplexing functionality. The OLT multiplexes the GEM frames onto the transmission medium using the GEM Port ID as a key to identify the GEM frames that belong to different downstream logical connections. Each ONT filters the downstream GEM frames based on their GEM Port IDs and processes only the GEM frames that belong to that ONT.

In the ITU's family upstream direction, the traffic multiplexing functionality is distributed, as ONTs transmit in burst mode [1]. The OLT grants upstream transmission opportunities (or upstream bandwidth allocations) to the ONTs, informing about them in the bandwidth maps (the traffic-bearing entities of ONTs are recipients of the upstream bandwidth allocations, identified by their Allocation IDs or Alloc-IDs). Within each upstream bandwidth allocation, the ONT uses the GEM Port ID as a multiplexing key to identify GEM frames that belong to different upstream logical connections.

IEEE family data units are Ethernet frames. In downstream direction, Ethernet frames are broadcasted, and ONTs selectively receive frames destined to them, matching Destination Address field. In upstream, ONTs share the common channel. As a single ONT may transmit during a time slot to avoid data collisions, upstream transmissions are scheduled and occasionally may collide and need a retransmission.

Specifically for NG-PON2, and additional to the use of TDM and TDMA (TDM Access), WDM is introduced. It comes with multiple so-called OLT TWDM channel terminations, each one associated with a specific TWDM channel. Within each individual TWDM channel, the existing TDM and TDMA principles apply. ONTs in NG-PON2 can be switched (handed over) between the TWDM channels by the OLT.

6.2.2.2 Features Needed in PONs

Protocols for PONs incorporate mechanisms exclusive of their kind (due to their point-to-multipoint nature in a tree-shape structure) and other common ones. ITU family includes them in the Transmission Convergence layer. IEEE family defines the Multi-Point MAC Control Protocol (MPCP).

The aspects PONs need to be solved are as follows:

- Discovering unknown ONTs in the PON, be it under first start-up conditions, or recovering after power failures, power down, or unexpected disruptions.
- Orchestrating ONTs upstream burst transmissions to avoid collisions and assign capacity according with SLAs.
- Coordinating the access of ONTs at different distances (propagation delay) from the OLT.
- Accommodating drift in the propagation delay (due, e.g. to temperature changes).
- Defining upstream bursts overhead (e.g. header) for quick detection and interpretation, allowing fast and accurate optical receiver calibration.
- Defining data and frames structures to maximize data rate-efficient use.
- Structuring payload to identify individual flows and manageable traffic classes in those flows.
- Protecting the PON and its users from non-authorized or faulty ONTs.
- Securing the association of OLT and ONTs (authentication, authorization, eavesdropping, etc.).
- Mapping different payload protocols to the transmission medium (Ethernet, management protocol, etc.).
- Measuring and improving (FEC mechanisms) links quality (optical parameters, error rates, etc.).
- Providing signaling and control channels for the PON links.
- Diagnosing ONTs remotely, using operations and management communications channels and protocols.

6.2.2.3 Dynamic Bandwidth Assignment

A static bandwidth assignment to the different ONTs in a PON is suboptimal. Thus, a bandwidth assignment that reacts adaptively to bursty traffic patterns will improve its capacity to allow more users and improve their service experience (e.g. allowing traffic peaks).

Dynamic bandwidth assignment refers to the process by which an OLT distributes upstream PON capacity between the ONTs. Using traffic buffering in ONTs, and depending on the ONT buffer occupancy inference mechanism, several DBA methods are supported by ITU standards:

- Status reporting DBA. It is based on explicit buffer occupancy reports solicited by the OLT to the ONTs.
- Traffic monitoring DBA. It is based on the OLT's observation of the idle GEM frame pattern and its comparison with the corresponding bandwidth maps.
- Cooperative DBA. Introduced from XG(S)-PON onwards, it is based on the application-level upstream scheduling information provided by the OLT-side external equipment (used, e.g. in transport for 3GPP systems base stations).

OLTs will support a combination of both the two first methods and may additionally support the third one. The specific efficiency and fairness criteria are open to implementers. Status reporting is optional for ONTs. Status reporting DBA method involves in-band signaling between the OLT and the ONTs.

IEEE standards do not define specific bandwidth assignment algorithms, and this is left open to implementers [8]. Bandwidth arbitration among ONTs for EPON and 10G-EPON is referred to as inter-ONT (or ONU), and the bandwidth allocation among queues in the same ONT is known as intra-ONT scheduling.

6.2.3 ITU's GPON Family

In addition to the recommendations in Table 6.1, described in this section, ITU-T G.988 was defined to specify ONU (ONT) management and control interface.

6.2.3.1 GPON

GPON is considered the de facto PON standard in ITU family, with networks covering typically distances between 20 and 40 km.

GPON uses TDM in a continuous downstream of fixed-length frames (125 μs) that is propagated to all OLTs, so that each ONT can select its own traffic. In the upstream direction, the OLT authorizes each ONT to transmit upstream collision-free TDMA bursts of traffic. Bursts are sized and spaced appropriately for the traffic and the QoS committed to each ONT. The upstream flow works with the 125 μs frame concept, marking a repetitive series of ticks to serve as reference points. Figure 6.4 shows the relevant framing structure.

GPON Transmission Convergence layer architecture defines the framing of the information and builds the key functions of the protocol:

- ONT registration once it is activated on the PON. OLTs need to be registered to each user, attain synchronization, and establish the physical layer Operation, Administration and Management (OAM) channel.
- ONT ranging (estimation of the transmission delay), with ranging request messages for this purpose.
- OLT – ONT security, through an encryption process (key exchange, etc.).
- Alarms and performance monitoring, with mechanisms to detect link failure and monitor the health and performance of the PON links.

GPON traffic encapsulation uses GEM, a variant of the Generic Framing Procedure (GFP) defined in ITU-T G.7041. GEM, for Ethernet traffic, removes the frame preamble and adds a 5 Byte

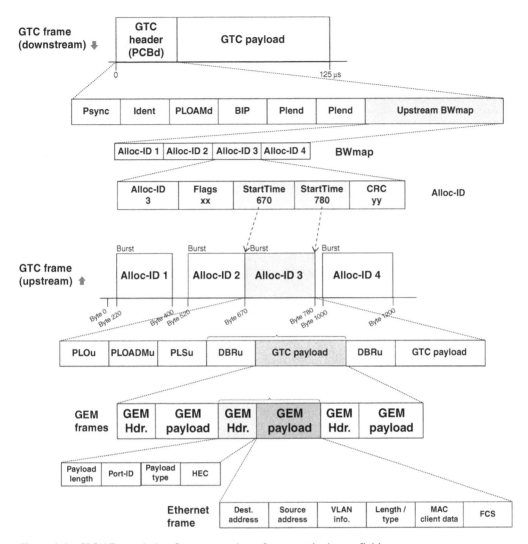

Figure 6.4 GPON Transmission Convergence layer frames and relevant fields.

header that, e.g. identifies an OLT's port (Port-ID). The Port-ID is an OLT number to identify a logical connection (different ONTs can be assigned a variety of Port-IDs, e.g. to map different traffic classes).

Several GEMs are mapped to the downstream flow, adding the so-called PCBd, to conform the GTC frame. This PCBd includes, among other fields, the PLOAMd and the upstream bandwidth map (BWmap). The PLOAMd field is used for OAM, downstream; it is complemented by the PLOAMu, in the upstream bursts.

The BWmap informs each OLT of the time in which they can have access to the channel, identifying its start and stop bytes in the upstream TDMA frame. Each OLT is identified by the Alloc-ID.

Upstream bursts send the information from each OLT in Transmission Containers (T-CONTs), in the positions indicated in the upstream bandwidth map sent from the OLT. Each T-CONT may represent a group of logical connections, and each OLT may have several Alloc-ID and thus send several chunks of information in its burst.

GTC frames are mapped into the physical layer. FEC parity bytes are inserted at the end of codewords, in downstream and upstream direction.

6.2.3.2 XG(S)-PON

XG-PON supports speeds of 10 Gbps downstream and 2.5 Gbps upstream, while its next evolution, XGS-PON provides symmetrical 10 Gbps both upstream and downstream using the same wavelengths.

Physical fiber and data formatting conventions are identical to GPON, and the wavelengths have been shifted, so that both GPON and XG(S)-PON can operate the same PON infrastructure, simultaneously.

6.2.3.3 NG-PON2

NG-PON2 utilizes WDM to configure multiple 10 Gbps wavelengths, both up and downstream, with the intention to deliver a symmetrical 40 Gbps service. Using different wavelengths to GPON and XG(S)-PON, NG-PON2 allows for the three technologies service coexistence (see Figure 6.5).

The OLT NG-PON2 concept consists of multiple OLT channel terminations connected via a wavelength multiplexer to the optical fiber. NG-PON2 system is designed to provide flexibility to configure trade-offs in speed, distance, and split ratio parameters to adapt to various applications. This time and wavelength division multiplexing passive optical network (the TWDM PON) allows each wavelength to be shared between multiple ONTs.

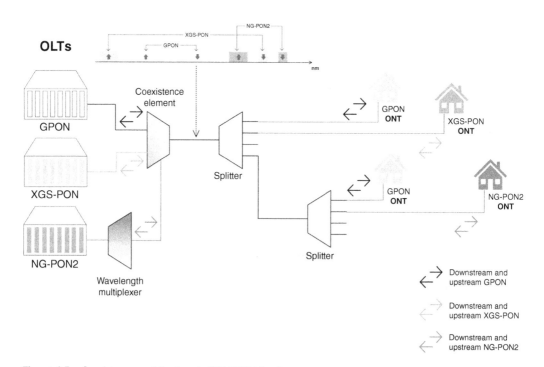

Figure 6.5 Coexistence architecture in ITU GPON family.

6.2.4 IEEE's EPON Family

In addition to the standards shown in Table 6.1, IEEE includes several evolutions of them in IEEE 802.3bk-2013, IEEE 802.3ca-2020, and IEEE 1904.1-2017 (SIEPON; for interoperability purposes).

6.2.4.1 EPON

EPON was developed for seamless compatibility with Ethernet devices. EPON is based on the IEEE 802.3 standard and needs no additional encapsulation or conversion protocols to connect to Ethernet-based networks. EPONs cover typically distances between 20 and 40 km.

EPON relies on an asymmetric point-to-multipoint architecture, making the most of standard Ethernet standard definition and mechanisms (switched Ethernet with source and destination address fields, CSMA/CD, etc.). However, the preamble of an EPON Ethernet frame differs from a regular one, e.g. in that it contains the Logical Link ID (LLID; a PON-level address) [4].

The MPCP is the protocol that, on top of the MAC, defines the different functions needed in EPON. Each ONT has an MPCP entity communicating with another entity in the OLT [8]. Thus, even if the PON is a point-to-multipoint, the MPCP becomes an ensemble of logical point-to-point links for upper layers, using the so-called P2P Emulation (P2PE). Each ONT is identified by a Logical Link Identification (LLID). The MPCP performs different functions, including auto-discovery, ONT registration, ranging, bandwidth polling and bandwidth assignment, with limited size control messages, namely GATE, REPORT, REGISTER_REQ, REGISTER, and REGISTER_ACK.

ONT discovery and registration depend on the reception of GATE messages in the ONT. These messages are periodically broadcasted from the OLT. Upon reception of such messages, ONTs wait for the discovery window to be opened and send REGISTER_REQ messages from other ONTs in the same process. Unicast GATE messages, with a unicast LLID and a unicast destination address, are sent to the ONT as part of the process. Once the ONT is registered, its access to the PON will be coordinated with the TDMA mechanism.

ONT ranging aims at calculating the Round-Trip Time (RTT) to understand the logical distances between ONTs and adjust synchronization to be able to transmit without collisions.

6.2.4.2 10G-EPON

EPON evolved into 10G-E-PON to provide symmetrical 10 Gbps upstream and downstream data rate. 10G-PON uses the same band as XG(S)-PON and can thus coexist with EPON in the same PON.

6.3 Power Line Communication Technologies

PLC technologies are a unique set of technologies that use the power system cables for telecommunication purposes. PLC technologies have been present in the history of electricity since its early stages, with different roles and applications depending on the segment of the grid where they are used.

The early use of PLC [9] is connected with the pioneer years of telecommunications, where any wireline medium was explored as a transmission medium. While the early wireline transmission started to consolidate, and radio was being explored, PLC demonstrated to be a reliable transmission mean in long-distance applications in HV power lines (high-power transmissions – around 50 W – and tens of kilometers covered using frequencies from 50 to 300 kHz). From then on, PLC applications in the power systems and telecommunications industry have shown an uneven evolution.

PLC systems rely heavily on the electricity grid infrastructure and on its structure since power lines and their interconnection determine the paths and the reach of PLC propagation. Its feasibility is constrained by the harsh nature of power lines (signal adaptation and noise) and the compatibility with other uses of radio frequencies in free space.

PLC is undoubtedly recognized as an enabler of the Smart Grid.

6.3.1 Mainstream Technologies and Standards

PLC uses power lines to transmit information in parallel to the high-power 50 or 60 Hz signal. From a telecommunication perspective, the system behaves as an FDM transmission. In this case, the high-power signal uses a carrier frequency different to the one used for data transmission. This is analogous to the xDSL connection and the voice transmission in a landline, where the voice is sent using a different carrier than the information.

Even though PLC technology may not be common for the general public (with the exception of the in-home domain), it is widely present in the utilities ecosystem since it has proven to provide services where other technologies offer deployment difficulties, poorer performance, or higher costs.

PLC is not a Core Network technology; it is an Access technology for specific segments of the grid.

6.3.1.1 PLC Technologies Evolution

From a historical perspective, the first patents for PLC technologies date back to the beginning of the nineteenth century, connected to PLC applications interested in meter reading (1897 as per [10] and 1903 as per [11]). However, the first implementations of PLC for operational purposes were seen to 1918 in Japan [12] for voice telephony over power lines at long distance. By 1927, PLC had been widely adopted in United States, Europe, and Asia in HV applications [13, 14], and the initial voice communications to reach distant PSs, evolved to include low-speed data transmission. This PLC application solved the difficulties to find telecommunication services in isolated areas of the grid. These services are still used in the grid with efficient PLC digital communications [9, 15], although their use is secondary to optical fiber.

Central load management systems were developed in the 1950s to control load peaks in the grid [16]. These systems insert a signal with frequency around the 250 Hz into the electrical wires. These systems are limited by the available data rate (around 1 bps) and its one-direction nature (broadcast) [17]. However, due to the high power of the injected signals, they reach a high coverage propagating through all grid levels, independently of the transformers interconnecting them.

The interest in advanced meter reading (AMR) systems appeared in the 1970s, with the need of two-way communications to retrieve consumption from individual meters in an LV network. From the early AMR interest, to the automatic metering infrastructure (AMI) consolidation in the Smart Grid, the requirements have led to a sophistication of the telecommunication needs that has been accompanied by the state-of-the-art evolution, which leads to higher data rates even in noise and power-limited scenarios.

The proliferation of PLC technology for AMR and AMI initiatives gave birth to the European Norm EN 50065-1 [18], released in 1991 by CENELEC in Europe, which organized the spectrum used in power lines for such applications in the range 3–148.5 kHz (from A to D bands; similar USA FCC provisions can be found in clause 15.113 of Title 47 of the Code of Federal Regulations from 9 to 490 kHz; and Japan as per ARIB STD-T84 from 10 to 450 kHz). The first commercial systems where based on FSK modulations, achieving some kbps up to a few hundred meters, with repeaters used to extend the coverage through a relaying function. The 2000s and 2010s have witnessed the evolution of these PLC-based AMI systems to increase the bit rate without increasing the used bandwidth, with standardized OFDM digital communication techniques [19, 20] in narrowband and broadband contexts.

6.3.1.2 Supported Services and Applicability Scenarios

The diversity of PLC technologies is high, probably comparable to that of radio solutions, as different parts of the spectrum and segments of the grid can be used in PLC. Thus, based on the grid part they use, some of them can be better categorized as backbone technologies while some others as local access. Moreover, the nature of the existing grid segments is different when it comes to the PLC channel (propagation, noise, etc.) and, consequently, its performance and availability.

In terms of service offering, one may differentiate Narrowband PLC (NB-PLC) and Broadband PLC (BB-PLC) technologies. However, there are other niche PLC technologies that have their particular application and proven implementations on the field:

- Ultra-Narrowband (UNB) PLC systems. These systems work in either Ultra-Low Frequency (0.3–3 kHz) or in the Super-Low Frequency (30–300 Hz), with very low data rates (roughly 100 bps) spanning tens or even one hundred kilometers. TWACS is used for medium to low-density remote areas [21].
- Power line carrier applications. A modern version of HV PLC technology, using the low part of the spectrum (typically below 1 MHz) to deliver moderate throughput (typically up to 2 Mbps) for PS connection for voice, data, and teleprotection purposes [22].

6.3.1.2.1 Narrowband PLC

NB-PLC technologies are referred to as the transmission within (loosely) the 3–500 kHz bandwidth. NB-PLC technologies happened earlier than BB-PLC and have two categories:

- Low Data Rate technologies (LDR). They are based on single-carrier designs that could only manage a few kbps. Their main applications are the Smart Metering and home automation. Communication techniques are based on FSK or spread-FSK modulation, such as the case of the IEC 61334-5; or an even simpler solution such as the one of X10, where a bit is transmitted every time the mains signal crosses 0 V.
- High Data Rate technologies (HDR). Higher data bit rates can be achieved by transmitting through several carriers simultaneously, i.e. multicarrier transmission, which allows of transmission rates going from hundreds of kbps to very few Mbps. Two of the pioneer solutions in this group were promoted by industrial alliances: the PoweRline Intelligent Metering Evolution Alliance (PRIME) and the G3-PLC Alliance, which developed the PRIME and G3-PLC solutions, respectively.

Signal injection in NB-PLC, when applied over the LV grid, is usually performed connecting between one phase and neutral cables or any two phases. This is natural in smart meters, as most of them are just connected to two wires, and it is beneficial from a performance perspective at the SS [23] to improve transmitted power and minimize the chance of receiving the worst case noise level.

6.3.1.2.2 Broadband PLC

BB-PLC (also known as BPL or Broadband PLC) is generally used in the range of 1.8–30 MHz (some references mention limits of 80 MHz or even 250 MHz).

This bandwidth allows for a transmission speed of almost 250 Mbps. The problem that BPL faces is the high signal attenuation in part of this frequency band. This limits the range of the PLC links.

For these reasons, this technology has traditionally targeted in-home LAN applications (short distances) or backbone communications to interconnect SSs using the MV power lines (low noise). New applications of BPL will happen in the LV segment as well, taking advantage of the evolution of multiple-input multiple-output (MIMO) techniques [24] to adapt to this harsher channel scenario. PLC transmission occurs over the three wires (even if the signal is injected in one of the phases, due to inductive and capacitive coupling effects), and the receiver can use the three different channels with MIMO techniques.

There are two main BPL standard solutions, namely IEEE 1901 and the ITU-T G.hn. There are other nonstandard solutions such as OPERA and HomePlug AV2 [25]; the first one is the only one massively deployed in the MV grid for Smart Grid purposes, while the second one was developed to improve the performance of BPL solutions in in-home devices. Other standards are: ISO/IEC 12139-1, developed and used in Korea and operating between 2.15 and 23.15 MHz using OFDM; IEEE 1901.1, operating up to 12 MHz; and finally, IEEE 1901a, promoted by Panasonic and evolved from IEEE 1901 specification.

6.3.1.2.3 Access to PLC Services

Most PLC standards implement communication layers 1–2 and provide interfaces for higher layer protocols. Common interfaces are serial, Ethernet and IP protocols, with a different degree of adaptation through what it is known as *convergence layers*.

6.3.1.3 Architecture

PLC can only be deployed using the grid architecture, and this fact limits the support of certain telecommunication needs. However, all grid devices are connected to it and may potentially be communicated through PLC.

The general considerations on PLC technologies worth mentioning are as follows:

- Broadband vs. Narrowband. Originally, PLC networks were designed to use low-frequency carriers (some tens of kHz). As technologies evolved, more bandwidth was pursued to provide higher transmission speeds. One barrier against extending PLC bandwidth was the existence of the AM radio broadcasting band around 530 and 1700 kHz. This fact separated the evolution of the technology into two different frequency bands: narrowband, up to 530 kHz; and broadband, in the range from 1.8 MHz onward.
- Voltage level. The original concept of using FDM to transmit data over power lines has been adapted to any voltage level. Grid and safety-related considerations need to be taken when coupling the information signal to the power lines at specific grid locations.
- Standardized versus proprietary PLC technologies. As with any other technology, there are systems that are marketed by a single provider and based on nonpublic specifications. These were for many decades the only available option for PLC users. This is not the case today, and hence, the adequacy of each PLC solution can be assessed in the different field environments where good vendor's implementation would outperform others.
- Separation of HV, MV, and LV grids. While the power system is a continuum (the 50/60 Hz signal propagates from Generation to Customers), this is not the case for all frequency signals. This apparent weakness of the system for telecommunications becomes a strong point as frequency ranges can be reused in the different domains without interference. Thus, transformer-separated grid segments can reuse spectrum ranges.

Each utility grid topology is unique. When PLC is used, the several combinations of grid segments with NB-PLC and BB-PLC ranges configure varied PLC network architectures that may be unique to each utility. More importantly, PLC becomes a complementary handy technology to be combined with other technologies.

To provide indications on how PLC architectures are configured, a reference sample model is provided. This architecture does not cover the HV domain (as this grid segment is today mostly served with non-PLC technologies) and may evolve with the potential evolution of BB-PLC into the LV Distribution grid.

The PLC architecture takes the reference of the electricity grid architecture in Chapter 1. When overlapping the PLC structure over the architecture shown in Figure 6.2 (Chapter 1), the PLC overlay is shown in Figure 6.6. The non-PLC telecommunications Core is represented at the bottom of the figure with the connectivity toward central systems where applications can be hosted. When PLC in LV is used for Smart Metering purposes, the MDMS will be connected through this Core connectivity. If the MV PLC is used for remote control in the SSs, the SCADA systems will reach the SSs through this telecommunication's Core.

The reference architecture uses BPL in MV [9]. The main reason for it is the requirements in terms of bandwidth, since the amount of data to be transmitted is significant in this aggregation part of the telecommunications network. When using PLC for this access backbone link, the communication signal coming from the Core is injected in the MV network with the appropriate signal couplers. Within the MV domain, devices typically have a Master/Slave relationship, with Repeaters being configured at PLC level, to allow the signal to travel further. Assuming, i.e. that IEEE 1901 [26] could be used to communicate several SSs in the MV network as in Figure 6.6, three different kinds of nodes would be present, namely the Head-End (HE), the Repeating Stations (RSs),

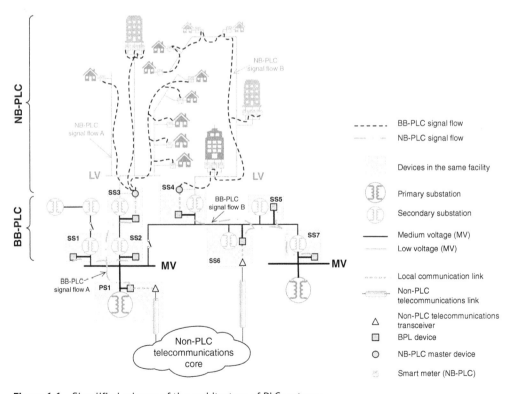

Figure 6.6 Simplified scheme of the architecture of PLC systems.

and the Network Termination Units (NTUs). NTUs represent the leaves of tree-shape topology and work as gateways to other networks connected at lower levels of the hierarchical grid Distribution infrastructure (LV networks, as mentioned later); RSs work as communications relays; and the HE coordinates the communications inside the PLC "cell," acting as a backhaul communications gateway to the Core as well.

The reference architecture for LV is commonly used with PLC for Smart Metering purposes. Once an SS has been reached (using PLC or other technology), the link that interconnects the Data Concentrators (DCs) of the Smart Metering system (see Chapter 5) with the smart meters placed at the customer premises through the LV network covers the *last mile*. The DC usually hosts an NB-PLC device with the role of a Master toward the PLC "slaves" inside the smart meters.

The location of DC may depend on the size of the LV network. European countries tend to have larger LV networks (i.e. with a higher number of customers per SS). In these cases, the DC is placed at the LV side of the transformer. Given the high attenuation and the possibility of noise sources in the LV network, the coverage of NB-PLC signal may be limited. This is why NB-PLC technologies have defined and developed relying mechanisms, further elaborated in Section 6.3.3.1.

6.3.2 Main Capabilities and Features

6.3.2.1 Common Transceiver Designs in PLC Systems

Since the appearance of the multicarrier solutions, the communications techniques follow a rather similar scheme. The purpose of this section is to present a generalized common design for a PLC transceiver (Figure 6.7).

PLC uses a multicarrier modulation with orthogonal carriers, making use of the advantages described in Chapter 3. Common to many OFDM implementation, not all carriers are used for data transmission; thus, zeros are inserted in the nonused frequencies. The reasons for this depend on electromagnetic compatibility (i.e. those frequencies may have been assigned to other services and need to be notched), high channel attenuation, etc. Additionally, some other carriers may be used for additional purposes different to the pure data transmission; this is the case of the pilot signals, i.e. modulated carriers with preestablished data known by the receiver for channel estimation.

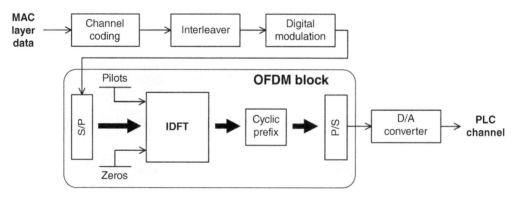

S/P = serial/parallel
P/S = parallel/serial

Figure 6.7 General transmitter scheme for multicarrier PLC systems.

To transmit digital information, the amplitude of each carrier is modulated using digital modulation techniques explained in Chapter 3.

Sometimes, the robustness achieved by the Digital Modulator is not enough. In such scenarios, the use of Channel Coding improves the performance in terms of the quality of the communication at the expense of reducing the effective data transmission rate. For this reason, some of the PLC technologies make use of Channel Coding mechanisms in combination with Interleaving.

The selection of the specific Channel Coding mechanism varies depending on the specific PLC technology; in some of them the use of this technique is optional or is chosen adaptively depending on channel conditions.

6.3.2.2 PLC Signal Coupling

Couplers are the elements that allow PLC signals to be superimposed onto and/or extracted from a power waveform, while also preventing the power waveform from entering the PLC transceiver. In contrast with other transmission media (such as radio), now there is a preexisting voltage signal that is difficult to manage and harness. Consequently, the design of a PLC coupler must be safe, since the voltage levels might be extremely high. This has an impact in the cost of couplers, with LV ones being cheaper than HV and MV ones.

PLC couplers must take into consideration impedance matching to achieve small coupling loses, flat frequency response to minimize signal distortions, efficient electrical protection against transients, and reduced costs and size.

Impedances in the grid change due to the time and frequency varying nature of the loads connected. The widest variation can be found in LV grid, where the lack of impedance matching will be equivalent to a reduced power transmission. On the contrary, HV and MV are less affected by this variability, as they are separated from the dynamics of the LV grid.

PLC couplers can be classified depending on [27] the following:

- Physical connection. They can be capacitive, inductive, resistive, and "antenna," being the first two ones the most prevalent alternatives.
- Grid voltage level. They can be prepared for HV, MV, or LV.
- Frequency band. They can be narrowband (up to 500 kHz) or broadband (from 1.8 MHz up to several MHz).
- Propagation mode. They can inject the signal in common or differential mode.
- Number of connections. They can connect to a variable number of conductors (three phases and neutral cables).

Coupling of PLC signals into power lines is a limiting factor to consider in the deployment of PLC systems, as a proper access to the power line is needed. In the case of HV and MV segments, this needs to be done through specially designed external coupling units that have a non-negligible size that needs to be accommodated in the points where the access is needed (e.g. the switchgear of a MV SS). In the case of LV solutions, while the coupling can be accommodated inside the PLC equipment, power lines need to be accessible to connect them.

Capacitive coupling is usually preferred to inductive coupling, especially in HV scenarios. Although inductive coupling is flexible and easy to install as it does not need physical contact with the power line, inductive coupler behavior may vary with line impedance to the point of limiting the performance of circuits. Capacitive couplers include coupling capacitors of the appropriate capacitance ranges (e.g. between 350 and 2000 pF for MV broadband applications; and higher for HV applications) installed in series with the power line.

6.3.3 Narrowband PLC Systems

NB-PLC technologies are among the ones that have gained a greater momentum in the last decade. The need to provide future-proof Smart Metering solutions stimulated the development of solutions that overcome existing NB-PLC state-of-the-art, incorporating technologies from the BB-PLC domain but adapted to the NB-PLC environment. These solutions evolved with the European Commission mandates M/441 [28] and M/490 [29], the Open Meter project [30, 31], and the interest of both IEEE and ITU to incorporate such telecommunication solutions to their body of standards for the Smart Grid.

6.3.3.1 ITU-T G.9904 (PRIME v1.3)

In 2009, a group of vendors formed the PoweRline Intelligent Metering Evolution (PRIME) Alliance [32], mainly led by the Spanish utility Iberdrola. The idea was to provide an open specification for a PLC communication system in the CENELEC A band that could target the need for communications in the last mile of the power networks. In terms of requirements, Iberdrola needed to provide a technical solution to the Spanish national mandate [33] to replace all its meter base in Spain with smart meters that could be remotely managed.

A few years after the Alliance was formed, the solution was standardized as ITU-T G.9904. PRIME includes the definition of the PHY and MAC layers together with a convergence layer to interact with higher-layer protocols.

6.3.3.1.1 ITU-T G.9904 (PRIME v1.3): PHY Layer

At PHY layer, PRIME communication is based on frames. Each PHY frame consists of three major blocks: preamble, header, and payload.

The Preamble is a linear chirp-based signal that is used for synchronization purposes. This is followed by the Header, which contains important information for interpreting the Payload (see Type A in Figure 6.8).

To transmit the information, PRIME uses an FFT-based OFDM transmission system with Channel Coding like the one described in Figure 6.7. A scheme for the actual PRIME transmitter is shown in Figure 6.9. The transceiver makes use of the spectrum between 42 and 89 kHz to insert 97 subcarriers per OFDM symbol. The transmission spectrum is directly implemented at the OFDM stage by setting 318 of the subcarriers to zero; thanks to this, no up-conversion stage is required afterward. In order to reduce ISI [34], the time sequence is extended with a Cyclic Prefix of 48 samples, leaving the whole OFDM symbol with 560 samples. Since the sampling period is $T_s = \dfrac{1}{250\,\text{kHz}} = 4\,\mu s$, the overall OFDM symbols take 2.24 ms to be transmitted. In addition to this, the frequency separation is $\Delta f = \dfrac{F_s}{512} \approx 488\,\text{Hz}$.

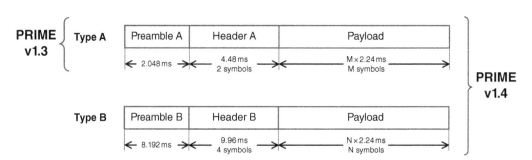

Figure 6.8 Frame types in PRIME v1.3 and v1.4.

6.3 Power Line Communication Technologies

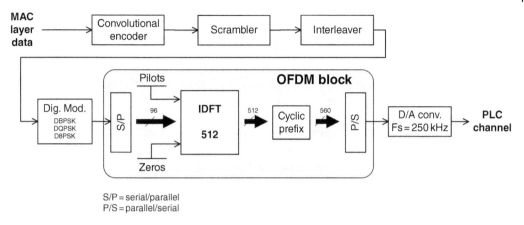

Figure 6.9 PRIME v1.3 transmission scheme as defined the standard.

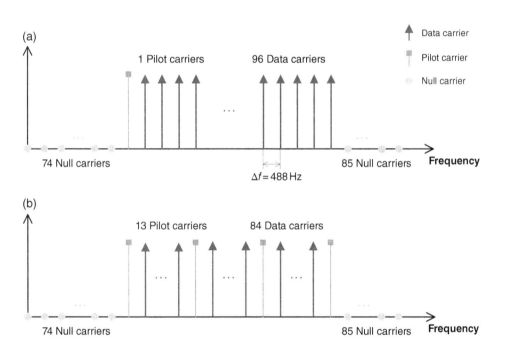

Figure 6.10 Subcarrier assignation. (a) Payload block. (b) Header block.

At the Digital Modulation stage, carriers can be modulated using different Differential Phase Shift Keying (D-PSK) techniques. However, all subcarriers in the same OFDM symbol must use the same modulation, although it may vary from OFDM symbol to OFDM symbol. For instance, while the modulation in the Payload block may be chosen to fit the characteristics of the channel, the modulation for the Header block is always set to DBPSK to increase its robustness, given the criticality of this block. Despite using 97 data subcarriers in the OFDM spectrum, the use of these carriers differs for the Header and the Payload block. In the Payload, the first carrier in each OFDM symbol is used to transmit the phase reference used in one of the possible digital modulations used in the other 96 (see Figure 6.10a). This mechanism is used to minimize the frequency shifts produced by the channel, since the demodulator is not based on the absolute phase of the received symbol but in the phase change with respect to the previous digital symbol. As a drawback, this procedure slightly decreases the data rate, as one of the carriers is not used to transmit information but to transmit the phase reference.

To increase its robustness, this carrier assignment is different for the Header block, as Figure 6.10b shows. In the case of the Header, 12 extra subcarriers are used as phase reference to compensate for scenarios where the channel frequency response changes drastically.

The transmitter cannot disable blocks of carriers in the transmission.

With respect to the FEC blocks, PRIME uses a Convolutional Encoder with coding rate of $R_c = 1/2$. The coding performance is enhanced by interleaving all bits within the same OFDM symbol. In terms of bits, this means 96, 192, or 288 bits, depending on the digital modulation in use.

The information encapsulated in the Header is always sent using the FEC modules, whereas, for the case of the Payload, FEC is optional. However, most implementations of PRIME use the system with the FEC enabled, to reduce Frame Error Rate (FER).

The combination of FEC ON/OFF and the different modulations defines the set of Communication Modes in PRIME. The trade-off between data rate and robustness is shown in Figure 6.11a for the six Communication Modes in PRIME.

Finally, the bit data rate for PRIME communication can be computed by dividing the number of bits present in one OFDM symbol by the time it takes to transmit this very symbol, as shown in Table 6.3.

6.3.3.1.2 ITU-T G.9904 (PRIME v1.3): MAC Layer

PRIME specification divides the DLL in two sublayers: the Medium Access Control (MAC) and the Logical Link Control (LLC).

PRIME defines two kinds of nodes in the MAC sublayer: Base Node (BN) and Service Nodes (SNs). The role of the BN is to manage the network resources and connections. There is only one BN per subnetwork and all the rest of the nodes (SNs) must register to it to be able to take part in the subnetwork. In Smart Metering deployments, the BN is typically implemented in the DC at the SS.

Given the physical characteristics of the power lines, the medium acts as a shared channel for all devices connected to it. The BN coordinates SNs so that every device has a chance to transmit. To do so, time is organized in MAC frames.

MAC frames are divided into three sections: Beacons (BCN) period, where information about the subnetwork is transmitted; Shared Contention Period (SCP), where all SNs in the network contend to access the channel; and Contention-Free Period (CFP), where only SNs that previously reserved a dedicated transmission time can use the channel. Moreover, Beacon frames can only be used by the BN and Switches (a state of a SN) existing in the subnetwork.

The length of each MAC frame is constant, while the length of the SCP and CFP varies dynamically depending on the network needs, and their duration is broadcasted in the beacon's information so that every device is synchronized. The structure of the three parts of the MAC frame is shown in Figure 6.12:

- Beacon. All beacon slots are placed at the beginning of a MAC frame. The first slot is reserved for the BN's beacon. The rest of the slots are assigned to the Switches present in the network. This assignment is done by the BN during the promotion process.
- SCP. It is a period where all devices in the network contend to access the channel to transmit. The length of the SCP is variable. Since collisions are likely to happen during this period, CSMA/CA is used.

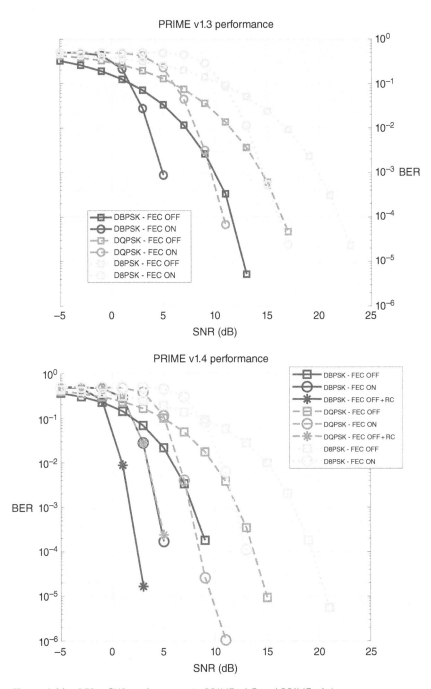

Figure 6.11 BER – SNR performance in PRIME v1.3 and PRIME v1.4.

Table 6.3 Data rates for the Communication Modes in PRIME.

Modulation	FEC ON (kbps)	FEC OFF (kbps)
DBPSK	20.48	40.96
DQPSK	40.96	81.92
D8PSK	61.44	122.89

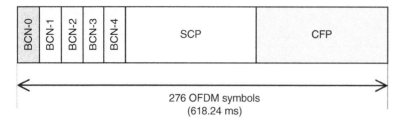

Figure 6.12 MAC frame scheme.

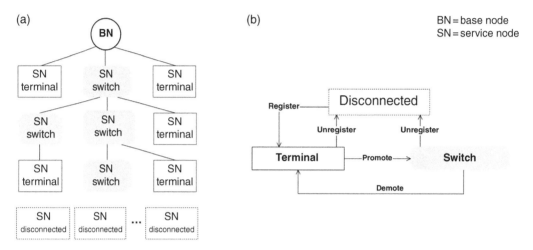

Figure 6.13 (a) Logical structure of a PRIME network. (b) Different states that a PRIME's SN node can reach.

- CFP. All devices have the possibility to request a CFP to the BN to transmit information without contending for the channel. If no request is done, the CFP is left empty. BN's criterion to accept CFP requests from the SNs is not specified in the standard, and it is left open to manufacturer's implementations.

MAC topology construction is based on SNs taking different roles or states in its relationship with the BN. Although the channel is physically shared, the logical topology of PRIME networks has a tree-like structure where SNs act as leaves or branch points, as shown in Figure 6.13a. When powered up, an SN starts in a Disconnected state. SNs change to Terminal by registering to the subnetwork's BN, and only when they are in Terminal state, they can be promoted to Switch to help the BNs to reach all SNs. A SN in Terminal state just provides connectivity to the upper layers; Switch nodes are also responsible for forwarding traffic to and from other SNs. The evolution of the different states and the name of their processes are displayed in Figure 6.13b.

The evolution of an SN to transition from Disconnected to Terminal is referred to as Registration (see Figure 6.14a). This is an automated process that is triggered by the SN when connecting to the network. After booting up, the SN starts listening for beacons being transmitted in the PLC channel. Once the SN detects and interprets one of the beacons, it has the necessary information to contact with the subnetwork's BN. A three-way handshake is initiated then where the SN shares important information with the BN and vice versa. Some of this information is the capabilities of the SN and the dynamic identification for the node within the network.

In contrast, the evolution from Terminal to Switch ("promotion") does not occur automatically. The promotion process can be directly initiated by the BN to build a specific logical network structure, or more often, it can be the result of a request from an SN listening to other SNs in need of support to reach the BN. This second case can be triggered by a situation where a given SN does not receive any beacon message. If, after a certain amount of time after booting-up, an SN does not see any beacon message, it shall send a Promotion Needed (PRN) signal. All registered SNs that listen to this message must send to the BN a PRO_REQ_S (Promotion Request from SN) message to ask for a promotion to Switch state to help the BN expand the reach of its subnetwork with beacons and selective traffic relaying. Among others, some of the information encapsulated in the promotion request is the quality of the received PRN message and an identifier of the disconnected node.

Upon the reception of the PRO_REQ_S message, the BN shall decide whether to accept the promotion or not. In case of several requests being received, it must also decide which one is to be accepted.

The BN accepts promotion request by replying with a PRO_REQ_B (Promotion Request from BN). Finally, the whole process is finished with a PRO_ACK from the newly created Switch node. From this moment on, this new Switch will include its contact information at a specific slot in the Beacon's part of the frame.

Figure 6.14b shows the previously described process from the Promotion Needed message to the transmission of beacons by the recently promoted Switch.

The BN performs network management functions using periodic keep-alive messages to poll all SNs and obtain an up-to-date status view of all connected SNs.

Finally, the LLC sublayer provides mechanisms to perform Segmentation and Reassembling (SAR) of the packets, flow control, and automatic repeat request (ARQ) functions. In the case of NB-PLC, SAR is needed since shorter-length packets allow for room to let other transmissions happen, and they will have a smaller likelihood of being affected by bursty noise.

MAC layer in PRIME also includes data security and encryption. It defines two security profiles (profile 0 and profile 1). In security profile 0, transmissions are performed without encryption. This

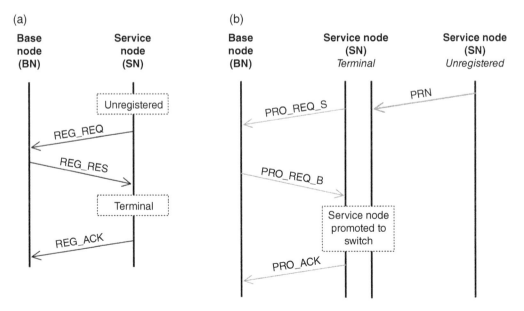

Figure 6.14 Evolution of Service Node states: (a) registration process; (b) promotion process.

may be of use when higher layers are implementing the encryption. By contrast, when using the security profile 1, both the data and its associated CRC are encrypted with a 128-bit Advanced Encryption Standard (AES) algorithm. This guarantees privacy in the communication, thanks to the use of the 128-bit AES; secure authentication, since the encryption key is only known by the corresponding service node and the base node; and data integrity, since the CRC is also encrypted.

6.3.3.2 Future ITU-T G.9904.1 (PRIME v1.4)

Even though ITU-T G.9904 (PRIME) is globally present in several countries, at the end of 2014, a new version of this technology was introduced, with the name of PRIME v1.4. To the date of writing this book, this version has not yet finished the publication process.

PRIME v1.4 introduces several new features and is backward compatible with version v1.3, so that both PRIME v1.3 and v1.4 devices can interoperate in the same PLC network. Its improved bit error rate performance can be seen in Figure 6.11b.

At PHY level, the new version introduces bandwidth expansion and a new (more robust) Communication Mode.

With respect to the bandwidth expansion, PRIME v1.4 includes the possibility of transmitting up to almost 500 kHz. The bandwidth is organized into eight so-called transmission channels. Figure 6.15 shows the structure of the channels. There is a guard band between channels of 7.3 kHz. PRIME v1.4 can use all the newly available bandwidth; however, devices can enable and/or disable specific channels. Channel number 1 conveniently corresponds to the frequency range used by PRIME v1.3 devices.

Regarding the new robust communication mode, PRIME v1.4 includes an extra combination in the transceiver by adding a Repetition Code (RC)[1] of 4 bits (see Figure 6.16). However, this new block works in combination with one of the frame types defined in this new version. PRIME v1.4 defines two types of frames:

- Type A. Essentially, the same frame as in PRIME v1.3. This is obviously needed for backward compatibility reasons.
- Type B. This new type of frame is longer than Type A (see Figure 6.8). The Preamble is four times longer; the Header includes two extra OFDM symbols; and the Payload may contain up to 252 OFDM symbols, in contrast with the 63 maximum symbols of Type A frames.

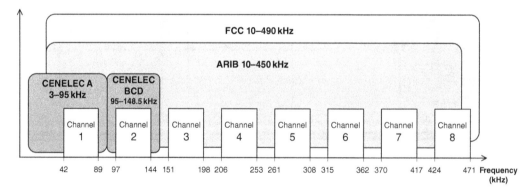

Figure 6.15 Channel structure in PRIME v1.4. Channel 1 refers to the spectrum associated to Figure 6.10.

---

1 The Repetition Code technique consists in sending data bits several times to increase the chances of better reception. The example discussed in Section 3.3.5.1 in Chapter 3 consists of a Repetition Code of 3.

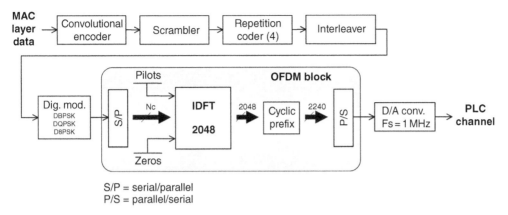

Figure 6.16 Transmitter structure for PRIME v1.4.

Table 6.4 Transmission rates in PRIME v1.4 in kbps. N_{ch} represents the number of channels in use.

Modulation	FEC ON + RC	FEC ON	FEC OFF
DBPSK	$5.22 \cdot N_{ch}$	$20.48 \cdot N_{ch}$	$40.96 \cdot N_{ch}$
DQPSK	$10.44 \cdot N_{ch}$	$40.96 \cdot N_{ch}$	$81.92 \cdot N_{ch}$
D8PSK		$61.44 \cdot N_{ch}$	$122.89 \cdot N_{ch}$

The new robust mode can only be used in combination with the DBPSK and DQPSK modulations and in Type B frames.

Additionally, as a backward compatibility mechanism, Type B frames are pre-appended in its Preamble and Header with a special content so that PRIME v1.3 devices would discard them.

The newly available communication channels and communication modes provide a new range of transmission rates in PRIME v1.4. Transmission rates (Table 6.4) depend on how many channels are in use. As all channels have the same carrier structure, the total transmission is proportional to the number of channels and the individual channel data rate.

The new specification also defines the inclusion of link quality information inside the packet header. This allows for a more up-to-date information on the link quality between pairs, to use the most appropriate communications mode. Additionally, to allow for longer data packets, a flexible MAC frame length has been defined, and a change in the MAC frame to support different durations for the beacon messages.

6.3.3.3 ITU-T G.9903 (G3-PLC)

In 2011, the G3-PLC Alliance, sponsored by Electricité Réseau Distribution France (ERDF; now Enedis), was formed to support, promote, and implement G3-PLC for Smart Grid applications. The technical specification defined by the Alliance became ITU-T G.9903 in 2011. In the same year, ITU-T G.9901 was released to cover the Power Spectral Density and the OFDM parameters used in G3-PLC [35].

6.3.3.3.1 ITU-T G.9903 (G3-PLC): PHY Layer

At PHY layer, G3-PLC communication is based on data frames consisting of three major blocks, namely Preamble, Frame Control Header (FCH), and Payload, as shown in Figure 6.17.

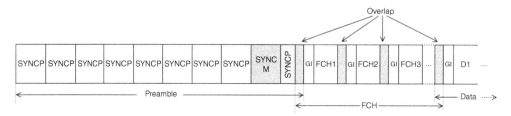

Figure 6.17 ITU-T G.9903's frame structure according to the standard.

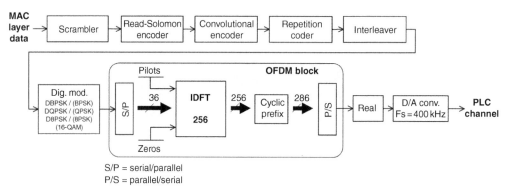

Figure 6.18 Scheme for a G3-PLC transmitter according to the standard.

The preamble consists of eight OFDM symbols (referred to as SYNCP) and one and a half OFDM symbol (SYNCM). Additionally, both the FCH and Payload sections may have different durations and modulation depending on the frequency band.

G3-PLC defines the usage of the available NB-PLC bands: CENELEC-A/C, FCC, and ARIB. Even though some transmission parameters are shared in the bands, an FFT-based and pulsed-shaped OFDM modulation is used together with several digital modulation options and Channel Coding as the transmitter in Figure 6.18 shows.

As in PRIME, the spectrum is built up in base band. Depending on the band plan, a different number of carriers are used for data transmission, as shown in Table 6.5. In contrast with PRIME, in G3-PLC some carriers can be disabled dynamically to help avoid noisy bands and create sub-domains.

OFDM symbols are pre-appended with 30 samples that, due to some overlapping intended for reducing out-of-band spectral leakage, are reduced to 22 samples.

Table 6.5 Number of data carriers used in G3-PLC.

Band	Data carriers
CENELEC-A	36
CENELEC-B	16
FCC	72
ARIB	54

Table 6.6 Maximum bit data rates for G3-PLC.

Modulation – mode	CENELEC A/B (kbps)	FCC (kbps)
BPSK – robust	5.77/2.30	15.58
DBPSK – normal	20.58/9.47	77.21
DQPSK – normal	35.40/16.12	166.47
D8PSK – normal	43.50/18.85	152.90

Several digital constellations (Figure 6.18) are available for the carriers, both differential and coherent modulation. The FEC mechanisms can also be configured in various ways. In fact, the FCH transmission uses different blocks than the Payload. The former uses a Convolutional Coder (CC) with $R_c = 1/2$ together with an RC of 6 bits; this combination is referred to as *Super Robust Mode*. The latter uses a three-layer coding scheme that consists of a Reed–Solomon (RS) encoder (with a 16 or 8 redundant bytes per block, depending on the configuration) and the same CC as described before, to compose the *Normal Mode*. This combination can be appended with a 4-bit RC block to form the *Robust Mode*. When using this *Robust Mode*, only DBPSK and BPSK digital modulations are available.

In all cases, an Interleaver is used to avoid burst errors. In contrast to PRIME, the Interleaver used in G3-PLC applies a two-dimensional shuffle of data between several OFDM symbols. Thanks to its longer size, interleaved bits are more scattered, leading to better correction performance, though increasing the decoder's complexity.

When analyzing the transceiver design, it seems that G3-PLC is set for a more robust design against noise when compared with PRIME. However, this extra robustness achieved thanks to channel coding technique has the direct consequence of reducing the data rate. Table 6.6 shows maximum achievable data rates, and Figure 6.19 shows the performance of the different modes defined in G3-PLC.

6.3.3.3.2 ITU-T G.9903 (G3-PLC): MAC Layer

At MAC layer, G3-PLC is strongly based on the IEEE 802.15.4 specification.

The frame structure of the messages follows the scheme shown in Figure 6.20. This structure defines two priority levels, high and normal. High-priority messages can only use the High Priority Contention Window (HPCW), whereas Normal-priority messages use the Normal Priority Contention Window (NPCW). Prior to these contention windows, a Contention-Free Slot (CFS) is defined.

The MAC layer is responsible for guaranteeing the integrity of the data transmitted. This is implemented via acknowledgment messages (or ACKs). ACKs are messages sent from the receiver of a frame to the transmitter to assert that the frame has been correctly received. Messages are acknowledged in the corresponding slot in Figure 6.20.

The transmitter may or may not request an ACK response. When messages are critical, they may be sent several times to increase the probability of correct reception. The recipient of the message must be able to distinguish and discard those redundant copies using the sequence number and the count of segments.

In the case of G3-PLC, the maximum size for the PHY PDU is 400 bytes. If a message does not fit, it is divided into as many segments as needed. When sending the message through the channel, the first segment uses the contention window corresponding to the message's priority.

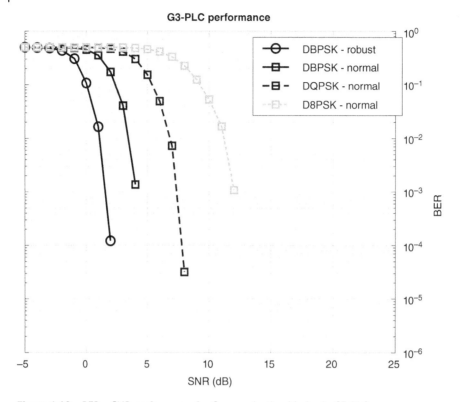

Figure 6.19 BER – SNR performance for Communication Modes in G3-PLC.

Figure 6.20 Superframe structure in G3-PLC.

The following segments corresponding to the same message always use the CFS, and this ensures that no other node is using the channel.

The quality of the communication link is also optimized by the MAC layer making use of the Tone-Map mechanism. The objective is to make sure that the most appropriate FEC mode and modulation are always used. Each node periodically sends to it neighbor devices a Tone-Map Request (TMRq) messages, indicating the need to update the FEC mode and modulation to be used. Upon reception of the message, the requested party performs an analysis of the quality of the signal received and replies with a Tone-Map Response (TMRs) message.

At a network level, G3-PLC organizes the network as a mesh topology where there is a Personal Area Network (PAN) Coordinator and several devices. With this network structure, virtual end-to-end connectivity is carried out using Lightweight On-Demand Ad-hoc Distance-vector routing protocol – Next Generation (LOADng) [36]. This protocol has low-memory requirements, capability of exploiting different routes to the destination, and is able to handle looped topologies. Consequently, messages are forwarded between devices following a one-to-one communication link until they reach their destination, as shown in Figure 6.21. Devices maintain a routing table with their neighbor's information that is updated at every reception.

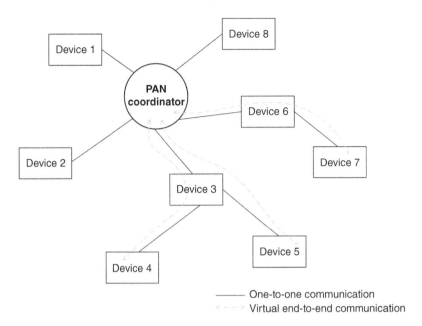

Figure 6.21 Logical topology in a G3-PLC network.

Encryption is also performed at MAC layer. Security is based on a symmetric key delivered by higher layers. A common Group Master Key (GMK) is shared between all nodes in the same PAN and is used to encrypt all messages in the network if the security mechanism is enabled. The GMK is delivered to the nodes in the network as part of their registration mechanism when they connect for the first time.

6.3.3.4 IEEE 1901.2

The IEEE 1902.2 was standardized in 2013, and some parts are influenced by both ITU-T G. 9904 (PRIME) and ITU-T G.9903 (G3-PLC)'s designs.

The IEEE 1901.2 specification contains the PHY and DLL description for an NB-PLC solution for the transmission of information using the 10–490 kHz bandwidth.

6.3.3.4.1 IEEE 1901.2: PHY Layer

The transceivers are configured with different parameters to adequate the transmitted signal to the corresponding band plan to be used (i.e. CENELEC or FCC).

The structure of an IEEE 1901.2 frame is very similar to the one used by G3-PLC: a Preamble followed by an FCH and the Payload.

The transmitter is based on an OFDM modulator that inserts several carriers on the corresponding transmission bandwidth. The scheme is very similar to G3-PLC's (Figure 6.18); however, it uses a pre-emphasis block prior to the OFDM modulation. This block uses feedback information from the receiving party to boost certain carriers to overcome highly attenuated frequency bands in the channel. Digital modulations available are identical to G3-PLC, and interleaving also spans over several OFDM symbols.

The RC block replicates the input bit R times at its output. R may be set to 1, 4, and 6 for the *normal*, *ROBO*, and *Super-ROBO* modes, respectively.

6.3.3.4.2 IEEE 1901.2: MAC Layer

In the case of the MAC layer, IEEE 1901.2 and G3-PLC are similar. Both solutions have adopted the same mechanisms for issues such as the medium access, the acknowledgment implementation, and SAR.

However, in IEEE 1901.2, the multi-tone mask mode is available for both the beacon-enabled and non-beacon-enabled modes.

6.3.3.5 ITU-T G.9902 (G.hnem)

The G.hnem specification is detailed in the ITU-T G.9902 for NB-PLC below 500 kHz, approved in October 2012.

G.hnem is also influenced by some of the main characteristics of the previously described PLC standards. The specification details the PHY and DLL for the transceiver's implementation.

There are no known field implementations of this standard.

6.3.3.5.1 ITU-T G.9902 (G.hnem): PHY Layer

Figure 6.22 shows the scheme of the G.hnem transceiver. Transmission is based on OFDM using up-conversion to shift the spectrum to the transmitting band. Interestingly, this approach is different to the other four already analyzed systems.

All data carriers are modulated using one of the available coherent digital modulations. These carriers can be enabled and disabled to cope with channel's disturbances. G.hnem differentiates from other solutions where digital modulation is always noncoherent and only coherent modulations are considered as optional. The use of a coherent modulation forces the need of an equalizer at the receiver to extract the channel's offset in the constellation, which implies a more complex receiver.

G.hnem includes FEC techniques in its normal communication mode. They consist of an inner and an outer block that implement a convolutional and Reed–Solomon encoder respectively. In this respect, the redundancy added by the Reed–Solomon encoder depends on the length of the transmitted message. More redundancy is added for longer streams of information. In contrast, the coding rate for the convolutional encoder does not depend on the message and can be set to either 1/2 or 2/3. G.hnem allows for two configurations of the Interleaver. One of them interleaves the bits within an AC cycle (either 50 or 60 Hz), whereas the other one performs what is called fragment interleave, which consists of interleaving over blocks of bits with variable lengths (longer than one OFDM symbol in all cases).

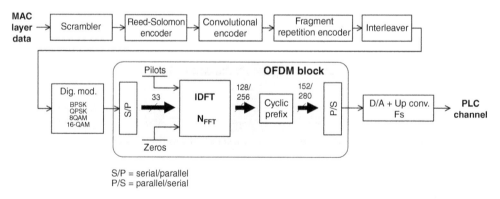

Figure 6.22 G.hnem transmitter scheme.

In addition, the Fragment Repetition Encoder (FRE) block is an option that is only used in the robust communication mode for G.hnem. The FRE provides repetitions of fragments with the repetition rate of R. Each fragment is copied R times. The FRE supports the values $R = \{1, 2, 4, 6, 12\}$.

6.3.3.5.2 ITU-T G.9902 (G.hnem): DLL Layer

The DLL in G.hnem is responsible for managing the medium access, acknowledgments, security, networking, and interfacing with upper layers.

The network architecture is organized in one or more domains. Each domain must contain a Domain Master (DM), in charge of admission, registration, and other operations. A domain is defined by a set of nodes registered to a specific DM. In addition, nodes in the same network but different domains can communicate using Inter-Domain Bridges (IDBs).

Medium access is implemented by a CSMA/CA algorithm with four priorities. The highest priority level (3) is reserved for beacons and emergency messages, whereas the rest of them (2 to 0) are used in data communications.

If enabled, medium access can be synchronous. In these cases, the DM sends periodical beacon messages to all domain nodes. Beacon information includes the duration of the CFPs and contention-based periods within the superframe.

6.3.4 Broadband PLC Systems

BB-PLC technologies make use of the spectrum starting at 1.8 MHz and offer a higher transmission data rate that comes at the expense of a higher attenuation that limits coverage. Thus, common applications today are in short-distance systems (e.g. in-home environments) and/or low-noise environments (e.g. MV grid).

6.3.4.1 IEEE 1901

IEEE 1901 is the standardized solution for BB-PLC supported by IEEE as a solution for both in-home (e.g. distances shorter than 100 m) and last-mile access (e.g. distances shorter than 1500 m).

IEEE 1901 standard, published in September 2010 and later upgraded in 2020, considers two different PHY specifications. Although both PHYs are based on OFDM, it can be built using an FFT or a wavelet mechanism [37].

At PHY level, the FFT-based implementation derives from a previous HomePlug AV [38] implementation, and it is backward compatible with that technology. Similarly, the wavelet-based OFDM is close to an HD-PLC Alliance [39] specification.

At MAC level, a single specification is defined. It interacts with both or any of the PHY possible layers through a so-called Physical Layer Convergence Protocol (PLCP).

An additional feature in the standards is the Inter-System Protocol (ISP) mechanism, formerly known as Inter-PHY Protocol (IPP), and defined in ITU-T G.9972. It guarantees the coexistence of both IEEE 1901 and ITU-T G.9960 devices.

6.3.4.1.1 IEEE 1901: FFT-OFDM PHY Layer

The block diagram for an IEEE 1901 transmitter is shown in Figure 6.23. Depending on the type of data from the MAC layer, the signal processing chain differs, using different FEC blocks.

In the common part of the transmission stage, data is modulated using a wide range of modulations to optimize the SNR channel conditions.

In the OFDM block, IEEE 1901 may work in two modes. The first mode operates in the 1.8–30 MHz (backward compatible with HomePlug AV 1.1) using 917 carriers for data transmission out

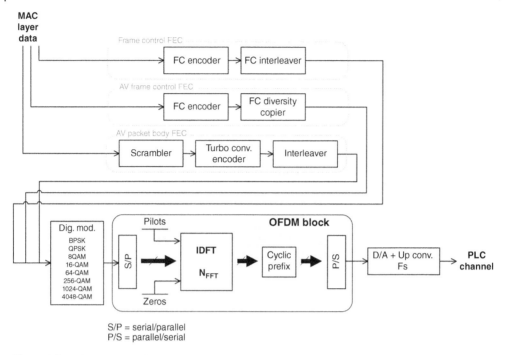

Figure 6.23 Transmitter's block diagram for an IEEE 1901 device.

of the 1155 available (the rest are masked out). The second mode extends the transmission to the 30–50 MHz band, while keeping the same carrier spacing as in the 1.8–30 MHz band (24.414 kHz), thus using in total of 1974 carriers for data transmission. In both cases, the duration of the OFDM symbol is kept to 40.96 μs. This is so since, when working in the first band, a 3072-point IFFT is used with a 75 MHz sampling frequency; when working in the second band, a 6144-point IFFT is used in combination with a 150 MHz sampling frequency. Both combinations produce a $\frac{3072}{75 \cdot 10^6} = \frac{6144}{150 \cdot 10^6} = 40.96\,\mu s$ symbol duration.

The cyclic prefix in IEEE 1901 is variable with the intention of adapting to the channel characteristics. Standardized values are {1.6, 3.92, 20.8, 2.56, 9.56, 11.56, 15.56, and 19.56} μs in order to optimize the performance in short delay-spread channels and avoid ISI in long delay-spread channels. The maximum transmission speed of IEEE 1901 is close to 500 Mbps $\left(1974 \cdot \log_2(4048) \cdot \frac{16}{18} \frac{1}{40.96+1.6}\right)$, assuming the most favorable conditions.

6.3.4.1.2 IEEE 1901: Wavelet-OFDM PHY Layer

The main difference of the wavelet-based OFDM implementation with the other PHY is that, instead of using an IFFT, the OFDM is implemented with a set of cosine-modulated filter banks that produces very low spectral leakage. When wavelets are used, there is no need to include the cyclic prefix extension to avoid ISI [40], achieving higher efficiency.

The transmitter uses 338 carriers to send information out of the total of 512 carriers in the 0–30 MHz band. These information carriers are placed in the 2–28 MHz band. It may seem that the use of a lower number of carriers with respect to the FFT-based version may lead to a lower transmission rate. However, this is not the case since Wavelet-OFDM is based on real carriers, in contrast

Table 6.7 Main parameters of IEEE 1901 PHY layer.

Frequency band	FFT-OFDM PHY		Wavelet-OFDM PHY	
	1.8–30 MHz	1.8–50 MHz	1.8–30 MHz	1.8–50 MHz
OFDM implementation	Fast Fourier Transform		Wavelet	
FFT points	3072	6144	512	1024
Sampling frequency	75 MHz	150 MHz	62.5 MHz	125 MHz
OFDM symbol duration	40.96 μs		8.192 μs	
Cyclic prefix duration	1.6, 3.92, 20.8, 2.56, 9.56, 11.56, 15.56, 19.56 μs		Does not apply	
Digital modulation	BPSK, QPSK, 8-QAM, 16-QAM, 64-QAM, 256-QAM, 1024-QAM or 4048-QAM		BPSK, 4-PAM, 8-PAM, 16-PAM, or 32-PAM	
Maximum transmission rate	229.5 Mbps	494 Mbps	193 Mbps	Around 500 Mbps

with the complex carriers used in FFT-OFDM. One consequence is that, out of the K available carriers, in the case of FFT-OFDM, only $K/2$ of them are effectively used, whereas in the case of Wavelet-OFDM, all of them can be digitally modulated.

Each real wavelet is modulated using one of the five possible digital constellations based on PSK and Pulse-Amplitude Modulation (PAM): BPSK, 4-PAM, 8-PAM, 16-PAM, or 32-PAM. These modulated bits are encoded using an outer/inner combination of channel coders, more specifically, a Reed–Solomon encoder followed by a convolutional encoder. There is an extra FEC block that can be used for extra redundancy. It is a Low-Density Parity-Check Code (LDPC), which benefits from a rather low coding/decoding complexity while providing good correcting capabilities.

This transmission scheme is able to deliver a maximum transmission rate of approximately 500 Mbps [37]. A list of the main transmission parameters for both PHY modes is shown in Table 6.7.

6.3.4.1.3 IEEE 1901: MAC Layer

At MAC level, IEEE 1901 follows a Master/Slave paradigm, where the Master coordinates the communication slots for the Slaves to control the overall QoS. The Master decides which Slaves can transmit in the two possible transmission periods: a CFP or a Contention Period (CP). The latter is based in CSMA/CA and is used for general-purpose devices. In contrast, the CFP is a TDMA-based period targeted at devices that have low-delay or low-jitter requirements. The standard differentiates three TDMA variants:

- Centralized TDMA, where the Master allocates slots for every path between a given node and the Head-End.
- Dynamic TDMA Polling, where the Master fixes the slots at the beginning of each transmission period based on the result of polling all the nodes in the network.
- Distributed TDMA, where the Master only assigns slots to its neighbors and delegates to them the responsibility of allocating time slots to their corresponding neighbors.

Different from other PLC solutions, network stations (i.e. slaves) can communicate directly with each other, instead of needing to relay on the network's Master.

The Master performs the association and authentication of network nodes. There are situations in which, due to noise or high attenuation, the Master node cannot communicate with all devices in the same physical network. In such scenarios, devices can be instantiated as Proxy Coordinators (PCo) and are responsible of relying on the messages to those Hidden Stations (HSTAs).

In some extreme scenarios where a device does not receive any beacon from the network's Master node or any PCo, the standard defines an Uncoordinated Mode. Devices running in this mode are responsible of maintaining its own beacon timing, since there are no external references to synchronize to.

6.3.4.2 ITU-T G.996x (G.hn)

ITU-T G.hn (Home Networking) group started in 2006 to elaborate a communication solution that could work under different legacy in-home physical media: telephone lines, coaxial and Cat5 cables, plastic-optic fibers, etc. Eventually PLC was added. The target was set to 1 Gbps transmission rate. This can be achieved thanks to the use of two "profiles": one using the 2–50 MHz band, and another one using the 2–100 MHz band.

ITU-T G.9960 and G.9961 specify the PHY and MAC layer of the G.hn technology, respectively. Several additional recommendations have been published afterward: ITU-T G.9962, ITU-T G.9963 and ITU-T G.9964.

G.hn standard defines different Profiles to accommodate the different scenarios with varied implementation complexity. It includes high-level devices such as residential gateways that are used to manage throughputs for in-home devices but also covers the definition of low-level apparatus that acts as sensor or appliance controller. It also needs to be parameterized (the OFDM) depending on the medium where the signal is being coupled.

6.3.4.2.1 ITU-T G.996x (G.hn): PHY Layer

At PHY layer, G.hn uses FFT-OFDM in combination with a variety of digital communication techniques to be able to adapt to different transmission scenarios. Other OFDM parameters, such as the carrier spacing and sampling frequency and the cyclic prefix length count with a predefined range of values; all of them have a power-of-2 relationship to simplify implementations. Additionally, G.9960 also defines a set of region-specific and medium-specific band plans, which are application-dependent and where different channel bandwidths can be selected. These configurations are schematically shown in Figure 6.24.

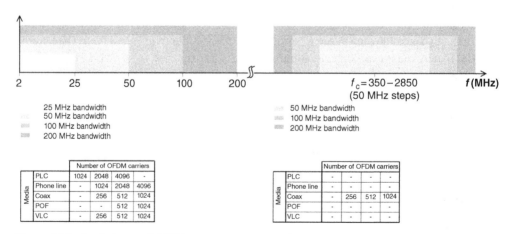

Figure 6.24 Band plans and OFDM parameters for ITU-T G.9960 (G.hn). *Source*: Adapted from [41].

Each of the OFDM carriers can be modulated using a range from 1 to 12 bit-long symbols, and each carrier can be modulated independently with a number of bits according to the SNR and attenuation at that specific frequency, as negotiated between transmitter and receiver.

ITU-T G.9960 uses some FEC mechanism. In this case, a specific subset of LDPC (Quasi-Cyclic Low-Density Parity Check or QC-LDPC) encodes with several possible coding rates: {1/2, 2/3, 16/18, 20/21}. A different set of parameters and frequency ranges are used to create a signaling structure able to communicate in extra-noisy environments that optimize the transmission, given some channel characteristics.

ITU-T G.9963 adds the possibility of using MIMO capabilities in the communication, to achieve a higher transmission rate and coverage.

6.3.4.2.2 ITU-T G.996x (G.hn): MAC Layer

From an architectural point of view, G.hn devices can be configured to work in up to 16 different domains in the same or in a different physical media. Each one of these domains counts with a DM that manages all devices (up to 250) in the domain in terms of resources allocation and priorities. Additionally, inter-domain communications for devices happen through IDBs. Finally, a Global Master (GM) node manages all domains and their interactions.

The media access in ITU-T G.9960 is coordinated by the DM with the MAC cycle, which can be synchronous to the mains cycle. This helps when dealing with periodically time-varying changes in the channel or noise sources coming from electrical devices or appliances connected to the same network.

Each MAC cycle is divided into several slots or transmission opportunities (TXOPs) designated by the DM. Since the TXOP structure is dynamic, the DM is responsible for setting it and informing all nodes in the domain. This is done in the Media Access Plan (MAP) frame. The assignment of TXOPs from the DM is done upon requests from the devices and mainly depends on the type of application nodes running. This approach ensures that critical applications are guaranteed a certain QoS.

The specification defines three kinds of TXOPs, depending on the application:

- Contention-Free TXOP (CFTXOP), where only the assigned node can transmit during that specific slot.
- Shared TXOP (STXOP), with managed time slots, implemented by a CSMA/CA access method.
- Contention-Based TXOP (CBTXOP), as a SCP where all nodes content to access the channel with some priority mechanism.

6.4 Applicability to Smart Grids

Optical fiber and PLC solutions complement each other in the access network. Optical fiber allows electric isolation in substations and electricity-related premises in general, and PLC technology is by default available in Smart Grids.

Specifically in the PLC case, there have been some non-Smart Grid PLC uses that have been instrumental to the eventual Smart Grid application of the same technology. These systems have enabled technology and ideas useful for utility applications.

The first example of PLC use for non-Smart Grid purposes took place at the turn of the last century for Internet Access. There were several approaches to this worldwide as, in a context of telecommunications deregulation, competition had to be fostered. Some of these approaches included the opening of the local loop, i.e. the copper pairs that reached all homes to provide the public telephony services through the traditional PSTN (Public Switched Telephone Network) or POTS (Plain Old Telephone System); others looked for alternative technologies and infrastructures different from local loops. HFC (Hybrid Fiber-Coaxial) networks, WLL (Wireless Local Loop), and plain optical fiber emerged as the most prevalent alternatives in the telecommunications industry and individually helped in the market deregulation. It was coincidental that deregulation came with the emergence of Internet access and the need of broadband access networks to fill this gap. Thus, PLC and specifically BPL emerged as another local loop alternative. Different countries and utilities worldwide developed different initiatives with a different degree of success in this domain. Despite the commercial clash and success (or not) of this alternative BPL access, the technology to enable the broadband transmission of PLC signals flourished [42] and was eventually adapted to work on utility services-only scenarios [43]. As a result, it has been successfully applied to Smart Grid deployments at a massive scale (Section 6.4.2) since then.

The second example of non-Smart Grid PLC use is the in-home communications domain. Once the BPL technology was proven at a wide scale, the opportunity to use it for in-home connectivity appeared. The increasing relevance of telecommunications and electronic elements local interconnection made BPL appear as an alternative to other wireless options (e.g. Wi-Fi), to be used in parts of the network where wireless (radio) has difficulties to operate (e.g. physical obstacles that may prevent signal propagation). This technology, matured in the in-home environment, will in turn be tested in Smart Grid scenarios, to solve some LV distribution grid challenges.

6.4.1 Passive vs. Active Optical Fiber Networks

PON has become popular for Internet residential access, both as a replacement for xDSL services and as an evolution of HFC networks. PONs allow last-mile optical-fiber-based solutions become economically efficient, where active optical fiber had not been an alternative in the past.

In a utility context, on the contrary, however, optical fiber access had traditionally been an alternative for utility access, to premises that could absorb the high investment associated to optical fiber infrastructure. This has traditionally been the case for PSs, but not for SSs in general. Indeed, the PS's relevance and its high asset value are different from those of an SS and may justify optical fiber access by default.

However, when telecommunications for SSs are designed, utilities tend to think on alternative solutions to optical fiber, even if PON alternatives are an option. Apart from the higher cost of optical fiber compared with, e.g. PLC solutions, the difference of the service targets between the SS case and the PON Internet residential case determine that the cost per serviced end point (SSs in the utility case, and customer homes in the Internet access case) is higher in the former. To understand this, we need to consider that each SS serves several tens of buildings, while when a TSP reaches a building, the potential number of end points is the number of existing homes. There are, nevertheless, success stories connected to the use of PONs in utilities [44].

With this landscape, the use of optical fiber in the utility access networks takes these different options:

- Active private optical fiber networks, reaching only some specific SSs where the investment is cost-efficient. Due to the fact that utility optical fiber cables carry a high number of fibers, and that target SSs may be located some kilometers apart one from the other, it is usual to find point-to-point active configurations, linking SSs with the PS or other appropriate extraction points.

- PON commercial service, provided by TSPs. This case leverages third-party existing infrastructure and may be convenient when urban Smart Grid cases are considered. The penetration of PONs in cities, and the fact that SSs are sometimes integrated in building basements, makes it possible. However, there are some requirements that must be met:
  - Secure power supply in the SS is commonly 48 Vcc.
  - SS access may be limited only to utility accredited contractors (for commissioning and maintenance purposes).
  - The CPE to be deployed at the SS must be prepared to support SS installation conditions, while being validated in the TSP network. The current PON CPE options available, with a few exceptions [45] will not be compliant.
  - CPE OAM procedures may need to be adapted, as the utility will need to manage its own network, according with all its safety, security, and operational procedures.

6.4.2 Broadband PLC over Medium Voltage for Secondary Substation Connectivity

The telecommunications access to SSs can be facilitated using PLC over the MV grid. As all SSs are connected to the grid and due to the good channel conditions of the MV grid, it is just natural that the MV power lines can be used as the way to communicate SSs.

The type of PLC technology that can more conveniently serve the needs of SS connectivity is BPL. BPL provides a throughput that can be used to connect the multiple Smart Grid services in SSs, including not only those within the SS, but also others that may be aggregated as a consequence of the LV connectivity in the SS. Additionally, BPL technologies tend to use a spectrum range that is different from the NB-PLC that is nowadays used in the LV grid typically for Smart Metering purposes.

The fact that BPL existing technologies work in the MHz range affects propagation distances and consequently, the applicability to field use cases depending on the grid morphology. However, 1 km broadband communication in underground power lines, and 2 km in overhead, can be covered easily.

The biggest rollout of MV BPL for Smart Grid purposes is in Spain. Iberdrola has reported to have deployed over 25 000 OPERA BPL modems over its MV grid for Smart Grid purposes [46, 47]. Specifically, Smart Metering and Automation services are aggregated in the BPL links, with an architecture that leverages a private and a commercially available connectivity of optical fiber, ADSL, and cellular (2G, 3G and 4G) radio [48].

The rollout in Spain is the result of the application of the second wave of research and development with OPERA technology. The rollout was performed over a 10-year period [49] and was simultaneously performed with the rest of the smart meter rollout mandated in Spain by the local Regulation, following European guidelines on the matter. Thus, it cannot be considered an ad-hoc PLC technological effort, but a "business-as-usual" Smart Grid rollout, where telecommunication technology is fully integrated with the rest of the Smart Grid operational activities. Along the years, operational processes matured, obtaining a high degree of operational efficiency based on a high level of technology resilience, and a proper network design with conservative rules, to guarantee that coverage ranges, throughputs, and latencies are aligned with service requirements. Compared with other telecommunication technologies, BPL over MV has demonstrated to have availability levels higher than those of commercial radio services, but lower than those of optical fiber. Throughputs and latencies are seen to follow the same rule [49].

The main highlights of this deployment are summarized as follows:

- Underground MV power lines offer a higher degree of performance predictability, as noise sources are more stable over time. However, overhead MV power lines can reach further (up to 2.3 km for a power spectral density of -50 dBm/Hz [49]).
- Coupling solutions have to be adapted to the specific features of the MV network, specifically to the type of switchgear (the connection point to the grid), and safety and asset lifetime need to be taken as the highest priority.
- Capacitive coupling solutions provide a more stable network, as the grid input impedance is more stable, independent of the grid operational state.
- Telecommunications network design and planning are performed before its rollout takes place. This design is based on a modular approach, based on an FDM split in two bands (2–7 and 8–18 MHz) and subsequent frequency planning; availability of the MV topology data; and understanding the signal propagation through new and legacy cables.
- Network designs leverage non-BPL connectivity possibilities, meaning that although the connectivity of the MV grid determines the connectivity paths among SSs and PS, the connectivity to extract telecommunication signals from cells can be deployed anywhere in the grid.
- Operational preproduction tests must show a minimum point-to-point throughput to guarantee that installation procedures have been performed correctly. A convenient threshold was fixed at 10 Mbps, although the worst-case real scenario is 20 Mbps [49].

BPL results for this field scenario have been reported in literature. Throughput results can be expected to realistically be between 35 and 60 Mbps [49], and latency ranges between 10 and 20 ms in each BPL hop [50].

BPL presents itself as a Layer 2 technology, capable of supporting Ethernet traffic. Full networks that take advantage of BPL connectivity must implement the higher layers. IP traffic-based networks are typically used, and classic networking mechanisms are used to provide resiliency and redundancy mechanisms. From a pure infrastructure perspective, the fact that BPL may be used for grid automation and control purposes implies that battery backup power supply may be needed. High-availability operation can be implemented with mechanisms such as the ones shown in [51], where the different element failures and remedies are explained.

6.4.3 High Data Rate Narrowband PLC over the Low Voltage Grid for Smart Metering

EDF (with Enedis, its DSO, France) and Iberdrola (Spain) Smart Metering deployments are perhaps the most outstanding and representative examples of recent massive PLC application. The story of both companies shares some parallelisms, but certainly notable differences.

EDF and Iberdrola were part of the group of European utilities pushing PLC solutions as alternatives for telecommunications in the telecommunications access market (the "local loop," [52]). Their engagement level with the use of PLC for such Internet Access was different, but both understood the possibilities (at both a technology and a market opportunity level) of PLC for the emerging Smart Grids and specifically for Smart Metering.

EDF and Iberdrola followed different roads to find their solution for PLC-based Smart Metering systems, which could be deployed massively. These paths were different for a variety of reasons, but an important one was the need to come up with a solution at a different stage.

Iberdrola, as the rest of Spanish utilities were urged by the Government to start a massive substitution of the residential installed meters' base (first in [33], and more specifically in [53]). At the time this happened in Spain, Iberdrola had tested an LDR NB-PLC solution, which proved difficult to install, was limited in operation, and was eventually inadequate for Smart Metering. HDR NB-PLC solutions had not been developed yet, but BPL techniques had been proven feasible over the LV grid. Meanwhile, France did not have such a Government mandate; EDF was engaged with what it known today as G1 meter, and although they understood that a more flexible telecommunications platform was needed for future Smart Metering deployments, their time constraints were different to Iberdrola's and the rest of the Spanish utilities.

From this moment on, two solutions that eventually would become industrial references and consolidated standards appeared. Iberdrola started its Smart Metering rollout with PRIME (ITU-T G.9904) in 2009 with a target of 11 million meters in 85 000 SSs that had to be installed by the end of 2018 [54]. EDF started a massive meter replacement in 2015, progressively transitioning from a G1 meter to G3 (ITU-T G.9903) smart meter (the first G3 smart meters were installed in low volumes in 2016 [55]), with the intention of completing the rollout of their 35 million meters in 740 000 SSs by the end of 2021 (95% committed penetration [56]). Refs. [57, 58] show the status and some interesting data of each project.

The telecommunications performance of HDR NB-PLC systems in LV, once the technology is fixed, depends on the topology of the grid, its noise conditions, and the number of smart meters connected to each DC in the SS. From the figures of the grids of this example, it can be seen that the average number of smart meters per SS is roughly 48 for EDF and 130 for Iberdrola, and thus the average expected results may show differences. Ref. [59] tries to stablish another classification based on the number of smart meters and its density (concentration per meter room) that correlates nicely with results. However, it is not only the pure telecommunication aspects that affect performance, there are other elements that have a high influence over the overall performance of this systems, from the meter reading perspective. Of all these elements, the application level and the objects retrieved from the smart meters play a fundamental role. In both rollouts, DLMS/COSEM is used.

Available performance data in literature is limited. Early data from EDF, from a 30 000 limited deployment showed collection rates of 98.6% meaning that 1.4% of the meters were not read over each day on average. Such high-level data is typical of the application-level view. Iberdrola has provided more low-level detailed data with PRIME, due to its preproduction tests. Ref. [59] shows the results of the first rollout of 100 000 smart meters in Spain and defines the term "availability" over a period of time when the tests are performed, to show how all smart meters can be read many times during the period of a day. Specifically, data for 3 SSs is shown, with 4, 189, and 262 smart meters. For the two largest ones, a rough reading time of 5 seconds for short cycles and 20 seconds for long cycles in each meter is reported. The worst case is that 180 meters can be accessed each hour with traffic transactions equivalent to the long cycles. In terms of latency, these systems offer a result close to the 200 ms of latency per hop on average [55].

Performance in field deployments can be optimized using the proper selection of design decisions to maximize the results on the field. The use of single-phase injection to maximize power transfer and minimize input noise; modulation scheme selections; ARQ with the appropriate number of repetitions; ACKs in non-dedicated packets, but piggybacked; SAR use, with appropriate size, etc. are aspects covered in [9, 50].

Performance can be controlled from the MDMS, at application level, but can also leverage telecommunication management protocols such as SNMP in BNs as reported by Iberdrola [60], to help troubleshooting of these pervasive network elements deployments. These BNs implement an MIB (Management Information Base) that provides PLC-plane telecommunications information. A SNMP-based NMS can retrieve MIB objects in the BNs periodically, with information about the network topology, channel occupation, etc., to perform statistical analysis and/or generate alarms on particular subnetworks. BN sniffing capabilities can also be incorporated.

The performance results obtained in these deployments can be improved with the use of PLC gateways. This kind of approach has been used by Iberdrola [61] in approximately 20 000 SSs. Gateways are devices that, integrating a BN with some other traditional non-PLC connectivity, can be connected remotely to the MDMS HE. Thus, the PLC gateway adapts (converts) the media and the protocol to seamlessly connect devices on a non-PLC media used as a WAN connection (e.g. Ethernet or wireless), with devices on PLC media. Even if this architecture has been deployed by Iberdrola to avoid installing DCs in SSs with a low meter count, the generalization of the concept can separate the number of smart meters connected to an SS, from the need to coordinate its PLC connectivity from the DC in the SS. Ideally, gateways could be installed closer to the smart meters and create structures with a lower quantity of connected meters, thus improving performance.

Last but not least, PRIME and G3 have been demonstrated as technologies capable of supporting other Smart Grid services and applications. Refs. [62, 63] have shown how LV feeder and phase information can be retrieved based on the properties of the PRIME PLC propagation, and Ref. [60] has shown how IP remote control traffic can work in PRIME PLC networks.

References

1 VIAVI Solutions Inc. https://www.viavisolutions.com/en-uk/passive-optical-network-pon (accessed 2 April 2021).
2 Farmer, J., Lane, B., Bourg, K., and Wang, W. (2016). FTTx Networks: Technology Implementation and Operation. Morgan Kaufmann.
3 Fiber to the Home Council Global Alliance (FCGA). http://www.ftthcouncil.info (accessed 2 April 2021).
4 Hood, D. and Trojer, E. (2012). Gigabit-Capable Passive Optical Networks. Wiley Online Library.
5 Full Service Access Network (FSAN). http://www.fsan.org (accessed 2 April 2021).
6 Ethernet Alliance. ethernetalliance.org (accessed 2 April 2021).
7 Metro Ethernet Forum. https://www.mef.net/about-mef (accessed 2 April 2021).
8 Ansari, N. and Zhang, J. (2013). Media Access Control and Resource Allocation: For Next Generation Passive Optical Networks. Springer Science & Business Media.
9 Sendin, A., Peña, I., and Angueira, P. (2014). Strategies for power line communications smart metering network deployment. *Energies* 7 (4): 2377–2420.
10 Routin, J. and Brown, C.E.L. (1897). Power line signalling electricity meters. *Br. Pat.* 24: 833.
11 Thordarson, C.H. (1905). Electric central-station recording mechanism for meters. Google Patents.
12 Marumo, N. (1920). Simultaneous transmission and reception in radio telephony. *Proc. Inst. Radio Eng.* 8 (3): 199–219.
13 Schwartz, M. (2009). History of communications: carrier-wave telephony over power lines: early history. *IEEE Commun. Mag.* 47 (1): 14–18.
14 Dostert, K. (2001). Powerline Communications. Prentice Hall PTR.
15 Remseier, S. and Spiess, H. (2006). Making power lines sing. *ABB Rev.* 2: 50–53.
16 Zahavi, J. and Feiler, D. (1980). The economics of load management by ripple control. *Energy Econ.* 2 (1): 5–13.
17 Dzung, D., Berganza, I., and Sendin, A. (2011). Evolution of powerline communications for smart distribution: from ripple control to OFDM. *2011 IEEE International Symposium on Power Line Communications and Its Applications*, pp. 474–478. doi: 10.1109/ISPLC.2011.5764444.

18 CENELEC (2011). EN 50065-1: Signalling on Low-Voltage Electrical Installations in the Frequency Range 3 kHz to 148,5 kHz. CENELEC.
19 United States Department of Energy (2010). *Broadband over Power Lines Could Accelerate the Transmission Smart Grid*.
20 Galli, S. (2007). Power line communications: technology and market perspectives. *Proceedings of the ANEEL Workshop Sobre Power Line Communications (PLC)*.
21 Mak, S.T. and Reed, D.L. (1982). TWACS, a new viable two-way automatic communication system for distribution networks. Part I: outbound communication. *IEEE Trans. Power Appar. Syst.* 8: 2941–2949.
22 SIEMENS. *PowerLink IP – The PLC Solution for Digital Transmission Grids*. https://new.siemens.com/global/en/products/energy/energy-automation-and-smart-grid/smart-communications/powerline-carrier-power-link.html (accessed 2 April 2021).
23 Sendin, A., Llano, A., Arzuaga, A., and Berganza, I. (2011). Strategies for PLC signal injection in electricity distribution grid transformers. *2011 IEEE International Symposium on Power Line Communications and Its Applications*, pp. 346–351. doi: 10.1109/ISPLC.2011.5764420.
24 Berger, L.T., Schwager, A., Pagani, P., and Schneider, D. (2017). MIMO Power Line Communications: Narrow and Broadband Standards, EMC, and Advanced Processing. CRC Press.
25 Homeplug Powerline Alliance (2004). *HomePlug 1.0 Technical White Paper*, vol. 30.
26 Goldfisher, S. and Tanabe, S. (2010). IEEE 1901 access system: an overview of its uniqueness and motivation. *IEEE Commun. Mag.* 48 (10): 150–157.
27 da Silva Costa, L.G., de Queiroz, A.C.M., Adebisi, B. et al. (2017). Coupling for power line communications: a survey. *J. Commun. Inf. Syst.* 32 (1): 8–22.
28 Enterprise and Industry Directorate General European Commission (2009). *M/441 EN Standardisation Mandate to CEN, CENELEC and ETSI in the Field of Measuring Instruments for the Development of an Open Architecture for Utility Meters Involving Communication Protocols Enabling Interoperability*.
29 S. G. M. European Commission, Directorate-General for Energy (2011). *M/490 Standardization Mandate to European Standardisation Organisations (ESOs) to Support European Smart Grid Deployment*.
30 Arcauz, N. (2009). The EU OPEN meter project. *ERGEG Workshop on Smart Metering*.
31 Arcauz, N. (2011). *Final Report Summary – OPEN METER (Open Public Extended Network Metering)*.
32 PRIME Alliance. www.prime-alliance.org (accessed 2 April 2021).
33 Boletín Oficial del Estado Español (2007). *Real Decreto 809/2006, de 30 de junio, por el que se revisa la tarifa eléctrica a partir del 1 de julio de 2006*. https://www.boe.es/eli/es/rd/2006/06/30/809/dof/spa/pdf (accessed 2 April 2021).
34 Proakis, J.G. and Salehi, M. (2013). Fundamentals of Communication Systems. Pearson Education.
35 Haidine, A., Tabone, A., and Muller, J. (2013). Deployment of power line communication by European utilities in advanced metering infrastructure. *International Symposium on Power Line Communications and Its Applications*, pp. 126–130. doi: https://doi.org/10.1109/ISPLC.2013.6525837.
36 Clausen, T., Yi, J., and Herberg, U. (2017). Lightweight on-demand ad hoc distance-vector routing-next generation (LOADng): protocol, extension, and applicability. *Comput. Networks* 126: 125–140.
37 Galli, S. and Logvinov, O. (2008). Recent developments in the standardization of power line communications within the IEEE. *IEEE Commun. Mag.* 46 (7): 64–71. https://doi.org/10.1109/MCOM.2008.4557044.

38 Afkhamie, K.H., Katar, S., Yonge, L., and Newman, R. (2005). An overview of the upcoming HomePlug AV standard. *International Symposium on Power Line Communications and Its Applications, 2005*, pp. 400–404.

39 HD-PLC. *IEEE 1901 HD-PLC Technical Overview*. https://hd-plc.org/hd-plc-technical-overview (accessed 27 April 2021).

40 Galli, S., Koga, H., and Kodama, N. (2008). Advanced signal processing for PLCs: Wavelet-OFDM. *2008 IEEE International Symposium on Power Line Communications and Its Applications*, pp. 187–192.

41 Martínez, M. and Iranzo, S. (2019). Present and future of ITU-T G.hn Standard. *IEEE International Symposium on Power Line Communications and Its Applications (ISPLC)*.

42 European Commission. *Open PLC European Research Alliance for New Generation PLC Integrated Network*. https://cordis.europa.eu/project/id/507667 (accessed 2 April 2021).

43 European Commission. *Open PLC European Research Alliance for New Generation PLC Integrated Network Phase 2*. https://cordis.europa.eu/project/id/026920 (accessed 2 April 2021).

44 Newton, L., Panyik-Dale, D., and Kershner, A. (2010). Chattanooga, Tenn Announces only 1 gigabit broadband service in U.S. for both residential and business customers. http://chattanoogagig.com/pdf/Chattanooga_GPON_EPB.pdf (accessed 27 April 2021).

45 Telnet Redes Inteligentes. *GPON ONT WaveAccess 512*. https://www.telnet-ri.es/en/gpon-ont-din-rail-form-waveaccess-512 (accessed 2 April 2021).

46 Utility Telecoms (2020). *Transforming Utility Telecom Departments to Drive the Migration to MPLS Networks and Maximise Grid Security, Reliability and Visibility*.

47 Sendin, A., Simon, J., Solaz, M. et al. (2015) MVBPL – reliable, future proof and cost efficient. *Proceedings of the 23rd International Conference on Electricity Distribution (CIRED), Lyon, France*, pp. 15–18.

48 Sendin, A., Simon, J., Urrutia, I., and Berganza, I. (2014). PLC deployment and architecture for Smart Grid applications in Iberdrola. *2014 18th IEEE International Symposium on Power Line Communications and Its Applications (ISPLC)*, pp. 173–178.

49 Valparis, J.A., Amezua, A., Sanchez, J.A. et al. (2017). Complete MV-BPL communications solution for large AMI and grid automation deployments. *CIRED-Open Access Proc. J.* 2017 (1): 78–82.

50 Sendin, A., Sanchez-Fornie, M.A., Berganza, I. et al. (2016). Telecommunication Networks for the Smart Grid. Artech House Publishers.

51 Solaz, M., Simon, J., Sendin, A. et al. (2014). High availability solution for medium voltage BPL communication networks. *18th IEEE International Symposium on Power Line Communications and Its Applications*, pp. 162–167.

52 PLC Utilities Alliance (2004). *Powerline an Alternative Technology in the Local Loop*.

53 Boletín Oficial del Estado Español (2007). *Real Decreto 1110/2007, de 24 de agosto, por el que se aprueba el Reglamento unificado de puntos de medida del sistema eléctrico*.

54 Iberdrola. *Star Project Website*. https://www.i-de.es/smart-grids/deployment-projects-areas/star-project (accessed 2 April 2021).

55 Lys, T. (2016). First feedback on the G3-PLC roll-out in France. *ERDF*, 15 p.

56 Lyon Lighthouse Project (2019). *Monitoring Strategy of the Lyon Lighthouse Project*.

57 i-DE (Iberdrola Group). *Smart Meters Approved*. https://www.i-de.es/smart-grids/smart-meter/smart-meters-approved (accessed 2 April 2021).

58 Enedis. *The Linky Meter, an Industrial Project*. https://www.enedis.fr/linky-un-projet-industriel (accessed 2 April 2021).

59 Sendin, A., Berganza, I., Kim, I.H. et al. (2012). Performance results from 100,000 + PRIME smart meters deployment in Spain. *IEEE Third International Conference on Smart Grid Communications (SmartGridComm)*, no. Lv, pp. 145–150.

60 Sendin, A., Arzuaga, T., Urrutia, I. et al. (2015). Adaptation of powerline communications-based smart metering deployments to the requirements of smart grids. *Energies* 8 (12): 13481–13507. https://doi.org/10.3390/en81212372.

61 Sendin, A., Gomez, J.S., Urrutia, I. et al. (2018). Large-scale PLC gateway-based architecture for smart metering deployments. *2018 IEEE International Symposium on Power Line Communications and Its Applications (ISPLC)*, pp. 1–6.

62 Navarro, E., Sendin, A., Llano, A. et al. (2014). LV feeder identification: prior requirement to operate LV grids with high penetration of DER and new loads. *CIGRE*, pp. 24–29.

63 Marrón, L., Osorio, X., and Sendin, A. (2013). Low voltage feeder identification for smart grids with standard narrowband PLC smart meters. *International Symposium on Power Line Communications and Its Applications*, pp. 120–125.

7

Wireless Cellular Technologies

7.1 Introduction

A cellular, or mobile, network is a telecommunication system where terminals are connected wirelessly and network coverage is organized in geographical areas called "cells." These cells are realized by means of a distributed assembly of base stations (hosting the antennas and radio processing equipment) and interconnected through the so-called mobile core network, allowing terminals to seamlessly move across cells with no service disruption.

Cellular technologies are extensively used to provide voice and data services to the general public through large, country-wide commercial networks and, more recently, increasingly used for the operational support in professional activities across different sectors and industries.

This chapter describes the services, applicability scenarios, network architectures, deployment considerations, and main capabilities and features that can be delivered by a cellular system. The focus is on 4G and 5G technologies, addressing the baseline concepts and commonalities among them as well as the main differences and enhancements brought by the latest 5G technologies.

7.2 Mainstream Technologies and Standards

7.2.1 Cellular Technologies Evolution

Since first cellular networks were introduced in the early 1980s, mobile communication services have undergone a tremendous development, to the extent of becoming the largest communication technology platform in human history with around 8000 million mobile subscriptions worldwide [1]. Technology innovation has been continuous during this period of roughly 50 years, with important technology leaps, commonly known as *generations*, following a ~10-year cycle (see Figure 7.1).

7.2.1.1 1G and 2G. Voice-centric, Circuit-switched Services

1G systems were fundamentally designed to provide narrowband, circuit-switched voice services as a wireless extension of the fixed telephony network. Multiple technologies, not compatible with each other, were deployed. A common denominator across them was the use of analog modulations and FDMA. Channel bandwidths used for individual voice connections in the different

---

Smart Grid Telecommunications: Fundamentals and Technologies in the 5G Era, First Edition. Alberto Sendin, Javier Matanza, and Ramon Ferrús.
© 2021 John Wiley & Sons, Inc. Published 2021 by John Wiley & Sons, Inc.

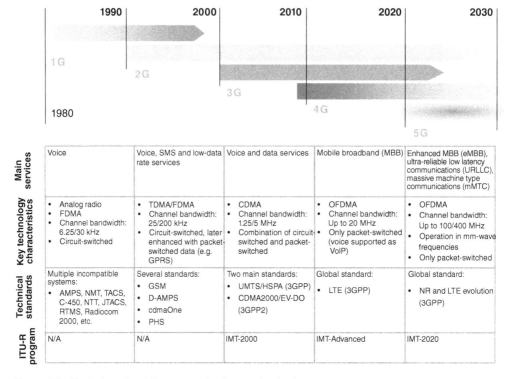

Figure 7.1 Evolution of mobile communications technologies.

systems were in the range of 6.25–30 kHz. 1G systems remained operational until late 1990s/early 2000s, when they were progressively displaced by a new wave of 2G systems built upon more resource-efficient and feature-rich digital technologies.

2G systems, while still designed for circuit-switched voice services, came also with support for basic data services such as Short Message Services (SMS) and Circuit-Switched Data (CSD) services, the latter allowing a mobile phone to be used as a dial-up data modem for Internet access at speeds of 10–20 kbps [2]. Progress toward improved compatibility through adoption of common standards received more attention in the transition to 2G systems, enabling roaming between networks and favoring economies of scale. At European level, GSM networks (a pan-European standard and the first digital cellular system) were deployed in the early 1990s. Other relevant 2G standards included Digital-AMPS (IS-54/136), developed in the US and mainly used there, CDMAOne (IS-95), another US standard used in Asia-Pacific, North and Latin America, and Personal Digital Cellular, developed and used exclusively in Japan. With the exception of CDMAOne, TDMA/FDMA technologies were at the core of 2G systems, with channel bandwidths in the order of tens to a few hundreds of kHz (e.g. GSM uses 200 kHz channels to handle up to eight simultaneous voice connections per radio channel).

Unlike 1G systems, many 2G systems remain operational as of today, especially GSM networks, since they are still central to operators for voice services. In this respect, it is worth noting that current 2G technologies count with numerous enhancements with respect to the first 2G systems deployed in the 1990s, especially with regard to the support of packet-switched data services such as the General Packet Radio Service (GPRS) and Enhanced Data rates for GSM Evolution (EDGE) features. GPRS is a GSM extension that allows for packet-switched services to be provided over the same TDMA radio structure used for voice services, adding a separate packet-switched core network. In turn, EDGE is primarily a GSM radio interface enhancement with more efficient modulation and coding schemes. Data rates achieved with GPRS/EDGE are typically in the range from tens to a few hundreds of kbps. GPRS and EDGE are often referred to as 2.5G.

7.2.1.2 3G. Paving the Way for Mobile Data Services

3G was an attempt to depart from the previous generations' voice-centric designs and transition toward a more flexible solution able to better accommodate both voice and high data rate services in the same system. CDMA became the technology of choice in 3G, motivated by, among others, its abilities to easily multiplex multi-rate connections over the same radio channel, to constructively exploit the multi-path propagation, and to support full frequency reuse deployments [3]. Two main standards were competing in the 3G era: Universal Mobile Telecommunications System (UMTS) and CDMA2000.

UMTS specifications were developed by the 3rd Generation Partnership Project (3GPP), a consortium established in 1998 by a number of Standards Development Organizations (SDOs) from Asia, Europe, and North America. UMTS capitalized on the core network protocols developed for GSM and introduced a new Radio Access Network (RAN) based on Wideband CDMA technology with 5 MHz channel bandwidth (the *wideband* term was adopted to differentiate UMTS technology from the CDMA technology used in the 2G CDMAOne/IS-95 standard, based on 1.25 MHz channels). Initial UMTS systems were able to deliver data rates of up to 384 kbps in outdoor scenarios. UMTS technology has progressively evolved since then, allowing nowadays to deliver data rates in the order of tens of Mbps by means of the so-called High Speed Packet Access (HSPA) capabilities.

CDMA2000 was a 3G standard promoted by 3GPP2, another consortium also established in 1998 under the purpose of evolving the 2G CDMAOne networks. Unlike UMTS for which a new RAN was specified from scratch, the radio part of the CDMA2000 standard was designed as an evolution of CDMAOne, allowing it to operate within the same frequency allocations and thus facilitating the migration to 3G. CDMA2000 specifications were also enhanced in subsequent releases under the commercial name of CDMA2000/1xEV-DO. However, 3GPP2's work in the CDMA2000/1xEV-DO standard was discontinued by 2013 as the whole mobile industry progressively converged around the Long-Term Evolution (LTE) standard developed by 3GPP for 4G systems.

Although technical standards have been developed within organizations such as 3GPP and 3GPP2, it is worth mentioning at this point the role taken by ITU-R in the whole process. In the 1990s, ITU-R launched the so-called International Mobile Telecommunications 2000 (IMT-2000) program, seeking to establish a global 3G standard recognized at ITU level. The idea was not to develop the technical specifications within ITU-R itself but define the requirements to be met by IMT-2000 systems, so that candidate specifications could be developed and submitted to ITU by national or regional SDOs (e.g. ETSI, ARIB, etc.). Indeed, 3GPP and 3GPP2 consortiums were created in response to the launch of the IMT-2000 program, and both UMTS/HSPA and CDMA2000/1xEV-DO radio interface technologies were submitted to ITU-R and recognized as IMT-2000 systems [4]. Following the approach started with the IMT-2000 program for 3G standardization, ITU-R has carried out similar programs for 4G and 5G, denoted as IMT-Advanced and IMT-2020, respectively.

7.2.1.3 4G. The First Global Standard for Mobile Broadband

The true mobile broadband revolution arrived with the 4G, which also popularized the smartphones and their rich application ecosystem.

In terms of standards, while initially there were some competing technologies (e.g. Mobile WiMAX, based on IEEE standards, and Ultra Mobile Broadband, as the evolutionary path of CDMA2000 systems set out by 3GPP2), the worldwide mobile industry aligned itself for the first time under a single global technology for mobile communication, known as LTE. LTE specifications are developed by 3GPP.

LTE uses OFDMA and relies on a shared radio resource allocation scheme that allows maximizing resource usage by dynamically scheduling, at millisecond timescales, transmissions from different terminals. LTE allows for reasonable receiver complexity also in combination with multiple antenna transmission and reception schemes. In contrast to previous generations, LTE shows a high level of spectrum flexibility, with selectable channel bandwidths from 1.4 MHz up to 20 MHz and support for FDD and TDD modes. Moreover, the whole LTE system is designed for packet data connectivity so that, all services, including voice services, are delivered over a packet-switched core.

The first release of the LTE specifications (3GPP Release 8) was completed in March 2009, and commercial operation of 4G networks began in late 2009. Subsequent LTE releases have followed, introducing additional functionality and capabilities. LTE Release 10, referred to as LTE-Advanced, was the one submitted and recognized as an IMT-Advanced standard at ITU level [5].

It is worth noting that the deployment of 4G networks during the 2010 decade did not result in the phasing out of 2G and 3G networks, which are still in operation today along with 4G. 2G and 3G technologies have remained central in all these years for the delivery of voice services as well as basic connectivity services (e.g. connections with low-cost GSM/GPRS modems used in diverse machine-to-machine applications, such as vending machines, SCADA, and smart metering systems, etc.). Indeed, utilities have the expectation that 2G services will be available up to 2030 to continue supporting the relatively high installed-base of metering devices deployed, e.g. in Europe, throughout their lifetime [6]. At the same time, announcements by operators about their plans on 2G and 3G network shutdowns are starting to be common [7].

7.2.1.4 5G. Expanding the Applicability Domain of Cellular Technologies

The 2020 decade is poised to be the timeframe for the adoption and consolidation of 5G systems. Among other enhancements, 5G systems come with a new radio access technology called New Radio (NR) that allows operation at, for the first time in cellular systems, millimeter wave (mm-wave) bands, in addition to the lower frequency bands used in 2G/3G/4G systems. Channel bandwidths of up to 400 MHz can be configured in mm-wave bands to achieve multi-Gbps speeds. NR is based on OFDMA and reuses many of the structures and features of LTE, whose evolution is also part of the roadmap of the 5G systems.

5G systems are not just intended to be a faster version of 4G/LTE systems. Remarkably, 5G technology development has also been centrally driven by the aim of extending the reach of cellular technologies to new application domains. Indeed, 5G often boasts for its ability to meet the development needs of so-called *verticals*, reflecting different sectors of the economy (e.g. energy, automotive, transportation, media and entertainment, smart cities, agriculture, manufacturing, e-health, etc.) that can leverage mobile telecommunications to improve their processes.

The first version of the 5G NR specifications was available by the end of 2017 (an early drop of 3GPP Release 15) to meet commercial requirements on early 5G deployments, with first commercial 5G networks going live in 2019. At the time of writing, the latest version of approved 5G specifications is Release 16, which has been already recognized as an IMT-2020 standard in November

2020 [8]. Networks and devices compliant to Release 16 are expected to hit the market in the 2022–2023 timeframe.

Evolution of cellular technologies does not stop here. In parallel to the continued evolution of the 5G specifications, several initiatives are already underway to start shaping the vision and key technologies of what could form the basis of the next generation of mobile technologies, the 6G, for 2030 onward [9].

7.2.2 Supported Services and Applicability Scenarios

7.2.2.1 Service Categories

The services and features delivered by the latest cellular technologies are commonly grouped as [10]:

- *Enhanced Mobile Broadband (eMBB) services*, which are the logical continuation and improvement of the mobile broadband services delivered in today's mobile commercial networks.
- *Massive Machine-Type Communication (mMTC) services*, evolving from classic M2M services; they enable the connectivity of low-complexity, low-cost devices, with specific features that enhance the link budget (and so the coverage in hard-to-reach locations) as well as the energy consumption requirements of these devices. This category is also often referred to as *massive Internet of Things* (IoT) and *cellular IoT* (CIoT) or placed under the umbrella of so-called *Low-Power Wide-Area* (LPWA) technologies.
- *Ultra-Reliable Low Latency Communications (URLLC) services*, able to deliver latencies in the millisecond scale or below, along with the necessary reliability mechanisms. This category is also commonly referred to as *Critical MTC*, *Critical IoT,* or *Industrial IoT* (IIoT).

7.2.2.2 Performance Indicators

Table 7.1 provides illustrative performance indicators for each service category based on the minimum technical performance requirements defined in the IMT-2020 program [11].

Further to the IMT-2020 technical performance requirements, a large number of use cases for the applicability of 5G technologies have been analyzed in 3GPP to define more detailed performance requirements needed to guide the development of 5G specifications. Use cases span from more traditional use cases involving the delivery of commercial mobile broadband services to very specific vertical applications in domains such as factories of the future, electric power distribution, medical facilities, vehicular communications, and railways. All these requirements are documented in 3GPP technical specifications (e.g. [12–15]), and an evaluation of the achievable performance against the IMT-2020 technical performance requirements for eMBB, mMTC, and URLLC usage scenarios is reported in [16].

7.2.2.3 Commercial Networks and Private Networks

The main applicability area of 2G, 3G, and 4G technologies has been predominantly in the commercial domain, with Mobile Network Operators (MNOs) deploying large country-wide commercial networks that offer voice and data communication services to the general public (commonly known as *public networks*). More recently, starting with 4G/LTE [17] and further incentivized by the new capabilities introduced in 5G, the deployment of private networks based on 3GPP cellular technologies is ramping up and expects to reach business volumes comparable to those of the commercial domain [18].

Table 7.1 Minimum technical performance set for IMT-2020/5G systems.

Metric	Service category	Requirement	Comments
Peak data rate	eMBB	Downlink: 20 Gbps Uplink: 10 Gbps	Defined for a single mobile terminal in ideal channel conditions
Peak spectral efficiency	eMBB	Downlink: 30 bit/s/Hz Uplink: 15 bit/s/Hz	Assumes the use of 8 and 4 spatial streams in downlink and uplink, respectively
User experienced data rate	eMBB	Downlink: 100 Mbps Uplink: 50 Mbps	Defined as the 5th percentile of the Cumulative Distribution Function (CDF) of the user throughput values during active time
User spectral efficiency	eMBB	Downlink: Up to 0.3 bit/s/Hz Uplink: Up to 0.21 bit/s/Hz	Defined as the 5th percentile of the CDF of the user throughput normalized by the channel bandwidth
Average spectral efficiency	eMBB	Downlink: Up to 9 bit/s/Hz/TRxP Uplink: Up to 6.75 bit/s/Hz/TRxP	Defined as the aggregate throughput of all users divided by the channel bandwidth and by the number of Transmission/Reception points (TRxP)
Area traffic capacity	eMBB	10 Mbps/m^2 (equivalent to 10 Tbps/km^2)	Defined for indoor hotspots
User plane latency	eMBB and URLLC	eMBB: 4 ms URLLC: 1 ms	Defined as the one-way latency over the radio access network for a terminal in active state. Assumes unloaded conditions and small IP packets with no payload
Control plane latency	eMBB and URLLC	20 ms (encouraged to consider 10 ms)	Refers to the transition time from a battery efficient state (e.g. idle state) to an active state
Connection density	mMTC	1 million devices per km^2	Assumes applications that generate low traffic volumes per device
Reliability	URLLC	10^{-5} packet error rate, equivalent to five nines reliability (99.999%)	Reliability relates to the capability of successfully transmitting a given amount of traffic within a predetermined time duration. Defined for a packet length of 32 bytes and 1 ms transmission time
Mobility	eMBB	Up to 500 km/h with normalized channel spectral efficiencies above 0.45 bit/s/Hz	500 km/h is mainly envisioned for high speed trains
Mobility interruption time	eMBB and URLLC	0 ms	Defined as the shortest time duration supported by the system during which a user terminal cannot exchange user plane packets with any base station during transitions between cells
Network energy efficiency	eMBB and mMTC	Up to 100x improvement over IMT-Advanced	No numerical values established. Covers both network energy and device energy efficiency
Bandwidth	eMBB	At least 100 MHz and up to 1 GHz in frequency bands above 6 GHz	Not necessarily continuous spectrum. It can be met through carrier aggregation features

A wide area of enterprises and organizations in sectors, such as energy, manufacturing, healthcare, shipping, and mining, may prefer a private network over a public network for several reasons [19]:

- Business orientation of MNOs, which may not be willing or able to provide the likely stricter SLAs needed to support the business or mission-critical communication needs of these organizations.
- Specific coverage requirements, which may require the deployment of dedicated radio equipment within the enterprises' premises (e.g. campus network, factory network) or in remote areas not properly served by public networks.
- Specific service requirements, for instance, in terms of supported features, bit rates, latency, and reliability levels necessary to meet the performance profiles of some applications (e.g. motion control in factory automation).
- High security requirements, using dedicated security credentials and the ability to retain sensitive operational data on-premises.
- Isolation from other networks, as a form of protection against malfunctions in a public mobile network.
- Accountability, making it easier to identify responsibility for availability, maintenance, and operation.

In contrast to previous generations, 5G standards come with features purposely designed for private networking.

In any case, private networking does not necessarily mean standalone, dedicated deployments totally decoupled from public networks. Instead, private networking can also be realized partly or totally through network slicing features on top of public mobile networks. More details on the features and implementation options for private networks are given in Section 7.4.7.

7.2.3 Spectrum

1G and 2G networks were mainly deployed in 800 and 900 MHz frequency bands. With 3G, the focus moved to the 2 GHz band and, progressively, several new bands have been made available for 4G at both lower and higher frequencies, spanning from below 450 MHz to around 6 GHz.

Importantly, the frequency bands assigned for mobile communication systems are typically technology neutral. This means that bands introduced or used for previous generations can be re-used for the new generations as well. In addition, 5G opens the door for the first time to the exploitation of spectrum bands above 24 GHz, where a higher amount of spectrum is available to achieve the envisioned multi-Gbps peak data rates and aggregated capacity at Tbps/km^2 scales.

7.2.3.1 Spectrum Harmonization. IMT Bands

The adoption of worldwide harmonized spectrum bands for the deployment of mobile communications systems is important to achieve economies of scale in the equipment as well as to facilitate interoperability and global roaming between networks.

International spectrum coordination is carried out in the form of an international treaty known as ITU Radio Regulations (ITU RR). ITU RR regulations, as covered in Chapter 2, establish the allocation of specific frequency bands for different radio services, including Mobile Services (MS). Since the launching of the IMT program in ITU-R, some of the spectrum bands attributed to MS are now explicitly recognized as IMT bands, which means that ITU members (national administrations) recognize these bands as *preferential bands* for the deployment of mobile communication

Table 7.2 IMT bands.

Frequency bands identified for IMT (MHz)	Bandwidth (MHz)	Regions
450–470	20	Global
470–698	228	R2, R3
694/698–960	262	R1, R2, R3
1427–1518	91	R1, R2, R3
1710–2025	315	Global
2110–2200	90	Global
2300–2400	100	Global
2500–2690	190	Global
3300–3400	100	R1, R2, R3
3400–3600	200	R1, R2, R3
3600–3700	100	R2
4800–4990	190	R2, R3
24 250–27 500	3250	Global
37 000–43 500	6500	Global
45 500–47 000	1500	Global
47 200–48 200	1000	Global
66 000–71 000	5000	Global

Sources: Adapted from Refs. [20, 21].

systems based on IMT standards (IMT-2000, IMT-Advanced, and IMT-2020). IMT bands (see Table 7.2) are defined with global or regional scope.

Most of the IMT bands below 6 GHz are currently used for 2G, 3G, and 4G systems. This represents a quite fragmented spectrum that, aggregated, totals <2 GHz of spectrum. WRC-19 introduced five additional frequency bands, all above 6 GHz, which raised the total amount of IMT spectrum to 17.25 GHz. This new spectrum allows for much larger contiguous spectrum allocations per operator and thus the use of wider channel bandwidths (up to hundreds of MHz). WRC-19 also identified specific bands to be studied as new IMT bands in next WRC-23 [21]:

- 3600–3800 MHz and 3300–3400 MHz (Region 2).
- 3300–3400 MHz (Region 1).
- 7025–7125 MHz (globally).
- 6425–7025 MHz (Region 1).
- 10.0–10.5 GHz (Region 2).

7.2.3.2 Frequency Bands Being Prioritized for 5G

The deployment of mobile systems typically relies on the use of a combination of spectrum in different frequency bands, commonly categorized as:

- *Low-band spectrum*, covering frequencies below 1 GHz. This spectrum shows the best propagation conditions and is well suited to deliver wide-area coverage across urban, suburban, and rural areas and help to support massive IoT services. However, the amount of low-band spectrum is rather limited, which restricts the capacity and data rates that can be achieved.

Table 7.3 Main frequency bands under consideration for 5G deployment.

Country/region	Low-band spectrum (MHz)	Mid-band spectrum (MHz)	High-band spectrum (MHz)
EU	700	3400–3800	24 250–27 500
US	600	2500	24 250–24 450
		(CBRS) 3550–3700	24 750–25 250
		3700–4200	25 250–27 250
		3450–3550	26 500–29 500
			37 600–38 600
			38 600–40 000
			47 200-48 200
			42 000–42 500
			31 800–33 000
South Korea	700	3420–3700	26 500–28 900
China	700	2600	24 750–27 500
		3300–3400	37 000–42 500
		3400–3600	
		3600–4200	
		4400–4500	
		4800–5000	
Japan	No band below 1 GHz	3600–4200	27 500–29 500
		4400–4900	

- *Mid-band spectrum*, covering frequencies from 1 to 6 GHz. This spectrum offers a good mixture of coverage and capacity benefits.
- *High-band spectrum,* covering mm-wave frequencies. This spectrum shows the most challenging propagation conditions, requiring practically LOS operation. In contrast, high-band spectrum can provide the necessary amount of spectrum needed to achieve multi-Gbps speeds and high traffic capacity in localized service areas.

Following this categorization, Table 7.3 summarizes the main frequency bands under consideration for 5G deployments in the European Union, the United States, South Korea, China, and Japan [22].

7.2.3.3 Spectrum Exploitation Models

The traditional and mainstream spectrum exploitation model followed in the cellular industry is based on the granting, via spectrum auctions, of licenses for exclusive spectrum rights of use over, typically, large geographical areas (e.g. country-wide). These licenses are mostly acquired by MNOs, which pay significant amounts of money.

In addition to the traditional model, other approaches are progressively gaining more relevance, as discussed in the following.

7.2.3.3.1 Allocation of Dedicated Spectrum for Vertical Industries

There are some verticals whose needs may not be adequately addressed by public networks and other spectrum solutions should be considered [23]. As discussed in Section 7.2.2.3, some verticals may wish to deploy their own private network for a variety of reasons, including cost, control, data security, and flexibility to upgrade/change technologies. In this case, access to dedicated spectrum for private networking may be realized through trading or leasing of spectrum owned by MNOs. However, this puts vertical users under the dependence of MNOs' legitimate interests, which sometimes may be conflicting with those of the vertical users.

Alternatively, specific authorization regimes can be established by regulators for some vertical sectors in the form of, e.g. individual licenses granted by the regulator to the vertical itself. Two situations can be distinguished with different impacts on this specific authorization regime:

- *Dedicated spectrum for vertical users with wide-area connectivity requirements.* This could be the case of verticals that are network infrastructure dependent over a large area. Examples are utilities, Public Protection and Disaster Relief (PPDR) organizations, and railways operators.
- *Dedicated spectrum for vertical users with local-area connectivity requirements.* This could be the case of a wider range of sectors that may require mainly on-site, localized connectivity, and in most occasions, indoor deployments. Examples are companies in sectors such as manufacturing, logistics, mining, and health.

Localized use of spectrum facilitates the direct allocation to industries, given that the same spectrum can be re-used at different locations and the spectrum can be allocated in mid or high bands, where higher spectrum amounts are available. In Europe, some countries like the Netherlands and Germany have allocated 100 MHz of spectrum for industrial use in the 3.4–3.8 GHz band [24], which is among the pioneering bands for 5G deployment.

7.2.3.3.2 Use of Shared Spectrum

Shared spectrum regimes are established when, for whatever reasons, different services and users are authorized to operate in a common frequency band under some level of coordination which can guarantee interference protection to the services/users with a high priority.

One clear exponent of a shared spectrum regime is the Citizens Broadband Radio Service (CBRS) developed in the United States for the exploitation of 150 MHz in the 3.5 GHz band [25]. CBRS provides a three-tiered priority spectrum access scheme where the highest priority is reserved for incumbent users (e.g. federal uses such as Department of Defence radar systems, which are the legacy users of the band). Incumbent users receive protection against harmful interference from any other authorized use in the band via an automated frequency coordination mechanism, known as a Spectrum Access System (SAS), which keeps track of the location and channels used by the incumbent users and coordinates the operation of the other users so that incumbents do not get interfered (i.e. users other than incumbents need to check channels availability from the SAS before using them).

In addition to the incumbent access, CBRS establishes two other access categories:

- *Tier 2 – Priority Access.* This access tier consists of Priority Access Licenses (PALs) which are 10-year renewable licenses granted on a county-by-county basis through competitive bidding. Each PAL authorizes for a 10 MHz channel within the 3550–3650 MHz band, and up to seven PALs may be licensed in any given county. PALs must protect and accept interference from incumbent users but receive protection from GAA users.
- *Tier 3 – General Authorized Access (GAA).* This tier is licensed-by-rule to permit open, flexible access to the band for the widest possible group of potential users. GAA users can operate throughout the 3550–3700 MHz band. GAA users must not cause harmful interference to incumbent users or PALs and must accept interference from these users. GAA users also have no expectation of interference protection from other GAA users.

Auctions to acquire PALs have been celebrated in the United States in 2020, with several utilities winning spectrum rights in this band for private networking [26]. Similar models have been defined at European level (e.g. Licensed Shared Access [LSA], for the 2.3–2.4 GHz band [27]), though unlike CBRS, no wide-scale commercial implementation has followed so far.

7.2.3.3.3 Unlicensed Spectrum Operation

Both LTE and NR interfaces support the operation over unlicensed spectrum (also referred to as license-exempt spectrum) such as the 5 GHz band currently used in Wi-Fi networks. In contrast to shared spectrum regimes, no explicit authorization or coordination is necessary for operation in unlicensed bands, where radio equipment operating is mainly mandated to obey some rules such as limiting the maximum transmit power and duty cycles and requiring the use of Listen-Before-Talk (LBT) mechanisms to guarantee a fair co-existence among all the possible users.

Support of unlicensed operation by cellular technologies brings benefits to both vertical users and MNOs. The former may deploy private networks in unlicensed spectrum. The latter can get access to a significant amount of spectrum at no cost and use it in combination with licensed spectrum (via Carrier Aggregation or Dual Connectivity) to offer higher data rates via parallel data transmission.

7.2.4 3GPP Standardization

3GPP is the organization that brings together the global wireless industry for the development of cellular technology standards. 3GPP members include organizational partners, market representation partners, and individual members. Organizational partners are SDOs from different regions and countries that jointly determine the general policy and strategy of the 3GPP. 3GPP currently unites SDOs from China, Europe, India, Japan, Korea, and the United States. Market representation partners are organizations that provide advice to 3GPP on market requirements (e.g. services, features, functionality). And, individual members are the companies and organizations that actually carry out the technical contributions and develop the 3GPP specifications. As of today, more than 700 companies across the global wireless industry, research institutions, and academia are contributing to 3GPP work in a collaborative manner.

3GPP specifications provide a complete description of a mobile communication system, including radio access, core network, and service capabilities. The 3GPP specifications are published as 3GPP Technical Specifications, which are the normative documents that define the characteristics of the standards, and Technical Reports, which contain the feasibility studies conducted to support the normative work. All these documents are publicly available in the 3GPP website [28]. Work in 3GPP includes from 2G to 5G specifications (while 3GPP was established for 3G, later on it also took over from ETSI the responsibility to maintain 2G/GSM specifications). Specifications are constantly improved over time and delivered in the form of sequential packages called *releases*, which are produced every 1.5 or 2 years. In each release, updates may be provided for each of the different technologies, though most efforts are usually placed on the enhancement and expansion of the newest ones.

From a marketing perspective, Figure 7.2 shows the naming and logos adopted to refer to the different releases delivered since 3GPP Release 8, which was the first to include the 4G LTE specifications. The basic LTE logo is used to refer to the technology specified up to Release 9. The LTE-Advanced logo covers the extensions and enhancements from Release 10 up to Release 12. The LTE-Advanced Professional logo was introduced for Releases 13 and 14 to account for the multiple extensions added to address markets other than the consumer market such as public safety

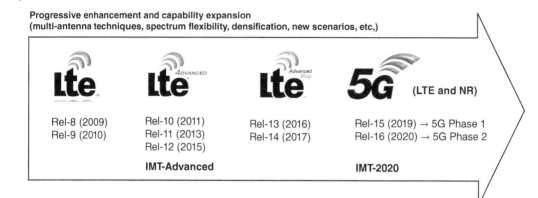

Figure 7.2 3GPP standards and releases.

communications and automotive. And, the newest logo for 5G has come with Release 15, which is actually being used for both NR and evolved LTE air interfaces.

3GPP Release 15 was finalized in June 2019, with a focus on eMBB services. Release 15 provided the first complete baseline specification for the NR air interface, a Next-Generation RAN (NG-RAN) and a new service-based 5G Core (5GC), including support for features such as network slicing and edge computing. On this basis, multiple new features and enhancements are being addressed in subsequent releases [29–31]. This is illustrated in Figure 7.3, grouping the new

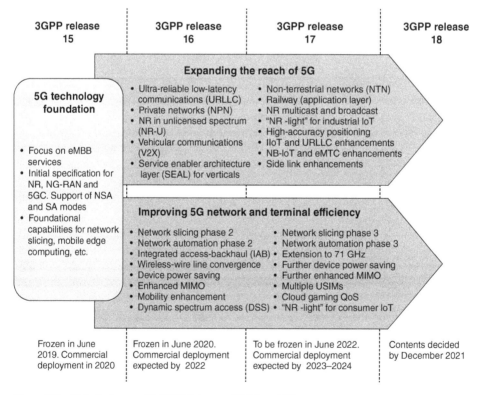

Figure 7.3 Main features with 3GPP Releases 15, 16, and 17.

features in two categories: those seeking to extend the applicability reach of 5G and those mainly intended to enhance the operation of 5G networks and terminals.

7.3 System Architecture

7.3.1 High-level Architecture of 4G/5G Systems

Cellular networks have evolved from voice-centric, circuit-switched networks with embedded service-level capabilities (e.g. built-in support for voice services), as it was the case for 1G/2G/3G systems, toward general-purpose, packet-switched networks for the delivery of *QoS-aware packet-based connectivity services*, that is, communication services for the transfer of packetized data streams (e.g. IP packets, Ethernet frames) with specific QoS constraints (e.g. guaranteed bit rate, latency, reliability), as it is the case for 4G/5G systems. Such connectivity services extend from mobile terminals on one side to external data networks interconnected to the mobile network, on the other. On top of the packet-based connectivity services, multiple applications and service layers with diverse QoS needs can be deployed.

Figure 7.4 illustrates the high-level architecture of a 4G/5G mobile network, which includes the following components:

- *User Equipment (UE):* This refers to any mobile terminal (e.g. smartphone, modem, IoT device) that gets connected to the mobile network via the radio interface (e.g. LTE and/or NR). The UE includes a Subscriber Identity Module (SIM), known as the SIM card, to securely store the subscriber data needed for network access (e.g. identities, security keys).
- *RAN:* Part of the mobile network that handles all aspects associated with radio transmission. The RAN consists of a large, geographically distributed assembly of *base stations*. A country-wide network may easily encompass thousands of base stations.
- *Core Network (CN):* Part of the mobile network in charge of the delivery of the connectivity services to UEs, including subscriber authentication, network access control, tracking of UE locations, session management, QoS configuration, security management, and traffic routing and forwarding toward/from the external data networks. Indeed, the CN constitutes the brain of

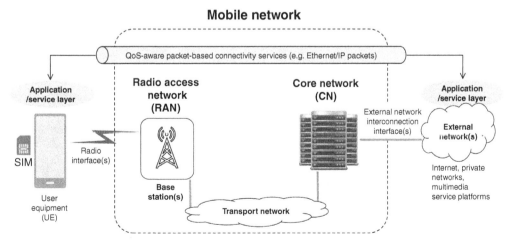

Figure 7.4 Illustrative architecture of a 4G/5G mobile communication system.

the mobile network for all functions except radio transmission. From an infrastructure point of view, the CN mainly consists of network functionality running in a few data centers.

- *Transport Network (TN):* This refers to the transport network layer (see Chapter 4) used to interconnect the locations where the RAN and CN functions are hosted (i.e. base station sites and data centers). From a functional perspective, the TN layer is decoupled from the RAN/CN layers. The part of the TN connecting the base stations is commonly referred to as the *backhaul network*.
- *External network(s)*: This refers to any external network made reachable to the UEs through the mobile network. An external network can be a public network, such as the Internet, or a private network, such as a corporate network or service platform for the delivery of e.g. voice services.

In terms of the overall quality of cellular service and associated subscriber experience, the RAN is the key piece of the mobile network infrastructure [32]. On the one hand, the RAN infrastructure is the ultimate edge of the network and, as such, it defines the achieved coverage and signal levels across the service area. On the other hand, the RAN infrastructure is directly responsible for efficient utilization of radio spectrum, which is an expensive and limited resource. All these aspects make the RAN to be the largest contributor to MNO's network CAPEX and OPEX, and the most dynamic and innovation-intensive network domain.

RAN base stations come in very different form factors (see Figure 7.5). One example is outdoor *macro cells* base stations, with radio transmitters of tens of Watts, antennas located at elevated points over the ground (e.g. 20–30 m), and achieving coverages of up to a few kilometers. Another example is outdoor *small cells*, transmitting only a few Watts and providing localized capacity for traffic hot spots with ranges limited to a few hundreds of meters. Moreover, specialized RAN deployments (e.g. distributed antenna systems, indoor small cells) are also commonly used to improve quality and capacity inside venues.

In addition to the RAN sites, a mobile network infrastructure includes a number of locations, commonly referred to as central offices, or more recently, data centers, where the CN functionality is deployed. Nowadays, a large part of the CN functionality is implemented as software applications running on general-purpose hardware (e.g. standard servers) instead of dedicated hardware as used in the past. This implementation approach is known as Network Function Virtualization (NFV) [33], which reduces the amount of hardware needed (several functions can run on the same server) and improves scalability and agility by allowing the software applications, referred to as Virtual Network Functions, to be instantiated, scaled, and moved across different servers as needed. NFV technology, after being widely adopted in the CN domain, it is also moving into the RAN domain, especially to implement the radio processing functions above the physical layer [32].

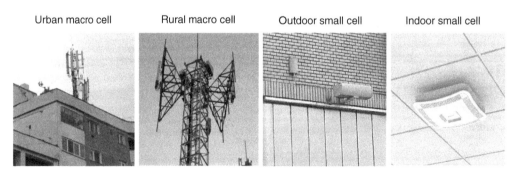

Figure 7.5 Illustration of different types of RAN base stations.

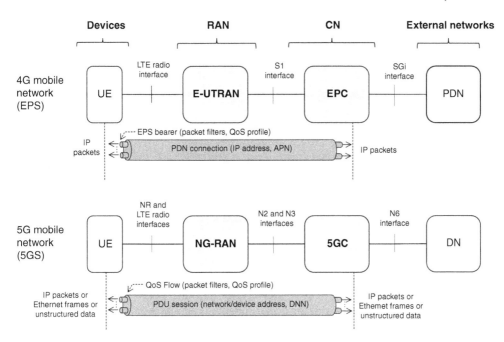

Figure 7.6 High-level functional view of 4G and 5G mobile networks.

7.3.1.1.1 Functional Architecture and End-to-end Connectivity Model

Figure 7.6 provides a functional representation of the architecture of 4G/5G networks, detailing the specific terminology used in each case to name the whole system, RAN, CN, and external networks:

- In 4G, the overall system is the Evolved Packet System (EPS), the RAN is the Evolved UMTS Terrestrial RAN (E-UTRAN), the CN is the Evolved Packet Core (EPC), and external networks are referred to as Packet Data Networks (PDNs).
- In 5G, the overall system is the 5G System (5GS), the RAN is the NG-RAN, the CN is the 5GC, and external networks are referred to as Data Networks (DNs).

Figure 7.6 also illustrates the concepts of *PDN connection* and *Protocol Data Unit (PDU) session*, which are the names used in 4G and 5G networks, respectively, to call the QoS-aware packet-based connectivity service between the UE and the external network.

In 4G, a PDN connection supports the exchange of IP packets (IPv4 and/or IPv6). It is associated with an Access Point Name (APN), which identifies the external PDN to which the UE is connected (e.g. Internet), and with an IP address, allocated to the UE for traffic exchange. Within the same PDN connection, one or more *EPS bearers* can be established for QoS differentiation (e.g. IP packets carrying voice frames prioritized over packets carrying web browsing traffic). An EPS bearer is associated with a set of packet filters for traffic classification (e.g. filters using specific IP destination/source addresses and/or TCP/UDP ports to determine which packets belong to each EPS bearer), and a QoS profile, which defines the expected traffic forwarding treatment for matching packets (e.g. priority, guaranteed rate, etc. – more details on the supported QoS profiles are given in Section 7.4.4).

In 5G, a PDU session, in addition to IP traffic, also allows for the exchange of Ethernet frames or unstructured data (note, however, that a given PDU session is only intended to handle one type of traffic). The external network is now identified by a DN name (DNN), which is equivalent to the

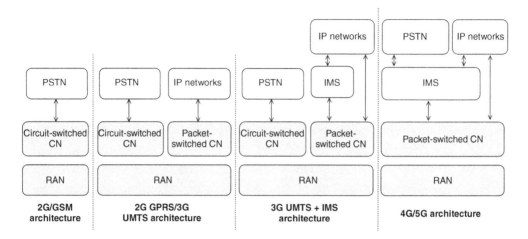

Figure 7.7 Evolution from circuit-switched-only to packet-switched-only architectures.

APN in 4G. To handle QoS, one or several *QoS Flows* can be activated within a PDU session. A QoS Flow is defined as the finest granularity for QoS differentiation, which is equivalent to the EPS Bearer concept in LTE.

7.3.1.1.2 Main Differences with Previous Generations Architectures

Figure 7.7 contrasts the functional architecture of 4G/5G systems with the architectures defined in previous 2G/3G systems, showing the progressive transition from circuit-switched only networks (e.g. GSM) mainly interconnected to the Public Switched Telephone Network (PSTN) to networks combining both circuit-switched and packet-switched cores (GSM/GPRS and UMTS) interconnected to both PSTN and IP networks, and finally to packet-switched-only networks (4G and 5G systems). In this process, 3GPP specified the IP Multimedia Subsystem (IMS), which is a standardized framework for the delivery of services previously supported as an integral part of the circuit-switched cores (e.g. voice services, video telephony) over the packet-switching cores.

7.3.2 Radio Access Network

7.3.2.1 E-UTRAN

7.3.2.1.1 Functional Architecture

The E-UTRAN (see Figure 7.8) consists of a collection of base stations named *evolved NodeBs* (eNBs). An eNB represents actually a logical node, not tied to any particular physical implementation (e.g. an eNB can be implemented as a macro cell site with multiple antennas or as a single small cell appliance for indoor coverage). An eNB interacts with UEs by means of the LTE-based radio interface, known as the Uu interface. Each eNB is connected to the EPC by means of the S1 interface, deployed over the backhaul network. Optionally, an eNB can also directly interact with other eNBs in their vicinity by means of the X2 interface, which is also deployed over the backhaul network.

7.3.2.1.2 LTE Cells and Main Operation Principles

A *cell* is the basic resource unit used by the base station (eNB in this case) to organize and support the transfer of information toward/from UEs. An LTE cell is implemented using a pair of RF carriers in case of FDD operation and using a single RF carrier in case of TDD operation. Through an

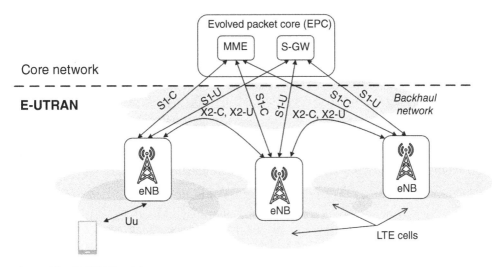

Figure 7.8 E-UTRAN architecture.

LTE cell, both control information (e.g. synchronization signals, system information, resource allocation information) and data traffic (e.g. IP packets) are exchanged. An eNB can handle one or several cells in the same or multiple frequency bands.

A cell is actually what a UE detects when looking for a cellular network (more specifically, the control information broadcast in the downlink RF carrier of the cell). Indeed, at a given location, a UE is likely to detect, and uniquely identify (using a "cell identifier" broadcast as part of the cell control information), multiple cells, which may be received with different signal power levels depending on the distance and the characteristics of the radio propagation between the UE and the eNBs.

An LTE cell is arranged (e.g. operating frequency band, transmit power, antennas) to be detected and used over a certain geographical area (i.e. the coverage area of the cell). Two examples of cell arrangements for a given eNB are illustrated in Figure 7.9. The example on the left shows an eNB

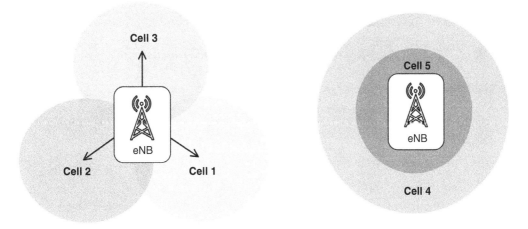

Figure 7.9 Examples of different cell configurations supported in an eNB.

handling three cells, each one radiated over a different area (i.e. sector) by using directional antennas. This is called a sectorized configuration, which is among the most common ones used in macro cell sites. In contrast, the example on the right shows the case of a non-sectorized eNB, using a single omnidirectional antenna but handling two cells with different coverage footprint. For instance, this could be the case of an indoor *small cell base station* with *Cell 4* operating in low-band spectrum, with better propagation characteristics and thus achieving a better coverage, and *Cell 5* using medium-band spectrum and offering a more limited range. More details on the time/frequency resources arrangement in LTE cells are covered later in Section 7.4.1.

Before delving into more details about the eNB functions and protocol stacks, some general principles and terminology related to how a mobile network operates need to be mentioned:

- As a baseline, a UE is typically connected to the RAN through a single cell, referred to as the serving cell (though, multi-cell connectivity is also supported in LTE).
- When a UE is switched on, it searches for cells (e.g. RF channel scanning, detection of cell synchronization signals), selects one (i.e. *cell selection procedure*), establishes a control connection with the eNB handling the cell (i.e. *Radio Resource Control, RRC, connection*), and registers for network services (i.e. *network registration procedure*).
- Once registered, connectivity services for traffic exchange can be established (e.g. activation of PDU connections and EPS bearers). This is called *session management*.
- Being registered, if traffic or signaling activity stops and there is no further information to be exchanged, the connectivity services are put on hold and the RRC connection is deactivated. This UE state is known as *idle mode* (i.e. no active RRC connection), in which the UE remains registered in the network but only listening to the information broadcasted through cell control channels (see Figure 7.10). In contrast, the UE state is said to be in *connected mode* whenever the RRC connection is active.
- In idle mode, a UE periodically monitors the signal level and quality of the serving cell as well as of other detected neighboring cells. If a neighboring cell gets better than the serving one, the UE may decide a cell change and start listening to a new cell (i.e. *cell re-selection procedure*).

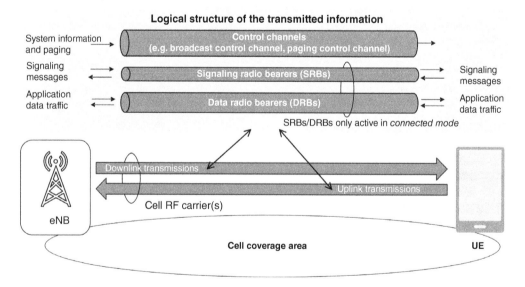

Figure 7.10 Types of information exchanged over LTE cells.

The selection of the new cell is based on measurements performed by the UE along with cell-selection configuration information sent through the cell control channels to guide the decisions made by UEs.

- The network keeps track of the location of UEs in idle mode. For this purpose, the cells are logically grouped in so-called Tracking Areas (TAs), with each cell belonging to a single TA. The TA identity to which a cell belongs is communicated as part of the system information in the cell control channels. A UE in idle mode could move between cells of the same TA without having to notify the network about cell changes. However, when a cell change leads to TA change, the UE shall inform the network about its new location (i.e. *TA update procedure*). In this way, the network always knows in which TA (i.e. in which group of cells) the terminal can be reached. This is called *mobility management* (and sometimes also referred to as *location management*).
- When a UE in idle mode needs to start sending either signaling messages (e.g. perform a TA update procedure) or data traffic (e.g. IP packets) or both, the RRC connection is first re-activated (switching the terminal state from idle to connected mode) and connectivity services are resumed. Over the radio interface, this results into the activation of the necessary information transfer services between the UE and eNB. Such transfer services are called *Signaling Radio Bearers* (SRBs), for signaling transfer, and *Data Radio Bearers* (DRBs), for application data transfer, as illustrated in Figure 7.10. Further details on how SRBs, DRBs, and cell control channels are realized and mapped onto the RF carrier resources are given in Section 7.4.1.8.
- When a UE is in connected mode, the decision to change the serving cell is always made by the network, in contrast to the cell (re-)selection in idle mode which is decided by the UE. The change of cell in connected mode may obey to different purposes (e.g. serving cell quality is deteriorating, cell becomes congested). In this respect, decision-making is assisted with measurements reported by the UE about the quality of the serving and other monitored cells. Remarkably, when the cell change is executed, this is done in a way that any on-going SRBs/DRBs are released from the old cell and automatically re-activated in the new cell, with no or minimal service interruption. This procedure is known as *handover* (or *hand-off* in some contexts).
- Last but not least, it is worth noting how transition from idle to connected mode takes place when the trigger is on the network side (e.g. new traffic addressed to the UE reaches the network when the terminal is in idle mode). In this case, the network shall first send a *paging message* to the UE through all the cells that belong to the TA where the UE is known to be located. Paging messages are sent through the cell control channels. After receiving a paging message in one of the cells of the TA, the UE proceeds to re-activate the RRC connection in that cell.

7.3.2.1.3 eNB Functions

The eNB is in charge of managing how radio resources are properly assigned and used by UEs in each of its handled LTE cells. This is known as Radio Resource Management (RRM), which encompasses functions such as:

- *Radio bearer control*. This relates to the setup and release of SRBs and DRBs. It includes admission control and congestion control (e.g. a new DRB can be rejected or an on-going DRB forced to terminate if the cell gets congested).
- *Dynamic radio resource management and data packet scheduling*. This is probably one of the most critical functions in the eNB. In LTE, transmissions to/from the multiple UEs are jointly multiplexed over a common pool of radio resources (referred to as *shared channel*) by using a *scheduler* function that decides which UE is served and on which time and frequency resources, at 1 ms timescale.

- *Mobility Control*. This function seeks to ensure that UEs are always connected through the best cells. This includes handover decision-making and adjustment of cell re-selection criteria. To assist mobility control decisions on the network side, UEs can be configured to report channel state measurements periodically or in an event-triggered manner.
- *Inter-Cell Interference Coordination* (ICIC). This function allows for interference mitigation between transmissions from different cells, facilitating frequency reuse (i.e. the same RF carrier can be used in neighboring cells with partial overlapping coverage). ICIC is especially relevant to improve the SINR experienced by UEs at cell-edges, which may receive comparable signal levels from surrounding cells.

Other key functions performed by an eNB in E-UTRAN are:

- Scheduling and transmission of *system information* and *paging messages*. System information includes cell and network identifiers, cell (re-)selection thresholds, random access configuration parameters, etc. System information is broadcasted at periodic intervals via the cell control channels. Along with system information, *paging messages* are also scheduled at specific intervals to trigger UEs switching from idle to connected mode.
- Security. Ciphering mechanisms are used between the eNB and UE to maintain privacy over the radio interface and protect the information against eavesdropping and alteration.
- Selection and routing from/to EPC network entities. The S1 interface of E-UTRAN allows the eNB to be connected to more than one EPC node, that is, with one or several MMEs and with one or several S-GWs. This feature is referred to as S1-flex.

7.3.2.1.4 Protocol Stacks

Figure 7.11 shows the protocol stacks used in the different interfaces of a 4G network. The interfaces related to the EPC nodes will be described further in a later section. In all the interfaces, the protocol stack is structured in a User Plane (UP) stack for data application transfer and a Control Plane (CP) stack for control signaling transfer.

In the radio interface, both UP and CP share the same lower layers, which consist of a Layer 1, or physical (PHY) layer, and a Layer 2, further split into three sublayers: Medium Access Control (MAC), Radio Link Control (RLC), and Packet Data Convergence Protocol (PDCP). In more details [34, 35]:

- The *PDCP sublayer* provides the security functions, including encryption and checksums for data integrity protection. It also provides in-sequence delivery, retransmission, and duplicate detection of PDCP packets during handover procedures and IP header compression capabilities (RFC 2507). The services offered by the PDCP to upper layers are DRBs in the UP and SRBs in the CP. There is one PDCP entity per DRB/SRB.
- The *RLC sublayer* is responsible for reliable data transmission. It provides segmentation and concatenation of PDCP packets into smaller RLC packets required at scheduling level. It supports Automatic Repeat reQuest (ARQ) mechanisms and removal of duplicate RLC packets. The services offered to the PDCP are called *RLC channels,* and like PDCP, there is one RLC entity per radio bearer.
- The *MAC sublayer* provides capabilities for traffic prioritization, multiplexing/de-multiplexing of RLC packets into MAC packets and fast retransmissions using a Hybrid ARQ (HARQ) mechanism. It also handles the scheduling decisions, which determine the radio resource allocation to the UEs in both uplink and downlink RF carriers. The MAC services offered to upper layers are called *logical channels*. There is one MAC entity per cell.

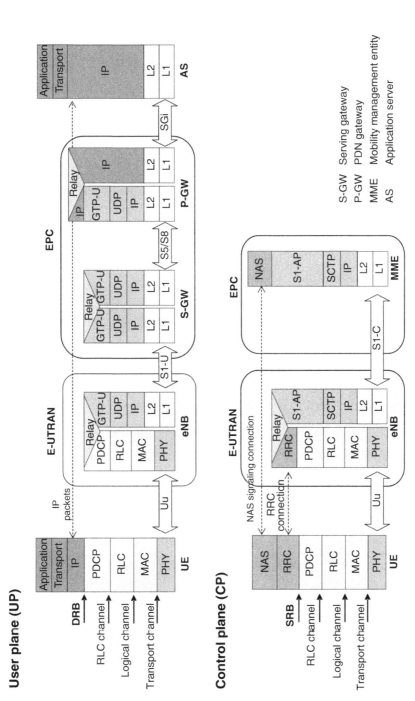

Figure 7.11 UP and CP protocol stacks in a 4G network.

- The *PHY layer* is responsible for channel coding, modulation, multi-antenna processing, resource element mapping, OFDM signal generation, and several other functions needed to support the transmission and reception of the LTE signals (e.g. channel estimation, equalization, etc.). The services provided to the MAC layer are called *transport channels*.

In the UP, the IP layer sits above the PDCP. As illustrated in Figure 7.11, the end-user IP packets flow between the UE and an Application Server (AS) in the external DN via the eNB, S-GW, and P-GW. Of note is that IP routing for these end-user IP packets may only take place at the UE and the P-GW. In contrast, these end-user IP packets are transferred transparently (i.e. without processing its IP headers) through the gNB and S-GW.

In the CP, above the PDCP, there is the RRC protocol used to handle all the radio-related signaling between UE and eNB, and the *Non-Access Stratum* (NAS) protocols, which handle network registration, session management and mobility management between the UE and the EPC. NAS protocols are encapsulated into RRC messages and sent transparently through the eNB, which only relays them between the UE and the EPC.

Unlike the radio interface where UP and CP share the same lower layers, the S1 interface is actually split into different interfaces: S1-U, which connects the eNB with the EPC node in charge of the UP (i.e. S-GW), and S1-C, which connects the eNB with the EPC node in charge of the CP (i.e. MME).

S1-U uses the GPRS Tunneling Protocol User Plane (GTP-U) to transfer the end-user IP packets. GTP-U is a protocol developed by 3GPP for GSM/GPRS networks and further adopted and extended to UMTS and LTE. The transfer service provided by GTP-U is based on encapsulation mechanisms. The GTP-U tunnel established between the eNB and S-GW is denoted as *S1 bearer*. There is a one-to-one correspondence between one S1 bearer and one DRB over the radio interface.

S1-C includes the S1 Application Part (S1-AP) protocol, which supports procedures needed to configure the eNB with the necessary information for UE access (e.g. subscription information transferred from the EPC such as identities and security keys). S1-C also includes procedures to handle handovers between eNBs that are not directly connected via X2 and for the MME to trigger paging procedures for terminals in idle mode. S1-AP is deployed over Stream Control Transmission Protocol (SCTP), which is a reliable connection-oriented transport protocol, similar to TCP but with enhanced features with regard to message-oriented transfer, multi-stream support, and multi-homing.

While not represented in Figure 7.11, the X2 interface (that can be used between eNBs) follows the same protocol structure as S1. Among others, X2 is used to support handover signaling and ICIC functions. It also allows transferring of end-user IP packets between eNBs during a handover execution to minimize packet loss. S1-C/S1-U/X2-C/X2-U are all interfaces designed to be deployed over IP-based transport networks.

7.3.2.2 NG-RAN

7.3.2.2.1 Functional Architecture

Figure 7.12 shows the architecture of the NG-RAN [36], which now includes two types of nodes: the *gNodeB* (gNB), which uses the 5G NR interface for interaction with UEs, and *next-generation eNodeB* (ng-eNB), which uses the evolved LTE interface (Release 15 onward).

Like in E-UTRAN, each gNB or ng-eNB handles one or several cells (NR cells in the case of the gNB, and LTE cells in the case of ng-eNB) and is responsible for all cell radio-related functions (e.g. RRM, RRC control connections). As the eNB in E-UTRAN, gNB and ng-eNB in NG-RAN are logical nodes and not a physical implementation, so that a possible implementation of a *5G base station* could embed both types of nodes as part of the same radio equipment (i.e. handle both LTE

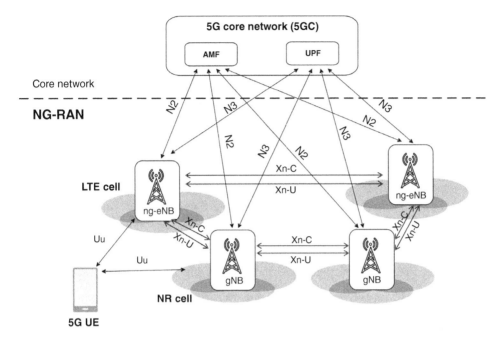

Figure 7.12 Next-Generation Radio Access Network (NG-RAN).

and NR cells). For simplicity, the term gNB will be used hereafter to refer to either gNB or ng-eNB, unless specific distinction between the two types of logical nodes has to be made.

As it can be seen in Figure 7.12, the gNB is connected to the 5GC by means of the N2 and N3 interfaces, more specifically, to the User Plane Function (UPF) of the 5GC by means of N3, and to the Access and Mobility Management Function (AMF) by means of N2. As in E-UTRAN, one gNB can be connected to multiple UPFs/AMFs for the purpose of load sharing and redundancy. And also like in E-UTRAN, there is an interface, Xn, split into Xn-C and Xn-U, for the direct interconnection of neighboring gNBs. All these network interfaces are deployed over the backhaul network, not represented in Figure 7.12.

7.3.2.2.2 NR Cells

Like in LTE, a NR cell is implemented using a pair of RF carriers (in case of FDD operation) or a single RF carrier (in case of TDD operation) and its operation follows the same basic principles explained in Section 7.3.2.1.2. However, NR cells:

- Support the operation in high-band spectrum (above 6 GHz), in addition to mid-band and low-band frequencies as in LTE.
- Allow a wider range of configurable channel bandwidths, from as low as 5 MHz up to 400 MHz, depending on the frequency band.
- Offer a more optimized and *clean* radio interface, with a set of different numerologies for scaled, optimal operation in different frequency ranges and fewer amounts of always-on signals (for enhanced energy efficiency and reduced interference).
- Enable *forward compatibility* by allowing a flexible mapping of the time/frequency resources in a cell (i.e. control channels and traffic channels do not have a pre-defined mapping onto the cell radio resources). Thus, future new signals and configurations not yet in the standard could be added to the NR carrier without impacting the operation of legacy terminals.

- Can support lower latencies than in LTE cells, below 1 ms.
- Allow for a more extensive use of beamforming and MIMO features, in both data and control channels (LTE does not support beamforming in the common control channels).

7.3.2.2.3 Protocol Stacks

Figure 7.13 illustrates the protocol stacks used in the different interfaces of a 5G network [37]. The interfaces related to the 5GC nodes will be further described in a later section. As in a 4G network, the protocol stack is structured in a UP and a CP.

At the radio interface, the protocol stack for NR follows practically the same structure as for LTE (see Section 7.3.2.1.4). In terms of layers, the only difference is the addition of a new Service Data Adaptation Protocol (SDAP) sublayer in the UP on top of the PDCP sublayer. The SDAP is responsible for mapping *QoS Flows* to DRBs according to their QoS requirements. This brings further flexibility in QoS management as compared to the approach used in the LTE radio interface, where a one-to-one mapping between an EPS Bearer (the equivalent in 4G to the QoS Flow) and a DRB is always considered. In NR, in contrast, several QoS Flows could be multiplexed into the same DRB if needed. Moreover, in terms of functionalities, the following differences can be noted [37, 38]:

- PDCP in NR introduces support for data duplication over different transmission paths (i.e. dual connectivity and split bearers). This allows improved reliability for URLLC applications. PDCP in NR also introduces integrity protection in the UP.
- The RLC in NR does not support in-sequence delivery of data to higher protocol layers. The reason is additional delay incurred by the reordering mechanism, which might be detrimental for services requiring very low latency. If needed, in-sequence delivery is provided by the PDCP layer in the NR interface.
- The header structure in the NR MAC layer has been changed to allow for more efficient support of low-latency processing than in LTE.
- The PHY layer in NR comes with a more flexible design (e.g. scalable OFDM numerologies) as well as multiple new capabilities (e.g. beamforming). More details are given in Section 7.4.2.
- RRC in NR adds a new state to the protocol machine, called RRC *inactive*, in addition to the *idle* and *connected* states already used in LTE RRC. This change intends to reduce the signaling load as well as the latency for the cases of frequent transmission of small packets.

In the network side, N2 and N3 interfaces are also similar to, respectively, S1-C and S1-U interfaces used in LTE. In N2, the Next-Generation Application Protocol (NGAP) is now used as the application layer protocol between the gNB and the 5GC node in charge of the CP (i.e. AMF). NGAP is transferred over SCTP, as S1-AP in S1-C. In N3, GTP tunnels are also used between the gNB and the 5GC core in charge of the UP (i.e. UPF). However, a new UP Protocol layer is added on top of GTP-U to multiplex QoS Flows into GTP-U tunnels. This layer is not present in LTE because a one-to-one mapping between an EPS bearer and a GTP-U tunnel is used.

7.3.2.2.4 Flexible Functional Splits of the gNB Functionality

NG-RAN specifications have also introduced standardized ways to split the gNB into several parts, defining the interfaces between these parts. This allows for more flexible and modular implementations of the NG-RAN as well as the introduction of NFV technologies for the implementation of the less computing intensive functions [39].

The split defined in 3GPP specifications considers the separation of the gNB into a Central Unit (CU), which handles the PDCP/RRC layers, and one or more Distributed Units (DUs), which handle the RLC/MAC/PHY. In addition, the CU can be further split into a CU-CP handling the RRC and a

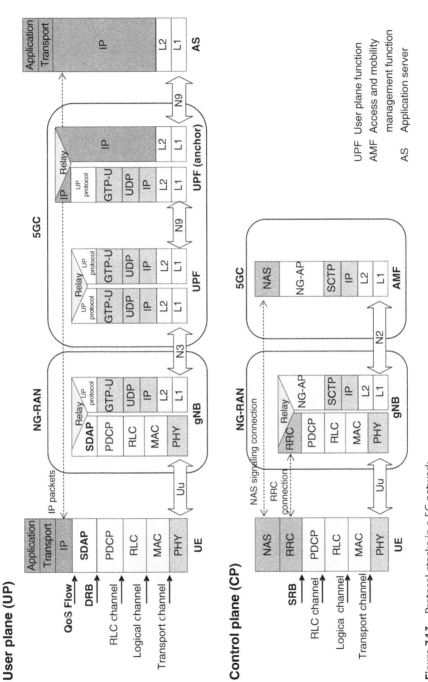

Figure 7.13 Protocol stacks in a 5G network.

CU-UP handling the PDCP. Other lower layer functional splits are currently under consideration in other related fora such as the Open RAN Alliance (O-RAN) [40] and Small Cell Forum [41].

7.3.3 Core Network

7.3.3.1 Evolved Packet Core

The EPC provides the core functionality to deliver packet-based connectivity services with diverse QoS needs over E-UTRAN [42, 43]. The EPC is responsible for:

- End-to-end session management, including the setup and release of PDN connections and EPS bearers.
- Mobility and subscriber data management, including network access control, subscriber authentication, authorization, and billing.
- Traffic routing and inter-working with the external PDNs.

7.3.3.1.1 EPC Nodes and Main Functions

As depicted in Figure 7.14, the EPC consists of several functional entities: Mobility Management Entity (MME), Home Subscriber Server (HSS), Serving Gateway (S-GW), Packet Data Network (PDN) Gateway (P-GW), and PCRF (Policy and Charging Rules Function).

The MME is the entity in charge of the CP of the EPC. Every UE that is registered in the LTE network has an assigned MME entity. The MME interacts with the UE via the NAS signaling protocols. Main functions are:

- Authentication and authorization of the users to access the network.
- Session management, including establishment of PDN connections and EPS bearers across the UP. The activation of DRB in the eNB or establishment of GTP tunnels between eNB, S-GW, and P-GW is commanded from the MME via S1-C and S11 interfaces.
- Mobility management of UEs in idle mode. This is based on the concept of TA described in Section 7.3.2.1.2. In particular, the MME is the entity on the network side that handles the information about the TAs where terminals are registered. If a terminal in idle mode has to be contacted from the network for any reason (e.g. new traffic to be sent to the UE), it is the MME

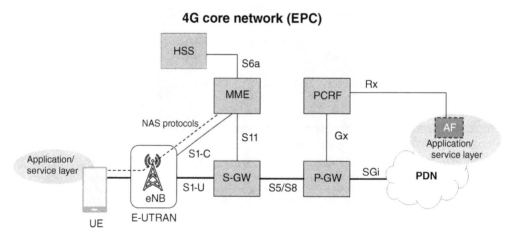

Figure 7.14 Evolved Packet Core (EPC) architecture.

which triggers the paging procedure toward all the eNBs handling the cells that belong to the TA where the UE is registered. LTE allows for multi-TA registration, avoiding unnecessary TA updates from UEs moving back and forth between cells at TA boundaries.

The HSS is the central database that contains, among others, user subscription-related information. The operation of the MME is assisted by the HSS, which is connected through the S6a interface.

The S-GW and P-GW are in charge of the UP of the EPC. In particular, the S-GW handles the UP connectivity with E-UTRAN via the S1-U interface and the P-GW connectivity with the external PDN via the so-called SGi interface. A UE registered in the network gets assigned a S-GW as the entry point to the EPC along with one or more P-GW to interact with the external networks. LTE supports multi-PDN connectivity, that is, the UE can be simultaneously registered to several PDNs. P-GW and S-GW communicate through the S5 interface (or through the S8 interface when they belong to different operators, e.g. if the UE is roaming). The S5/S8 interfaces are deployed over an IP-based transport network. From this perspective, the S-GW takes care of:

- User traffic routing to/from one or more P-GWs. User plane data transfer between S-GW and P-GW is performed by means of GTP tunnels.
- Buffering the downlink IP packets of the UEs that are in idle mode and initiate the network-triggered service request procedure toward the MME when new downlink traffic is detected.
- Transport level packet marking (i.e. QoS marking) in uplink and downlink so that the QoS differentiation can be applied in the underlying transport network.
- Serving as an anchor point for handovers between eNBs.

On the other hand, the main functions associated with the P-GW include:

- Assignment of IP addresses to the UEs.
- QoS enforcement mechanisms such as rate control and transport level packet marking for downlink traffic.
- Anchor point for inter-system mobility management (between LTE and non-3GPP networks).
- Interaction with the PCRF for policy enforcement.

The PCRF network entity is part of the Policy and Charging Control (PCC) system, providing operators with advanced tools for service-aware QoS and charging control between the network and external application platforms. The PCRF (see Figure 7.14) terminates an interface named Rx over which external application servers (e.g. residing on service delivery platforms such as IMS) can send service-related information, including resource requirements for the associated IP flow(s). In this regard, the term Application Function (AF) is the generic term used to refer to the functional entity that interacts with applications or services that require dynamic PCC. With the information received on the Rx interface as well considering subscription-based information available in the network, the PCRF makes decisions in the form of PCC rules (including the QoS and charging control rules) that are sent to the LTE network over the Gx interface for enforcement in the P-GW.

Of note is that all above described EPC entities are logical functions that can be implemented in a single physical node, distributed across multiple nodes, or executed on a cloud platform.

7.3.3.1.2 *Protocols Used Within the EPC*
All the interfaces used within the EPC are based on three protocols: Diameter, GTP-U, and GTP Control Plane (GTP-C).

The Diameter protocol (RFC 6733) is an IETF protocol that was developed for network access and IP mobility AAA (Authentication, Authorization, and Accounting). The Diameter protocol was extended by 3GPP to support specific commands and corresponding information elements needed in the context of mobile networks. Among others, Diameter protocol is used in the interfaces S6a, Gx, and Rx.

GTP-U (described in Section 7.3.2.1.4) is used for data transfer via tunneling and encapsulation mechanisms in the S5/S8 interfaces between S-GW and P-GW. In contrast, GTP-C is a signaling protocol used to manage the establishment of these GTP-U tunnels and transfer of other necessary information between EPC network nodes. In particular, GTP-C is used in S11 and S5/S8 interfaces.

7.3.3.2 5G Core Network

5GC is a platform conceived to provide QoS-aware packet-based connectivity services between UE and external DNs. From this perspective, 5GC builds upon the EPC principles and brings in new enhancement areas [44–46]:

- New design based on a Service-Based Architecture (SBA), in contrast to the traditional approach of building a system as a set of networked entities with entity-to-entity defined interfaces and "traditional" network protocols (e.g. GTP, Diameter).
- Further functional split of the CP and UP, resulting in the possibility to use general-purpose designs and implementations of the underlying switching/routing appliances in the UP.
- Native support of network slicing, enabling the customization of the 5GC to diverse applications and customers' needs.
- RAN-agnostic design of 5GC, allowing a 5GC system to also provide connectivity services over non-3GPP wireless and wireline access technologies.

7.3.3.2.1 Service-Based Architecture and UP/CP Decoupling

The adoption of an SBA design is a central novelty of the 5GC. An SBA is a system architecture in which the system functionality is achieved by a set of functional blocks, referred to as Network Functions (NFs), which are specified from the perspective of the services that they provide. A NF service has to be understood as one type of capability exposed by a NF (e.g. access to subscription data). A NF service may support one or more NF service operations (e.g. get subscription data, modify subscription data, etc.). In this way, procedures (i.e. the interactions between the NFs) are fully defined in terms of the service(s) that each NF provides, irrespective of which other NF consumes them. This contrasts with the traditional specification of system architectures based on the definition of point-to-point interfaces between any pair of interacting system components, which usually leads to some degree of repetition within the specifications when the same signaling procedure is used across multiple interfaces. Thus, in an SBA definition, point-to-point interfaces are replaced by a common bus which interconnects all the NFs.

The SBA representation of a 5G system is illustrated in Figure 7.15. The SBA is actually applicable to the CP section of the 5GC, where 5GC CP NFs are depicted with their associated Service-Based Interfaces (SBIs), labeled as N_{amf}, N_{smf}, etc. For the implementation of the SBIs, a Representational State Transfer (REST) architecture design and the Hypertext Transfer Protocol version 2 (HTTP/2) have been adopted [47], in contrast to the use of network protocols such as GTP-C and Diameter that were central for signaling purposes within the EPC.

Moreover, as it can be seen in Figure 7.15, there is a clear separation between the NFs responsible of the CP and the UP. This separation, which is a realization of the so-called Software Defined Network (SDN) concept [48], allows for centralized control and programmability of the forwarding capabilities in the User Plane Functions (UPF). It also enables independent scalability and

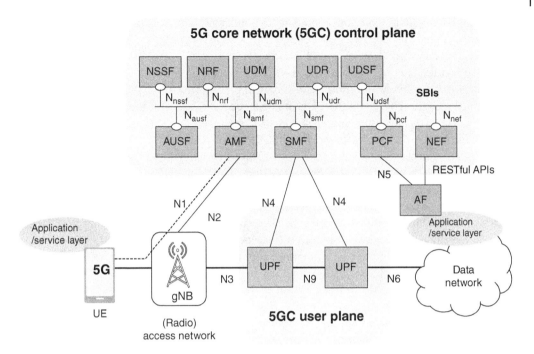

Figure 7.15 5G core network architecture (service-based representation).

evolution of the CP and UP (e.g. allocating more capacity to the CP without affecting the UP or vice versa) and more flexible deployments (e.g. centralized location for CP functions and distributed implementation of the UPF).

7.3.3.2.2 5GC Network Functions

Within the UP, the UPF is responsible for routing and forwarding UP packets between the NG-RAN and the external DN, in addition to other duties such as QoS enforcing (e.g. rate limiting) and traffic measurements. The UPF is connected to DNs via the N6 interface, which is similar to the SGi interface in EPC. On the RAN side, the UPF uses the N3 interface (described in Section 7.3.2.2.3). The UPF shall be regarded as a distributed functionality that may involve multiple switches with Layer 2 and Layer 3 forwarding capabilities located at different points within the mobile network and connected through the N9 interface. The behavior of the UPF is controlled by the Session Management Function (SMF) via the N4 interface. N4 is implemented by means of the so-called Packet Forwarding Control Protocol (PFCP), which follows the same principles as the OpenFlow protocol in SDN [48]. In this way, UPF does not embed specific mobile network functionality (as it was the case of S-GW and P-GW in LTE) and general-purpose forwarding appliances with PFCP support can be used.

Within the CP, NFs include: Access and Mobility Management Function (AMF), Session Management Function (SMF), Policy Control Function (PCF), Application Function (AF), Network Slice Selection Function (NSSF), Authentication Server Function (AUSF), and Unified Data Management (UDM):

- The AMF is responsible for network registration management, UE-5GC session management, mobility management, and reachability management (to ensure that a UE can be paged where there is a need to establish a mobile-terminated connection). An AMF serves as the single-entry

point for a UE for all its communication. The AMF interacts with the UE through the N1 interface (i.e. NAS protocols) and with other 5GC CP functions via the SBIs. The AMF also interacts with the access network through the N2 interface.
- The SMF handles, among other functions, IP address allocation for the UE, control of policy enforcement, downlink notification management (e.g. triggering of the paging procedure when downlink data have arrived and the UE is in idle mode) and general session-management functions (e.g. setup, modification, and release of PDU sessions). The SMF interacts with the UPF through the N4 interface and with the rest of 5GC NFs via SBIs.
- The PCF is responsible for providing policies associated with mobility management and session management (e.g. network selection policies, restricted tracking areas, authorized QoS based on network status, etc.).
- The AF acts as an application server that allows external applications (e.g. IMS applications) to interact with the PCF to influence on, e.g. UPF selection and traffic routing toward specific DNs. AFs are already used in the EPC architecture. Depending on the level of trust, AFs can have direct access to the 5GC (using a N5 interface connected to the PCF) or rely on the NEF, discussed below.
- The NSSF supports the AMF in the selection of the Network Slice during the UE registration procedure so that the service request from a UE can be properly accommodated by considering the UE's subscription and any specific parameters.
- The AUSF supports both subscriber and network authentication. The services provided by the AUSF are used by the AMF to support procedures such as UE registration. The AUSF uses the UDM function to retrieve the subscription information needed to handle the authentication functions.
- The UDM manages subscriber data such as the permanent and temporary identifiers (e.g. International Mobile Subscriber Identity [IMSI], Globally Unique Temporary Identifier [GUTI]), authorized QoS, authentication vectors generated by the AUSF, etc. It is similar to the HSS within the 4G EPC.

Additionally, the 5GC introduces some new functions that are a consequence of the adoption of an SBA architecture. These are the Network Function Repository Function (NRF) and Network Exposure Function (NEF):

- The NRF allows NFs to register their services and discover the services registered by other NFs. For instance, when selecting a Network Slice, the NRF can be used to discover the specific NFs that support the operation of the selected Network Slice.
- The NEF provides an interface for capability exposure, i.e. making 5GC functionalities available to external applications. For instance, network monitoring and event reporting services (e.g. loss of connectivity, location, etc.) can be exposed as well as services for aiding in the establishment of a communication initiated on the network side. NEF is regarded as one of the key pieces for the integration of 5G systems with third party applications needed in vertical industries. Like SBIs, capability exposure is based on the use of RESTful Application Program Interfaces (APIs) and aligns with the Common API Framework (CAPIF) specified within 3GPP for Northbound API development (more details on Section 7.3.4.2).

The 5G system architecture allows NFs to store their contexts and data in separate data storage functions. In particular, the Unified Data Repository (UDR) is used to store data related to the UDM, PCF, and NEF, while an Unstructured Data Storage Function (UDSF) is defined to store data belonging to any other NFs. Storing data at a central location help NFs to operate in a "stateless" manner. Stateless operation means that NFs do not rely on making transitions between states. Instead, every transaction is treated as new transaction. This allows multiple instances of a specific NF to be running in parallel so that one NF instance can take over from another without needing to transfer any information regarding to the current state. This is key to improve NF resilience and load balancing.

7.3.3.2.3 Support of Network Slicing

3GPP introduces the concept of network slicing, where different logical networks are supported within the 5G system to cater for distinct services and/or customer requirements for a given service. For example, one network slice can be set up to support mobile broadband applications with full mobility support, similar to what is provided by LTE, and another slice can be set up to support a specific non-mobile, latency-critical industry-automation application. Network slices may run on the same underlying physical core and radio networks, but, from the end-user application perspective, they appear as independent networks. Indeed, some specific 5GC NFs can even be individually instantiated for a particular network slice. In many aspects, it is similar to configuring multiple virtual machines on the same physical computer.

A single network slice is identified by its Single Network Slice Selection Assistance Information (S-NSSAI). The S-NSSAI(s) are included in the subscription information. A UE can be registered with up to eight S-NSSAI simultaneously. For each S-NSSAI registration, one or multiple PDU sessions can be established.

7.3.3.2.4 RAN-agnostic Core Network

The 5GC has been developed to be able to operate with Access Networks (ANs) other than the NG-RAN. This is facilitated by the design adopted in the N2/N3 interfaces and the QoS model based on QoS Flows, which are not coupled with any specific access technology.

One example of AN that can be supported is a Wi-Fi-based access network. Another example is a fixed broadband access network, paving the way for the convergence between mobile and fixed services.

7.3.3.3 Transitioning from 4G to 5G

For most mobile operators, the introduction of 5G systems will be a migration from their existing 4G network deployments to a combined 4G-5G network. In this respect, 5G specifications have been defined to support different combinations of core and radio access technologies that allow network operators to gradually evolve their infrastructure from 4G to 5G [39].

One of these combinations is the so-called Non-Stand-Alone (NSA) configuration, which has been adopted in the majority of the first deployments of 5G networks. Under NSA, a gNB is connected directly to an evolved 4G/EPC core and dual connectivity features, referred to as Evolved LTE-NR Dual Connectivity (EN-DC), are used to allow 4G/5G devices to be simultaneously connected to an LTE and a NR cell. With EN-DC, CP signaling (i.e. RRC and NAS signaling) is handled through the LTE cell (referred to as the Master cell) and the NR cell (referred to as the secondary cell) is used only for the UP data transfer.

Dynamic Spectrum Sharing (DSS) is another capability intended to facilitate the gradual introduction of 5G NR in frequency bands used for 4G LTE. With DSS, specific LTE subframes are not used for LTE transmissions to allow the insertion of 5G NR signal components. This enables both legacy LTE and new 5G/NR terminals to be served in the same channel in a 4G band and dynamically allocating resources to 4G and 5G terminals based on traffic demand.

7.3.4 Service Platforms

Both 4G and 5G networks are designed to provide "raw" connectivity services between mobile terminals and packet-based data networks. Thus, to support services such as voice, instant messaging, Push-to-Talk (PTT), etc., additional components to handle the necessary service delivery capabilities (e.g. session-level signaling for call establishment, etc.) shall be deployed on top of the mobile connectivity network.

One of these platforms is the IP Multimedia Subsystem (IMS), standardized as part of 3GPP systems, which is used to deliver voice services over 4G/5G networks.

7.3.4.1 IMS and Voice Services over 4G/5G

IMS was standardized to allow mobile network operators to progressively migrate from circuit-switched voice to packet-switched voice and pave the way for the introduction of a richer set of multimedia services based on and built upon Internet applications (e.g. videoconferencing, multimedia messaging, etc.) [49]. IMS is conceived as an access-type agnostic overlay to the mobile network, initially defined as part of the 3G/UMTS specifications and later evolved to include support for LTE and NR access [50, 51]. IMS has been re-designed under Release 16 to follow the SBA of the 5GC [52].

IMS provides functions and common procedures for end-to-end session control (e.g. voice call setup and release), policy and bearer control (e.g. selection of the QoS parameters used in the underlying network connectivity services), and charging for end-user services.

The use of IMS for the delivery of voice services over 4G/5G networks is commonly referred to as Voice over LTE (VoLTE) when the access technology is LTE and as Voice over NR (VoNR) when the access technology is NR. In the upper layers, the VoLTE/VoNR implementation relies on the use of a set of Internet-based protocols, including the Session Initiation Protocol (SIP), which is used for session management (e.g. service registration, activation, etc.), and Real-Time Transmission Protocol (RTP), used to encapsulate the voice packets. Supported codecs are Adaptive Multi-Rate (AMR), AMR Wideband (AMR-WB), and the Enhanced Voice Service (EVC) codec for High-Definition (HD) voice service offers. The mandatory set of functionalities that has been agreed for the UE, the access and core network, and the IMS functionalities for VoLTE/VoNR is specified in the Permanent Reference Document (PRD) IR.92 [53] published and maintained by the Global System for Mobile Association (GSMA).

7.3.4.2 5G Service Frameworks and Application Enablers

In addition to the IMS, 3GPP is also specifying service frameworks and application enablers to allow third-party applications to effectively develop and deploy new services over 3GPP networks (e.g. vertical applications) [54–56]. This includes the following components: Mission-Critical Services (MCX), Common API Framework (CAPIF), Service Enabler Architecture Layer (SEAL), and Edge Application Architecture (EDGEAPP).

MCX brings a number of features for broadband mission critical communications. MCX services include Mission Critical Push-to-Talk (MCPTT), Mission Critical Video, and Mission Critical Data (MCData). The MCX standard counts with strong support within the PPDR sector (MCVideo), with operational deployments already on-going in various countries (e.g. FirstNet in United States [57]).

CAPIF provides a unified Northbound API framework across 3GPP NFs to ensure that there is a single and harmonized approach for API development. CAPIF offers a single point of entry for third-party applications (CAPIF API invokers), with the ability to uniformly discover and securely access the underlying 3GPP capabilities through API exposure.

SEAL provides a "common services layer" including a set of core service capabilities such as group management, location management, device configuration, network resource, identify and key management functionalities that are common to industry verticals. This avoids redefining the individual services for each vertical industry and facilitates lowering deployment costs for operators, reducing adoption barriers, and improving time-to-market.

Last but not least, the EDGEAPP framework is intended to enable native support for hosting edge computing applications, while consolidating edge computing standardization in 3GPP.

7.3.5 Main System Procedures

Procedures define how different components of the mobile network interact to provide a given function. At the highest level, 5G mobile system procedures include [58]:

- Registration management procedures, used to register or deregister a UE with the network and establish the user context.
- Connection management procedures, used to establish and release the CP signaling connection between the UE and the AMF.
- Session management procedures, used to establish, modify, or release PDU sessions and QoS Flows.
- Mobility management procedures, used to keep track of the current location of UEs in idle mode.
- Handover procedures, used to change the serving cell of a UE in connected mode during an active communication session.

Some relevant examples of system procedures are described in the following, starting with the initial network registration when the UE is powered-on and then moving to communication session setup and the realization of a handover.

7.3.5.1 Network Registration

A UE needs to register with the network to get authorized to receive services, to enable mobility tracking and to enable reachability. The user has no access to the network until this step is successfully completed. The network registration is generally performed when the terminal is switched on.

Figure 7.16 shows a common message chart for the network registration procedure. When the UE is powered on, it needs to detect a suitable cell to camp on and decode the system information broadcast in that cell (Step 1). Once this is done, a random access procedure is initiated, leading to the establishment of the RRC connection between the UE and the gNB (Step 2). Some further details about the cell search and random access processes are given in Section 7.4.1.9.

Once the RRC connection is activated, NAS signaling exchange between the UE and the 5GC can happen. In particular, in this case, the UE initiates the registration procedure by sending the *Registration Request* message (Step 3). This message includes multiple fields such as UE identifiers (e.g. GUTI) and registration type (e.g. initial registration, registration for emergency call) as well as it may contain any requested Network Slice identifier (e.g. S-NSSAI). With this information, the gNB selects the AMF to which the *Registration Message* has to be forwarded (Step 4).

Once the selected AMF receives the registration request (Step 5), authentication and security procedures are triggered (Step 6). This involves some message exchanges between the UE, AMF, AUSF, UDM, and gNB for UE-network mutual authentication and activation of encryption and integrity mechanisms for NAS and RRC signaling. After a successful authentication/security

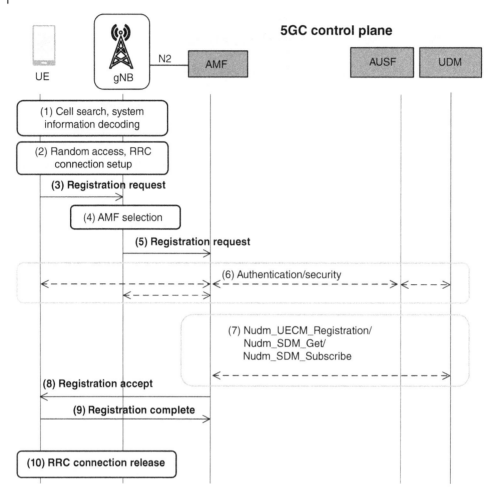

Figure 7.16 Network registration procedure description.

procedure, if the AMF is not the same that handled the last registration procedure with this UE or does not have subscription data for the UE, the AMF has to register with the UDM (e.g. using the *Nudm_UECM_Registration* service), retrieve subscription data (e.g. using *Nudm_SDM_Get* service), and request to be notified of subsequent changes in the subscription data (using *Nudm_SDM_Subscribe* service) (Step 7).

Once subscription data are updated in the AMF, a *Registration Accept* message is sent to the UE (Step 8), which acknowledges it sending a *Registration Complete* message (Step 9). At this point, if no further transaction is needed, the RRC connection may be deactivated (Step 10) and the UE transitions into idle state.

7.3.5.2 Service Request

Once a UE is registered in the network but has moved into idle state, the re-establishment of the CP connection with the AMF is carried out by means of the Service Request procedure. This procedure can be either UE-triggered or network-triggered. The former is used in order to send uplink signaling messages, user data, or as a response to a network paging request. On the other hand, the network-triggered procedure is used when it is the network which needs to signal with a UE in order to send N1 signaling to UE or deliver mobile terminating user data. A network-triggered

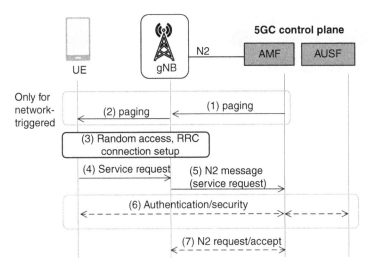

Figure 7.17 Service request procedure description.

procedure is illustrated in Figure 7.17, which only differs from the UE-triggered in the initial paging steps initiated by the AMF and leading to the paging notification transmissions over the air interface by the gNB (the paging message may be actually transmitted across the multiple cells belonging to the TA where the terminal is registered). As it can be seen in Figure 7.17, after detecting the paging, the UE starts the random access and establishes the RRC connection (Step 3), which is used to send a *Service Request* message. This message is forwarded by the gNB to the AMF where the UE is registered (Step 5) by means of N2 signaling. Optionally, authentication and security procedures may be also triggered at this point. If successful, the AMF requests the gNB to establish a context to handle the UE by means of the *N2 Request* message, which is acknowledged with an *N2 Accept* message (Step 7). At this point, the control connection with the AMF is active and subsequent transactions can be started (e.g. PDU Session establishment). The Service Request procedure is also used to re-activate the UP connectivity.

7.3.5.3 PDU Session Establishment

Figure 7.18 shows a message chart for PDU session establishment. The starting point assumes that the UE has an active CP connection with the AMF. The procedure is triggered by the UE sending a *PDU session establishment request* message, which includes information such as the DNN (used to identify the external DN) and the S-NSSAI. Upon reception of the request, the AMF selects the SMF that will handle the PDU session establishment and proceeds to create a context in that entity (e.g. using the *Nsmf_PDUSession_CreateSMContext* service) (Step 2). To that end, the SMF may retrieve registration and subscription data from the UDM (Step 3), acknowledge the context creation request (Step 4), establish a policy association with a PCF (Step 5), and trigger the session establishment on the UP by means of the N4 interface (Step 6).

Once the N4 session has been established in the UPF, the SMF transfers the details of the PDU session being activated (e.g. allocated IP address, QoS parameters, tunnels information) to the AMF by means of the *Namf_Communication_N1N2MessageTransfer* service (Step 7). This information is used by the AMF to request the creation of the PDU session context in the gNB (Step 8) and transfer the *PDU Session Establishment Accept* message to the UE (Step 9), which is going to be allocated with the necessary radio resources by the gNB (e.g. establishment of the DRBs in Step 10). At this point, first uplink data may be already transferred. Finally, to allow also for downlink

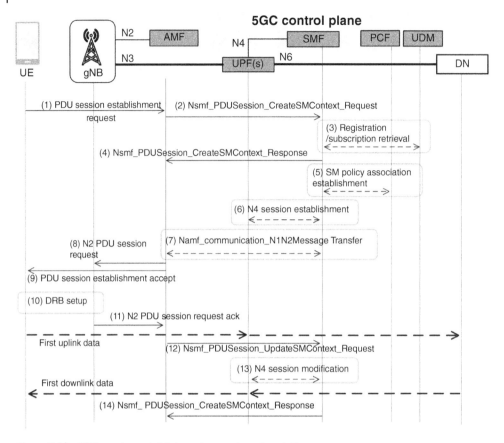

Figure 7.18 PDU session establishment procedure description.

transfer, the session configuration in the UPF must be updated with the details of the gNB handling the connection, which is done in Steps 12–14.

7.3.5.4 Handover

The Handover (HO) procedure allows switching the UE connection from one cell to another without or with minimal service disruption. A basic HO message chart is illustrated in Figure 7.19 for the case that the UE transitions between two cells handled by different gNBs interconnected with the Xn interface and with no need to re-allocate the AMF and UPF entities within the 5GC as a result of the change of gNB. As starting point, it is considered that the UE is connected to the *source* gNB and uplink and downlink data traffic exchange is ongoing. The *source* gNB serving the connection is the one responsible to make the HO decision (Step 2), which can be motivated by the radio measurements reported by the UE or other reasons (e.g. cell congestion). The HO decision includes the determination of the new candidate cell, which is this case is handled by *target* gNB. Hence, in Step 3, the *source* gNB requests the HO to the *target* gNB through the Xn interface and the target gNB checks if the connection can be admitted. If the connection can be admitted in the new cell, the HO Request is positively acknowledged and the source gNB sends a RRC reconfiguration message to the UE in order to execute the cell change (Step 6). At this point, downlink traffic data still reaching at the source gNB can be forwarded to the target gNB (Step 7), while the UE is detaching from the old cell and getting access through the new cell (Step 8). Once the RRC radio connection and the DRBs are restored in the new cell (Step 9), uplink and downlink data exchange can continue through

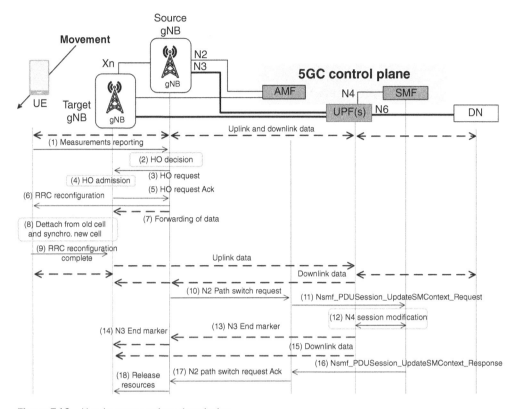

Figure 7.19 Handover procedure description.

the target gNB. However, while the uplink data path is already properly changed at this point, the downlink data path still flows through the source gNB. Therefore, additional signaling is needed basically to modify the PDU session in the SMF so that the GTP tunnel between the UPF and the source gNB is replaced by a new tunnel between the same UPF and the target gNB (Steps 10–17). The procedure finalizes by the target gNB sending a *Release Resources* message to the source gNB, confirming success of the HO and triggering the release of resources in the source gNB.

7.4 Main Capabilities and Features

This section provides more insight into the main capabilities and features of the 4G/LTE and 5G/NR radio interfaces and networks. Commonalities and main differences between LTE and NR are highlighted. It is worth noting that the description mainly focuses on the capabilities and features associated with eMBB and URLLC services. mMTC technologies (NB-IoT, LTE-M) are covered in Chapter 8.

7.4.1 LTE Radio Interface

The LTE radio interface can be configured with several channel bandwidths, from 1.4 to 20 MHz, and work in either FDD or TDD operation mode. In the downlink, the transmission scheme is Cyclic Prefix OFDM (CP-OFDM) with a subcarrier spacing of 15 kHz. The multiple access scheme is OFDMA. In the uplink, a Discrete Fourier Transform (DFT) Spread OFDM (DFT-S-OFDM)

Table 7.4 LTE operating bands.

LTE band	Name	Region	Uplink (MHz)	Downlink (MHz)	BW (MHz)	Duplex mode	Min–max channel BW (MHz)
1	2100	Global	1920–1980	2110–2170	60+60	FDD	5–20
2	1900 PCS	NAR	1850–1910	1930–1990	60+60	FDD	1.4–20
3	1800+	Global	1710–1785	1805–1880	75+75	FDD	1.4–20
...			...	...		...	
7	2600	EMEA	2500–2570	2620–2690	70+70	FDD	5–20
...			...	...		...	
20	800 DD	EMEA	832–862	791–821	30+30	FDD	5–20
...			...	...		...	
42	TD 3500		3400–3600	3400–3600	200	TDD	5–20
43	TD 3700		3600–3800	3600–3800	200	TDD	5–20
...			...	...		...	
46	TD unlicensed	Global	5150–5925	N/A	775	TDD	10, 20
...			...	...		...	
88	410+	EMEA	412–417	422–427	5	FDD	1.4–5

Source: Based on Ref. [59].

scheme is used, which shows less peak-to-average power ratio than CP-OFDM and allows terminals to transmit with a higher average power without increasing the amplifier complexity. The multiple access scheme on the uplink is referred to as Single Carrier FDMA.

7.4.1.1 Operating Bands
LTE specifications include support for more than 60 operating bands, which are defined to meet the operational and technical conditions established by global or regional regulations (e.g. maximum transmission power, out-of-band emissions limits, etc.). Table 7.4 shows some of these bands, including the operating band defined in the lowest frequency range (i.e. B88) and the one in the highest frequency range (i.e. B46). The most common bands used in European countries for LTE are B1, B3, B7 and B20.

7.4.1.2 Time-frequency Resource Grid
The structure of an LTE signal (Figure 7.20) can be viewed as a two-dimensional resource grid, with a subcarrier axis (frequency dimension) and an ODFM symbol axis (time dimension). In the time dimension, the transmitted signal is organized into frames, sub-frames, and slots. A frame has a 10 ms duration and is split into 10 subframes of 1 ms duration each. The subframe is the time granularity at which users(s) can be scheduled to transmit/receive in LTE, which is known as the Transmission Time Interval (TTI). A further division of the subframe results in the slot definition, which lasts for 0.5 ms. Over this time structure, each slot has room for the transmission of seven or six OFDM symbols, depending on whether a normal or extended cyclic prefix is used. The normal cyclic prefix is of 4.7 μs, suitable for most deployments, while the extended cyclic prefix of 16.7 μs can be more appropriate for highly dispersive environments. Therefore, the duration of the OFDM symbol is 71.4 or 83.4 μs, which equals to the inverse of the subcarrier spacing ($1/(15\,\text{kHz}) = 66.7\,\mu s$) plus the corresponding cyclic prefix.

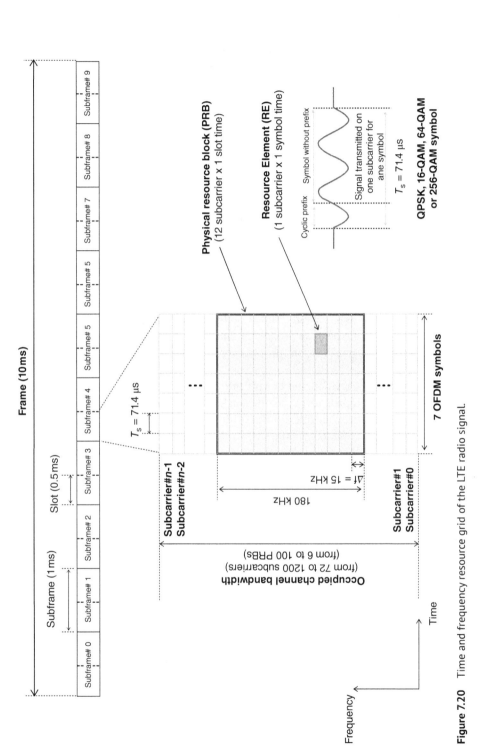

Figure 7.20 Time and frequency resource grid of the LTE radio signal.

Table 7.5 Channel bandwidth configurations in LTE.

Channelization	1.4 MHz	3 MHz	5 MHz	10 MHz	15 MHz	20 MHz
Number of PRBs (N_{PRB})	6	15	25	50	75	100
Subcarriers (Ks = 12 · N_{PRB} + 1)	73	181	301	601	901	1201
Occupied bandwidth B = Ks·Δf (MHz)	1.095	2.715	4.515	9.015	13.515	18.015
FFT size (Ns)	128	256	512	1024	1536	2048
Sampling rate fm = Ns · Δf (Msamples/s)	1.92	3.84	7.68	15.36	23.04	30.72

In the frequency dimension, the total number of occupied subcarriers depends on the LTE channelization being used. LTE defines the following set of transmission bandwidths (see Table 7.5): 1.4, 3, 5, 10, 15, and 20 MHz, which results in 72, 180, 300, 600, 900, and 1200 occupied subcarriers (i.e. subcarriers used for data and reference signals, not counting subcarriers left for, e.g. guard bands). The resource unit formed by one symbol time and one subcarrier is called Resource Element (RE). A single RE can accommodate a single modulation symbol (QPSK, 16QAM, 64QAM, or 256QAM). Within a slot, subcarriers are grouped in blocks of 12, forming what is known as the Physical Resource Block (PRB). A PRB contains 84 REs when normal cyclic prefix is used. PRBs are the central unit for resource allocation within LTE cells (to be more precise, the smallest resource unit that can be scheduled to a device within a subframe is one PRB pair mapped over two consecutive slots).

7.4.1.3 Scheduling, Link Adaptation, and Power Control

Data transmission in LTE is primarily scheduled on a dynamic basis in both uplink and downlink. For each 1 ms subframe, the scheduler in the eNB controls which PRBs are assigned to which devices, considering the QoS settings of the DRBs and information on the amount of buffered data pending for transmission. The scheduling decisions are provided to the UE through the Physical Downlink Control Channel (PDCCH).

LTE allows for channel-dependent scheduling [34]. In the downlink, Channel-State Information (CSI) is reported by UEs, reflecting the instantaneous downlink channel quality (i.e. Channel Quality Indicator, CQI) in both the time and frequency domains. In the uplink, the channel-state information is estimated by the eNB itself, based on Sounding Reference Signals (SRS) transmitted from UEs. In this way, the scheduler can allocate PRBs considering instantaneous channel conditions (e.g. avoiding the allocation of resources to terminals that may be momentarily experiencing a deep fade).

LTE scheduling also allows for different data rates to be selected in each resource allocation by adjusting the code rate of the Turbo Code as well as the modulation scheme from QPSK up to 256-QAM. The possible combinations are specified as a set of Modulation and Coding Schemes (MCSs). The selection of the MCS is known as *link adaptation* (or *rate control*). In the downlink, CQI reports sent by the terminals are used for MCS selection.

In addition to link adaptation, power control is also a key feature for LTE in the uplink. Uplink power control for LTE ensures that the transmit power used by devices for different uplink physical channels and signals is sufficient to allow for proper demodulation and, at the same time, do not cause unnecessary interference to other transmissions in the same or other cells. Adjusting the received power directly impacts on the MCSs that can be used in the uplink. LTE uplink power control is a combination of an open-loop mechanism, implying that the device transmit power depends on estimates of the downlink path loss, and a closed-loop mechanism, implying that the network can, in addition, directly adjust the device transmit power by means of explicit power-control commands.

7.4.1.4 Fast Retransmissions and Minimum Latency

In addition to the ARQ mechanism at RLC, LTE uses HARQ at the MAC layer. The key difference between HARQ and ARQ is that the former is able to buffer the received data if errors are detected, request a re-transmission, and combine the buffered data with the re-transmitted data (e.g. soft-combining) prior to channel decoding. In contrast, RLC ARQ discards the erroneous data and requests for re-transmissions. Moreover, thanks to the 1 ms timescale scheduling, the HARQ mechanism allows for feedback on success or failure of the transmitted data to be provided very fast, using a simple stop-and-wait mechanism. The total duration elapsed between the transmission of data in a single 1 ms PDSCH/PUSCH subframe until the reception of the corresponding HARQ acknowledgement is 8 ms. This value actually represents the minimum latency achievable over the radio interface. Moreover, the fact that the data sent in one 1 ms subframe is not acknowledged until 8 ms later leads to the need of having at least eight HARQ processes running in parallel for a UE to be able to fill the stop-and-wait gaps and reach the full capacity of the channel.

7.4.1.5 Multiple-antenna Transmission and Reception

LTE uses multi-antenna transmission and reception for *spatial diversity*, *beamforming*, and *spatial multiplexing*. Spatial diversity reduces the impact of fading by using multiple antennas at the transmitter, the receiver, or both. Beamforming allows directing the transmitted signal toward the UE. Both spatial diversity and beamforming are intended to improve the received SNR. In contrast, spatial multiplexing techniques, commonly known as MIMO techniques, are used to send multiple layers of data (i.e. spatial streams) simultaneously using the same time-frequency resources, thus increasing the achieved data rate. Spatial multiplexing can be used when the channel shows good SNR.

Figure 7.21 illustrates the key processing blocks in the case of a MIMO 2×2 [35]. It can be observed that the data to be transmitted are split in two layers and so-called *pre-coding* processing is applied to determine the symbol to be transmitted on each of two parallel OFDM transmission

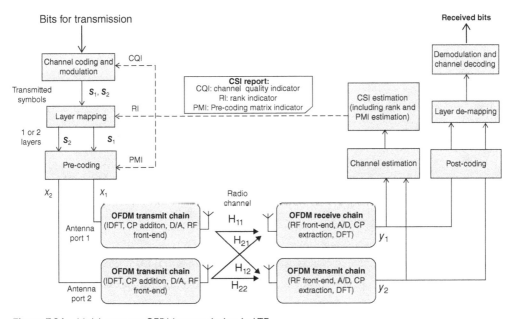

Figure 7.21 Multi-antenna OFDM transmission in LTE.

chains, referred to as antenna ports. At the receiver side, the pre-coding and layer mapping operations are just reversed. The example illustrated is known as *closed-loop spatial multiplexing Transmission Mode* (TM) and corresponds to the case that both the number of layers and the pre-coding matrices are adjusted dynamically based on receiver feedback as part of the CSI reporting, which includes in this case, in addition to the CQI, a Rank Indicator (RI), which measures the decorrelation between the layers, and a Pre-coding Matrix Indicator (PMI), used to select the pre-coding coefficients from a set of pre-defined options. Open-loop spatial multiplexing TM is also supported, as well as other TMs such as *transmit diversity*. The configuration of the different TMs, including switching between TMs, can be adjusted dynamically by the MAC scheduler in the same manner as MCS is selected to properly match channel conditions.

LTE supports MIMO configurations with up to eight layers, which requires to have at least eight transmitting and eight receiving antennas. This sort of MIMO configurations involving a single terminal is known as Single User MIMO (SU-MIMO). Alternatively, by combining the spatial properties of the channel with the appropriate interference-suppressing receiver processing, LTE also allows multiple devices to transmit and/or receive on the same time-frequency resource (i.e. multiple "layers" transmitted in parallel but, unlike SU-MIMO, each layer is associated with a different device). This feature is referred to as Multiuser MIMO (MU-MIMO).

LTE has also introduced Coordinated Multipoint (CoMP) techniques [34], which allow for coordinated transmission/reception from antennas belonging to different sites or sectors (often referred to as Network MIMO).

7.4.1.6 Carrier Aggregation and Dual Connectivity

Carrier Aggregation is a feature that enables a UE to simultaneously use multiple cells, denoted as Component Carriers (CCs). The aggregation can include CCs in the same frequency band (intra-band Carrier Aggregation), either contiguous or non-contiguous, as well as CCs located in different frequency bands (inter-band Carrier Aggregation). In this way, Carrier Aggregation allows increasing the transmission bandwidth (and consequently the peak data rates) that otherwise would be limited due to fragmented spectrum available in low- and mid-bands. Carrier Aggregation requires all the cells/carriers to be operated under the same scheduler. This restricts its use to cells handled by the same eNB.

LTE also supports Dual Connectivity for simultaneous connection to more than one cell. While conceptually similar to Carrier Aggregation, the fundamental difference is that the cells being combined in Dual Connectivity can be operated by distinct schedulers, located in different eNBs, and they may even belong to different radio-access technologies (e.g. EN-DC feature for Dual Connectivity with LTE and NR cells).

7.4.1.7 Physical Signals and Physical Channels

The REs in the time-frequency resource grid are arranged in different sets, forming what is known as *physical signals* and *physical channels*. Without loss of generality, the mapping of physical signals and physical channels to the LTE time-frequency resource grid is illustrated in Figure 7.22 for a 1.4 MHz downlink channel, which consists of six PRBs in the frequency domain (i.e. 1.08 MHz of occupied bandwidth). The time domain shows a frame interval (10 ms).

Physical signals are pre-defined sequences of symbols necessary to assist the operation of the PHY layer. In particular, LTE defines:

- *Primary Synchronization Signal* (PSS) and *Secondary Synchronization Signal* (SSS), which are used by UEs for cell search and downlink timing and frequency synchronization

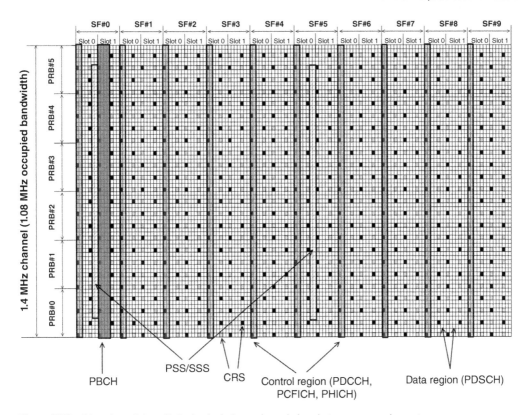

Figure 7.22 Mapping of downlink physical channels and signals to resource elements.

acquisition. These signals are transmitted every frame (10 ms) in Subframe #0 (SF #0) and SF #5, mapped always into the six central PRBs, irrespective of the channel bandwidth (i.e. for a channel of 20 MHz, the mapping of PSS/SSS will be exactly the same as the one shown in Figure 7.22).
- *Cell-specific Reference Signals* (CRSs), used for downlink channel estimation and channel quality measurements. The CRSs are arranged in the time-frequency domain so that they are uniformly spaced, allowing correct interpolation of the channel. Their number and specific location depend on different factors such as the number of antenna ports being used.

Physical signals do not carry information. In contrast, physical channels are arrangements of REs used for transmission of different types of information from upper layers. The following physical channels are defined in LTE:

- *Physical Broadcast CHannel* (PBCH), which carries system information messages. It is always located in SF #0, just next to the PSS/SSS and also limited to six PRBs in the frequency domain, irrespective of the channel bandwidth.
- *Physical Downlink Control CHannel* (PDCCH), which is used to send Downlink Control Information (DCI) to the UE such as PDSCH scheduling assignments. It is sent in the beginning of each SF, in the so-called control region, which can be configured to use two or three OFDM symbols. Unlike PSS/SSS/PBCH, the control region always spans the whole channel bandwidth.

- *Physical Control Format Indicator CHannel* (PCFICH), used to inform the UE about the number of OFDM symbols used for the PDCCHs. It is mapped to the control region.
- *Physical Hybrid ARQ Indicator CHannel* (PHICH), used to send the hybrid-ARQ ACK/NAKs. It is mapped to the control region.
- *Physical Downlink Shared CHannel* (PDSCH), which is used for upper layer data transmission. It is spread over the full resource grid, in the so-called data region. Usable REs for PDSCH within each PRB are those not used for other signals or channels.

Focusing now on the uplink, the following physical channels and signals are defined:

- *Physical Uplink Shared CHannel* (PUSCH), used for uplink data transmission.
- *Physical Uplink Control CHannel* (PUCCH), used to transmit Uplink Control Information (UCI) such as ACK/NACK of downlink transmissions, downlink CSI reports, and uplink scheduling requests.
- *Physical Random Access CHannel* (PRACH), used by terminals for initial access and uplink timing synchronization with the network.
- *DeModulation Reference Signal* (DM-RS), it is a physical signal used for channel estimation, enabling coherent detection of PUSCH and PUCCH transmissions.
- *Sounding Reference Signals* (SRSs), it is a physical signal transmitted by UEs in a periodic manner that is used by the eNB for sounding the channel quality and assists the operation of the scheduling algorithms. They are not associated with any uplink data transmission. SRSs are transmitted in the last symbol of a subframe.

Figure 7.23 shows how uplink physical channels and signals are mapped to the time-frequency resource grid for a 3 MHz channel. The outermost parts of the band are reserved for PUCCH. The network determines how many resources are allocated to the PUCCH, and the configuration is broadcasted to the UE via PBCH. The rest of the band is mainly used by the PUSCH and is allocated to individual mobiles in units of PRBs within each subframe. The base station also reserves certain resource blocks for random access transmissions on the PRACH. The PRACH has a

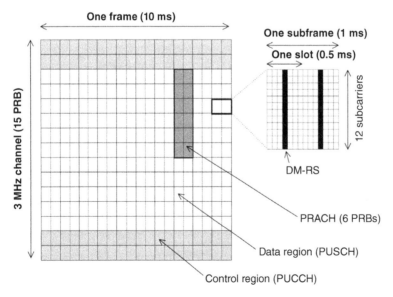

Figure 7.23 Mapping of uplink physical channels and signals to resource elements.

bandwidth of six PRBs (i.e. 1.08 MHz) and a duration from one to three subframes. The location of the PRACH in the resource grid as well as the number of PRACH opportunities over time is configured by the base station.

7.4.1.8 Mapping Between Physical, Transport, and Logical Channels

Figure 7.24 shows the mapping between the physical channels explained in the previous section and the concepts of transport channel, logical channel, SRB, and DRB introduced in Section 7.3.2.1.

The transfer capabilities of the physical channels are offered by the PHY layer to the MAC layer in the form of transport channels. A transport channel is defined by how and with which characteristics the information is transmitted. Data on a transport channel are organized into Transport Blocks (TBs), which are sent each TTI. Associated with each TB is a Transport Format (TF),

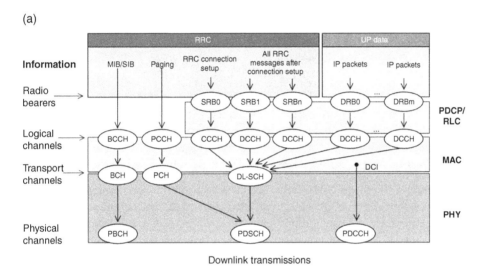

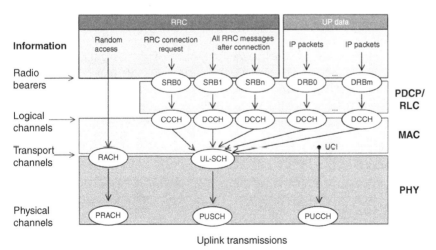

Figure 7.24 Mapping between DRBs/SRBs, logical, transport, and physical channels.

specifying how the TB is to be transmitted (e.g. selected MCS). The following transport-channel types are defined for LTE:

- *The Broadcast CHannel* (BCH), which has a fixed transport format and is directly mapped over the PBCH and broadcast across the whole of the cell. It is used for transmission of parts of the BCCH logical channel.
- *The Paging CHannel* (PCH), which is used for transmission of paging information from the PCCH logical channel. It is mapped over the PDSCH. The PCH supports Discontinuous Reception (DRX) to allow the device to save battery power by waking up to receive the PCH only at predefined time instants.
- *The DownLink Shared CHannel* (DL-SCH), which is the main transport channel used for transmission of downlink. It is mapped over the PDSCH. It supports features such as dynamic link adaptation and channel-dependent scheduling in the time and frequency domains, hybrid ARQ with soft combining, and spatial multiplexing.
- *The UpLink Shared CHannel* (UL-SCH), which is the uplink counterpart to the DL-SCH, that is, used for transmission of uplink data. It is mapped on the PUSCH.
- *The Random-Access CHannel* (RACH), which is also defined as a transport channel in the uplink, although it does not carry TBs. This is mapped to the PRACH.

In turn, the MAC sublayer uses the transfer capabilities offered by the PHY layer transport channels to send different types of information from upper layers. The transfer services offered by the MAC are in the form of the logical channels. A logical channel is defined by the information it carries. Logical channels are classified based on whether they carry UP or CP information and based on whether they are allocated to a specific mobile (dedicated) or used by more than one (common):

- *The Broadcast Control CHannel* (BCCH), which carries system information messages that allow mobiles to learn how the cell is configured (e.g. cell bandwidth, random-access transmission configuration, network identities, cell barring information, neighboring cells, etc.). This information is organized in a number of System Information Block (SIB) messages and a Master Information Block (MIB), the latter providing the scheduling details (e.g. periodicity) of the different SIBs.
- *The Paging Control CHannel* (PCCH), which carries the paging messages to address UEs in idle mode.
- *The Common Control CHannel* (CCCH), which carries messages on a special SRB, called SRB0, which is active by default for UEs that are moving from idle to connected mode in the procedure of RRC connection establishment.
- *The Dedicated Control CHannel* (DCCH), which carries the large majority of signaling messages. Several SRBs could be activated to handle e.g. signaling messages with different priorities.
- *The Dedicated Traffic CHannel* (DTCH), which carries data from DRBs. One or multiple DRBs can be established per terminal. Each DRB has a QoS profile and transfers the IP packets associated with an EPS bearer.

7.4.1.9 Radio Access Procedures

Complementing the description of the system procedures in Section 7.3.5.1, this section focuses on some key procedures carried out over the radio interface. These are *cell search*, *acquisition of system information*, *random access,* and *paging*.

The cell search procedure allows for acquisition of frequency and symbol synchronization, acquisition of frame timing, and determination of the physical-layer cell identity of the cell. In order to synchronize to the carrier frequency, the UE needs to correct any erroneous frequency offsets that are present due to local oscillator inaccuracy. In addition, due to the presence of multiple cells, the

UE needs to distinguish a particular cell on the basis of cell identifiers. Cell search is supported by means of the PSS/SSS/PBCH. Once downlink synchronization is achieved, the device can initiate the acquisition of system information by decoding the MIB/SIBs sent by the network.

Random access is mainly used for initial access when establishing a radio link (e.g. to request an RRC connection setup) but also for several other purposes such as assigning a unique identity within the RAN to the device (Cell Radio-Network Temporary Identifier, C-RNTI), establishing uplink synchronization with the new cell in a handover execution and as a scheduling request if no dedicated scheduling-requests resources have been configured in the PUCCH. Random access is a four-step procedure. Once the device is downlink synchronized, the device transmits a random-access preamble in the PRACH (Step 1), allowing the eNB to estimate the transmission timing of the device. Upon detection of the preamble, eNB sends a Random Access Response (RAR) via PDCCH (Step 2), which contains timing advance information and initial assignment of resources for PUSCH use. After adjusting uplink timing, the device can send an RRC message (e.g. RRC connection request) (Step 3), which is responded with another RRC message (e.g. RRC connection setup) via the PDSCH (Step 4). This last step also resolves any contention due to multiple devices trying to access the system using the same random-access resource.

When a terminal is in idle mode, paging is used for network-initiated connection setup. In LTE, the same mechanism as for "normal" downlink data transmission on the PDSCH is used and the mobile device monitors the PDCCH for downlink scheduling assignments related to paging. However, as a continuous monitoring of paging in the PDCCH during the idle mode would have serious implications on device battery lifetime, LTE supports mechanisms to establish paging cycles and allowing DRX operation in terminals. These mechanisms allow the network to configure in which subframes a device should wake up and listen for paging messages. The paging message includes the identity of the device(s) being paged, and a device not finding its identity will discard the received information and sleep according to the paging cycle. With DRX, paging occasions occur with a periodicity from 320 ms up to 2.56 seconds. The adjustment of the DRX cycle is a trade-off between responsiveness of network-triggered procedures and energy consumption on the terminal side.

7.4.2 5G NR Interface

The 5G NR interface builds upon and enhances many of the features introduced in subsequent releases for LTE. 5G NR has been designed from the start for:

- Support of multi-Gbps data rates, through the consideration of increased signal bandwidth and new frequency bands. This results in the need of, e.g. new procedures for mm-wave operation (e.g. beam management) as well as new channel coding schemes. In particular, channel coding in NR is based on LDPC codes, which are more attractive from an implementation perspective than Turbo Codes as used in LTE, because they offer a lower complexity especially at higher code rates.
- Reduced latency, tightening, e.g. UE processing time, and reducing the overall HARQ roundtrip latency compared to LTE.
- TDD-driven design, given that most of the new 5G bands are TDD. In 3G, FDD and TDD systems were fundamentally different. In 4G, TDD was an afterthought following FDD design with minimum changes. In contrast, 5G NR was designed first for TDD with FDD as a special case of the flexible frame structure.
- Minimizing always-on physical signals (e.g. reference signals) from the network. This has a direct impact on power savings at the network, reduced interference for partial loading scenarios, at the expense of higher UE complexity and battery power consumption.

- Limited bandwidth operation, with the introduction of the *BandWidth Part* (BWP) concept that allows devices to operate in a restricted portion of a wider bandwidth channel.
- Flexible design of the organization of the radio interface grid so that new access features could be introduced in a region of the channel resources with no impact on legacy transmissions (i.e. forward compatibility).
- Coexistence with LTE by means of DSS with LTE carriers.

5G NR also takes much background from LTE:

- Waveform. 5G NR is also based in CP-OFDM, which now is also mandatory for terminals in the uplink. In contrast to LTE, multiple numerologies are specified in NR (e.g. supporting subcarrier spacing from 15 kHz, as in LTE, to 240 kHz).
- Physical channel structure is almost identical to LTE, with some adaptations. For example, 5G NR specifies PSS/SSS, PBCH, PDCCH, and PDSCH, though it has removed PHICH and PCFICH and made some changes in the reference signals (e.g. no cell-specific reference signals). In the uplink, the same channels as LTE are still used: PRACH, PUCCH, PUSCH, and SRS.
- Radio connectivity operation and procedures. As LTE, 5G NR supports features such as Carrier Aggregation and Dual Connectivity and a lot of NR procedures are very similar to LTE (e.g. initial cell acquisition, random access procedure in four steps, uplink power control, etc.). A two-step random access procedure has also been introduced for 5G NR in Release 16.

A comparison of key features between NR and LTE is given in Table 7.6. The most novel aspects in 5G NR are detailed in Sections 7.4.2.1–7.4.2.6.

7.4.2.1 Flexible Waveform and Numerologies

5G NR allows for scalable OFDM numerology where the subcarrier spacing is no longer fixed to 15 kHz but selectable as $\Delta f = 2^n \cdot 15\,\text{kHz}$ for $n = 0, 1, 2, 3$, and 4, i.e. ranging from 15 to 240 kHz. The selection of the subcarrier spacing determines the frequency-domain bandwidth as well as the time-domain duration of a single RE, inversely proportional to the subcarrier spacing (Ts = $1/\Delta f + T_{CP}$, where T_{CP} is the cyclic prefix duration). Figure 7.25 shows the time and frequency dimensions of the 5G NR carrier.

In the time domain, frame and subframe structure follows that of LTE. However, in contrast to LTE, the division of one subframe into slots depends on the numerology, keeping fixed the number of OFDM symbols per slot to 14 (matching the 14 symbols within an LTE subframe). Hence, for $\Delta f = 15\,\text{kHz}$, there is a single slot per subframe; for $\Delta f = 30\,\text{kHz}$, there are two slots per subframe; for $\Delta f = 60\,\text{kHz}$, four slots; and so on.

In the frequency domain, NR also defines similar concepts to the RE and PRB used in LTE. In particular, the definition of RE is identical (i.e. one modulated symbol in one subcarrier) though now the time duration depends on the numerology. In contrast, the NR definition of PRB, now renamed to Radio Block (RB), differs from the LTE definition. In NR, a RB is defined as a one-dimensional measure spanning 12 REs in the frequency domain only, while LTE defines a PRB as a two-dimensional structure of 12 subcarriers in the frequency domain and one slot in the time domain. One reason for defining RBs in the frequency domain only in NR is the flexibility in time duration for different transmissions, whereas in LTE, transmissions always occupy a complete slot.

This range of numerologies allows 5G operation to be optimized for different deployment scenarios and end-user applications. For example, lower subcarrier spacing is beneficial to accommodate larger delay spreads and so larger cell ranges. In contrast, larger subcarrier spacing provides more robustness against increased oscillator phase noise which is an issue in mm-wave bands. Hence, lower frequency bands can use 15, 30, and 60 kHz subcarrier spacing, while higher frequency bands use 60,

7.4 Main Capabilities and Features

Table 7.6 Comparison between key NR and LTE radio interface features.

Specification	LTE	5G NR
Maximum bandwidth (per Component Carrier [CC])	20 MHz	50 MHz @ 15 kHz 100 MHz @ 30 kHz 200 MHz @ 60 kHz 400 MHz @ 120 kHz
Max. number of CCs	5	16
Subcarrier spacing	15 kHz	$2^n \cdot 15$ kHz ($n = 0\ldots4$)
Waveform	Downlink: CP-OFDM Uplink: DFT-S-OFDM	Downlink: CP-OFDM Uplink: CP-OFDM and DFT-S-OFDM
Max. number of subcarriers	1200	3300
Subframe length	1 ms	1 ms
Latency on air interface (HARQ roundtrip time)	8 ms	<1 ms
Slot length	7 symbols in 0.5 ms	14 symbols in $1/2^n$·ms ($n = 0\ldots4$) Also 2, 4 and 7 symbols for mini-slots
Channel coding	Data: Turbo Codes Control: TBCC	Data: LPDC Control: Polar Codes
Common control channels and initial access	No beamforming	Beamforming
MIMO	8 layers	8 layers
Reference signals	Cell-specific RS and UE specific DM-RS	Front-loaded DM-RS (UE-specific)
Duplexing	FDD; static TDD	TDD (dynamic/static); FDD

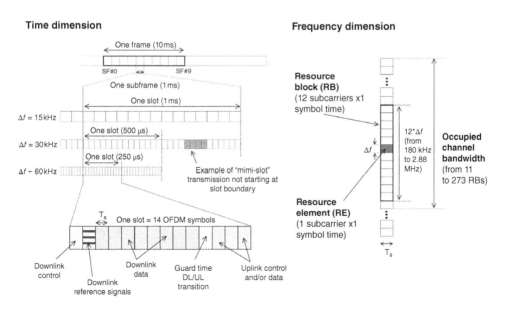

Figure 7.25 Scalable numerologies and time-frequency resource structure in 5G NR.

120, and 240 kHz. Numerologies can also be combined in the same 5G NR channel so that the time boundaries of the symbols are maintained across the different numerologies.

The NR subframe structure also allows for dynamic assignments of the OFDM symbol link direction and control within the same subframe, as also illustrated in Figure 7.25. This is known as dynamic TDD, allowing the network to dynamically balance uplink and downlink traffic requirements and include control and acknowledgement all in the same subframe.

7.4.2.2 Reduced Latency

5G NR interface allows air interface latency to be reduced below 1 ms. Several mechanisms contribute to achieve this [37, 38]:

- Use of front-loaded reference signals and control signaling, as shown in Figure 7.25. This allows a device to start processing the received data immediately without prior buffering, thereby minimizing the decoding delay.
- Enabling *pre-emption* and *mini-slot* transmissions, allowing the transfer of urgent data to a second device in resources previously allocated to a first device along with the possibility for transmission over a fraction of a slot using only two, four, or seven symbols.
- Use of *self-contained* subframes, allowing HARQ responses to be sent in the same subframe that data are received (e.g. uplink control part within the slot that carries the downlink data, as illustrated in Figure 7.25).
- Header structures in MAC and RLC re-designed to enable processing without knowing the amount of data to transmit. This is especially important in the uplink direction as the device may only have a few OFDM symbols after receiving the uplink grant until the transmission should take place.
- *Code-Block Group (CBG)* in HARQ. This is a new feature that allows NR HARQ mechanism to support retransmissions of only part of the TB received with errors, making it more efficient than having to send the whole TB. CBG is also useful when handling pre-emption since errors in the pre-empted device will be limited to some OFDM symbols only.
- *Operation without a dynamic grant*. Although dynamic scheduling is the basic operation of NR, operation without a dynamic grant can be configured as well. In this case, the device is configured in advance with resources that can be used for uplink data transmission (or downlink data reception). Once a device has data available, it can immediately begin uplink transmission without going through the scheduling request-grant cycle.

7.4.2.3 Bandwidth Parts

LTE was designed under the assumption that all devices shall be capable of using the maximum carrier bandwidth of 20 MHz. However, given the much higher maximum bandwidth in 5G NR (up to 400 MHz), 5G NR introduced the BWP feature, allowing devices to operate only in a fraction of the full bandwidth. This has many advantages such as the possibility to serve less-complex devices in wide channels as well as a reduction in the power consumption of terminals by avoiding them to have to monitor control channels across the full bandwidth (as it is the case in LTE for PDCCH/PHICH/PCFICH channels).

A UE can be configured with up to four BWPs per carrier, though only one BWP can be active at a time. The gNB can dynamically switch the active BWP.

7.4.2.4 Flexible Placement of the Control Channels

5G NR defines a more flexible time-frequency structure of downlink control channels where PDCCHs are transmitted in one or more Control Resource Sets (CORESETs). Unlike LTE where the full carrier bandwidth is used, CORESETs can be configured to occupy only part of the carrier bandwidth. This is needed in order to handle devices with different bandwidth capabilities (see BWP concept in Section 7.4.2.3) and also facilitates forward compatibility.

A CORESET can occur at any position within a slot and anywhere in the frequency range of the carrier (see Figure 7.26). The size and location of a CORESET is semi-statically configured by the network (a CORESET can have a duration of up to three OFDM symbols in the time domain and is defined in multiples of six RBs up to the carrier bandwidth in the frequency domain). A UE can only receive CORESETs in its active BWP. The reason for configuring CORESETs on the cell level and not per BWP is to facilitate reuse of CORESETs between BWPs. The first CORESET, CORESET 0, is provided by the MIB as part of the configuration of the initial BWP to be able to receive the remaining system information and additional configuration information from the network. After connection setup, a device can be configured with multiple, potentially overlapping, CORESETs.

Another difference with LTE is the placement of the PSS, SSS, and PBCH channels. In LTE, these are always located at the center of the carrier and transmitted once every 5 ms (see Section 7.4.1.7). Thus, by dwelling on each possible carrier frequency during at least 5 ms, a UE is guaranteed to receive at least one PSS/SSS/PBCH transmission if a carrier exists at the specific frequency. Without any prior knowledge, a device must search all possible carrier frequencies, which in LTE are defined over a carrier raster of 100 kHz. In contrast, in NR, the PSS/SSS/PBCH channels, arranged in a so-called Synchronization Signal Block (SSB) as depicted in Figure 7.26, are transmitted, by default, once every 20 ms in order to enable higher NR network energy performance. On this basis, in order not to negatively impact on the cell search time due to the longer period between consecutive SSBs, NR has specified a coarser frequency raster for SSB, referred to as the synchronization raster (e.g. 1200 kHz for frequencies below 3 GHz; though there is a specific formula to compute the raster). This implies that the possible frequency-domain positions of the SSB could be significantly sparser, compared to the possible positions of an NR carrier (the carrier raster), and not always match with the NR carrier center.

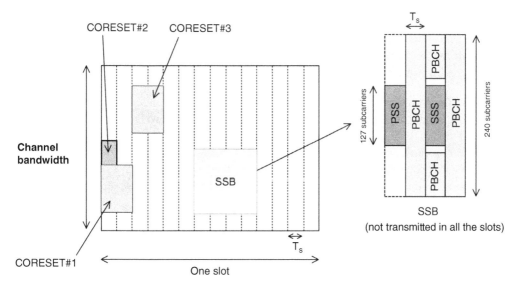

Figure 7.26 Mapping of control resources (CORESET) in NR.

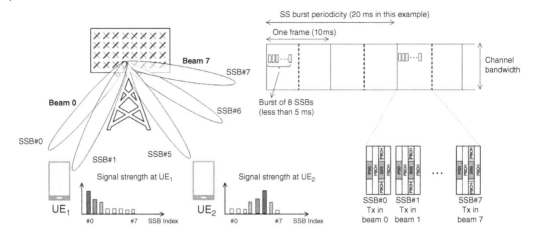

Figure 7.27 Beamforming in control channels.

7.4.2.5 Massive MIMO and Beamforming

5G NR has been designed from the very beginning to provide extensive support for large-scale antenna arrays (with potentially hundreds of antenna elements), often referred to as *massive MIMO* [37, 38]. As in LTE, multi-antenna transmission and reception are exploited for diversity, beamforming, and spatial multiplexing. The use of multi-antenna transmission and reception techniques enables significant gains in both coverage (especially when moving to high operating bands) and capacity. Furthermore, TDD facilitates the implementation of massive MIMO because of the channel reciprocity properties.

A major difference compared to LTE is the support of beamforming also in the control channels, which has required a different reference signal design with each control channel having its own dedicated reference signal. When beamforming is used, SSBs are transmitted in different beams at different times, arranged in SS *Bursts* as illustrated in Figure 7.27. A SS Burst may include as many as 4, 8, or 64 SSBs, depending on the frequency band (typically, 4 SSBs for frequencies below 3 GHz, 64 for frequencies above 6 GHz, and 8 for mid-band frequencies), and is always contained within a 5 ms time window. This allows the UE to identify the appropriate beam pair for receiving as well as transmitting. Indeed, in the uplink, when beamforming is used, the PRACH time/frequency occasions are also associated with different time indices of the SSBs within a SS burst set. Therefore, when a UE detects the SSB in one beam, it will only transmit the preamble in the PRACH occasions of that beam.

7.4.2.6 New Operating Bands

The frequency bands supported in the 5G NR interface come today in two Frequency Ranges (FRs): FR1, for low- and mid-band spectrum, and FR2, for the mm-wave spectrum. As per Release 16 [60], there are over 40 bands defined in FR1 and 5 in FR2. NR bands are named with a prefix "n" and the number corresponds to that of the LTE bands when there is overlapping.

In FR1, most of the bands currently specified for NR coincide with those already defined for LTE. In FR2, Table 7.7 shows the new supported bands.

7.4.3 Edge Computing Support

Edge computing allows a UE to access services that are hosted in appliances close to the serving base station. This approach helps to improve both end-user experience (reduced latency) and network efficiency (no need to move up/down traffic that is generated and consumed locally).

7.4 Main Capabilities and Features

Table 7.7 NR operating bands in FR2.

NR band	Name	Region	Frequency range (MHz)	BW (MHz)	Duplex mode	SCS (kHz)	Min–max channel BW (MHz)
n257	28 GHz	Global	26 500–29 500	3000	TDD	60, 120	50–400
n258	26 GHz	Global	24 250–27 500	3250	TDD	60, 120	50–400
n259	41 GHz	Global	39 500–43 500	4000	TDD	60, 120	50–400
n260	39 GHz	Global	37 000–40 000	3000	TDD	60, 120	50–400
n261	28 GHz US	NAR	27 500–28 350	850	TDD	60, 120	50–400

Source: Based on Ref. [60].

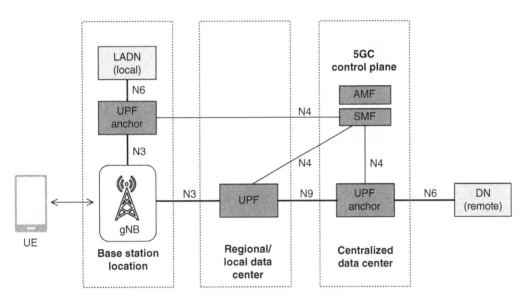

Figure 7.28 Illustration of edge computing support in 5G systems.

To support edge computing, a UPF and a DN or Local Area Data Network (LADN) can be placed at a location that is geographically closer to the base station (e.g. co-located with the same base station or in a regional data center). In contrast to a DN such as the public Internet or a private network that can be accessed from anywhere, a LADN is a DN specified to provide services across a limited number of TAs. A UE can request LADN information when registering in the mobile network and establish a PDU session with the LADN as long as it remains inside the associated TAs.

Figure 7.28 shows a possible architecture using a UPF and LADN co-sited with the base station [37]. This example assumes that separate PDU sessions are established for connection to the remote DN and the LADN, so that two Anchor UPFs are used (an Anchor UPF is responsible for mapping each downlink packet onto a specific QoS Flow belonging to a specific PDU Session). The example also shows an intermediate UPF used to provide routing between the local and remote networks. All the UPFs are under the control of the 5GC CP functions, which may remain located in a (distant) centralized data center.

In addition to the above network-level features, 3GPP, along with the ETSI Industry Specification Group (ISG) for Multi-Access Edge Computing (MEC), is also working on the definition of APIs for the integration of edge computing applications with 5G systems [61].

7.4.4 QoS Parameters and Characteristics

The support of QoS is a central piece in the specification of the connectivity models of both 4G and 5G systems. As explained in Section 7.3.1.1.1, an EPS Bearer in 4G systems or a QoS Flow in 5G systems is always associated with a QoS profile.

Focusing on 5G QoS, three categories of QoS Flows are defined:

- *Guaranteed Bit Rate* (GBR) QoS Flows, which ensures the delivery of a certain data rate. These QoS Flows are typically used for time-sensitive applications such as voice and video calls.
- *Non-Guaranteed Bit Rate* (Non-GBR) QoS Flows, with no assurances on the data rate provided. This is typically used for non-time-sensitive applications (e.g. web browsing) and application layer signaling (e.g. IMS signaling).
- *Delay Critical GBR Flows*, which provide significantly lower latencies than GBR QoS Flows. These QoS Flows are intended to support so-called Time-Sensitive Communications (TSC) for the delivery of deterministic and/or isochronous communication services, which may require strict bounds on latency, loss, packet delay variation (jitter), and reliability and where end systems and relay/transmit nodes can be time synchronized.

Each QoS Flow is associated with a set of *QoS parameters*. The most relevant are [37]:

- *5G QoS Identifier* (5QI). This is a pointer to a set of QoS characteristics that determine the behavior of the QoS Flow in terms of latency, error rate, and priority level (see Table 7.8).
- *Allocation Retention and Priority* (ARP). It defines the relative importance of the QoS Flow for admission control purposes in case of resource limitations. Pre-emption is supported so that a less priority QoS Flow could be released to make room for a new QoS Flow with higher priority in a congestion situation.
- Guaranteed Flow Bit Rate (GFBR), Maximum Flow Bit Rate (MFBR), and Maximum Packet Loss Rate. These parameters only apply to GBR QoS Flows.
- Session-Aggregate Maximum Bit Rate (AMBR) and UE-AMBR. These parameters only apply to Non-GBR QoS Flows. Session-AMBR limits the aggregate bit rate for all non-GBR QoS Flows of a given PDU session. UE-AMBR limits the aggregate bit rate for all non-GBR QoS Flows of a UE.

7.4.5 Network Slicing

The support of network slicing features in 5G systems allows operators to customize their network for different applications and customers.

Network slices can differ in functionality (e.g. air interface capabilities, mobility tracking features, security services, location services), in performance requirements (e.g. latency, availability, reliability, data rates), or they can serve only specific users (e.g. PPDR organizations, corporate customers, industrial users). One network, uniquely identified by a Public Land Mobile Network (PLMN) identifier, can support one or several network slices, each uniquely identified by a S-NSSAI identifier.

The implementation of a S-NSSAI includes resources from both the RAN and the CN, as illustrated in Figure 7.29. On the RAN side, a S-NSSAI can be associated with a particular configuration

Table 7.8 Standardized QoS characteristics associated with each 5QI.

5QI	Resource type	Default priority level[a]	Packet delay budget (ms)	Packet error rate	Default maximum data burst size	Default averaging window	Example services
1	GBR	20	100	10^{-2}	Not applicable	2 seconds	Conversational voice
2		40	150	10^{-3}			Live video streaming
3		30	50				Real-time gaming, Vehicle to Everything (V2X) messages
4		50	300	10^{-6}			Buffered video streaming
65		7	75	10^{-2}			Mission critical PTT voice
67		15	100	10^{-3}			Mission critical video
75		25	50	10^{-2}			V2X messages
5	Non-GBR	10	100	10^{-6}	Not applicable	Not applicable	IMS signaling
6		60	300				Video (buffered streaming), TCP-based applications
7		70	100	10^{-3}			Voice, video (live streaming), Interactive gaming
8		80	300	10^{-6}			Video (buffered streaming), TCP-based applications
9		90					Sharing, progressive video
69		5	60				Mission critical delay sensitive signaling
70		55	200				Mission critical data
79		65	50	10^{-2}			V2X messages
80		68	10	10^{-6}			Low latency eMBB, Augmented Reality
82	Delay critical GBR	19	10	10^{-4}	255 bytes	2 seconds	Discrete Automation
83		22	10		1358 bytes		
84		24	30	10^{-5}	1354 bytes		Intelligent Transport Systems
85		21	5		255 bytes		Automation for electricity distribution

[a] Priority level: a low numerical value corresponds to a high priority.

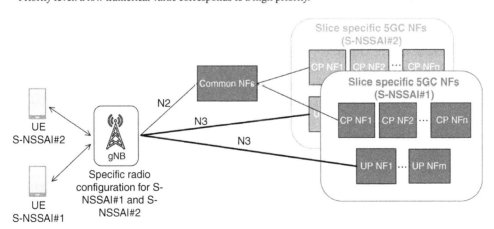

Figure 7.29 Implementation of several network slices.

Table 7.9 Standardized SST values for network slices.

Slice/service type	SST value	Characteristics
eMBB	1	Slice suitable for the handling of 5G enhanced Mobile Broadband
URLLC	2	Slice suitable for the handling of ultra-reliable low latency communications
MIoT	3	Slice suitable for the handling of massive IoT
V2X	4	Slice suitable for the handling of V2X services

of the radio protocol stack and RRM policies (e.g. allocation of radio resource utilization quotas per S-NSSAI [62]). On the CN side, a S-NSSAI may be implemented with a combination of dedicated NFs and common/shared NFs, that is, NFs that handle several S-NSSAIs. Indeed, a NF that is always common to all the network slice instances serving a given UE is the AMF (i.e. the AMF instance serving a UE logically belongs to each of the S-NSSAIs that the UE is registered). Network slicing builds upon technologies such as NFV for automating the lifecycle management of the NF instances (e.g. creation, scaling, termination of NF instances).

A S-NSSAI is a concatenation of a Slice/Service Type (SST) and a Slice Differentiator (SD). The SST refers to the expected network slice behavior in terms of the services and features it supports. 3GPP has standardized the SST values presented in Table 7.9 [46]. These values address the main use cases for 5G and are especially relevant for global interoperability in roaming situations. Non-standardized SST values can also be used. In contrast, the SD is optional but can be used to differentiate between network slices that have the same SST value but are offered to distinct subscriber groups.

7.4.6 Operation in Unlicensed Spectrum

Different extensions and features are available to deploy LTE and NR in unlicensed spectrum [63, 64].

Beginning with LTE, the cellular industry looked to unlicensed spectrum, typically associated with Wi-Fi, as a way to augment the capacity provided by licensed frequencies. Different technologies were developed, such as LTE–Unlicensed (LTE-U), LTE Wireless-LAN Aggregation (LWA), and Licensed Assisted Access (LAA). Among them, LAA has consolidated as the primary choice. LAA allows an eNB to jointly operate an anchor channel of licensed spectrum with multiple channels of unlicensed 5 GHz spectrum. Fair coexistence in the unlicensed channel with other systems (e.g. Wi-Fi networks) is provided by LBT capabilities to ensure channels are clear before transmission (LBT is mandatory in the EU and Japan).

Despite the progress achieved with LAA, 3GPP did not specify a standalone mode for LTE operation in unlicensed spectrum. However, this has been addressed by the MulteFire Alliance [65]. The MulteFire Alliance specifications are built upon 3GPP specifications, only adding those elements necessary for standalone operation in unlicensed channels. The first set of specifications, MulteFire 1.0 release, was centered on supporting basic LTE services, with the primary goal of delivering solutions such as enhanced cordless phone services within enterprises and public and private venues. In contrast, the latest specification, MulteFire 1.1 release, has been expanded to address the industrial IoT market with enhancements to support unlicensed NB-IoT and eMTC (further described in Chapter 8). Spectrum bands supported include 5, 2.4 and 1.9 GHz.

However, the approach of 3GPP with regard to the support of standalone operation in unlicensed bands has changed in 5G. In this context, 3GPP Release 16 has specified the operation of the NR interface in unlicensed spectrum, named NR-Unlicensed (NR-U). In addition to standalone operation, NR-U also supports the combination of licensed spectrum carriers and unlicensed spectrum carriers via Carrier Aggregation or Dual Connectivity, as a continuation within NR of the LAA feature already introduced with LTE.

The initial focus of NR-U is on unlicensed spectrum in the existing 5 GHz band as well as in the new 6 GHz band under consideration at regulatory level in some countries (e.g. in the United States, the FCC in the United States has already authorized the use of 1200 MHz of spectrum in the 6 GHz band for Wi-Fi and 5G NR-U). Support for unlicensed spectrum in the mm-wave range (e.g. 60 GHz) is expected in future releases.

7.4.7 Private Networks

The 3GPP focus on industry expansion (i.e. *verticals*) involved solutions to provide high-performance wireless connectivity with 5G private networks, referred to as Non-Public Networks (NPN) in the 3GPP specifications.

Key features introduced in 5G with the primary purpose of addressing this market include: operation in unlicensed spectrum (see previous section); URLLC features; TSC features, including deterministic networking and integration with IEEE 802.1 Time-Sensitive Networking (TSN) systems; 5G-LAN services, which provide similar functionalities to a LAN within a 5G network; and, last but not least, positioning services [66–68].

While 5G private networks may be deployed in a variety of configurations, these can be categorized under two basic forms, illustrated in Figure 7.30: Stand-alone NPN (SNPN) and Public Network Integrated NPN (PNI-NPN).

SNPN refers to the deployment of a private network as a dedicated, independent network, with all network functions (e.g. NG-RAN equipment, 5GC with the subscribers' database, service platforms) likely located within the enterprise premises, and not relying on any functions provided by a public network (i.e. a PLMN network). To support SNPN deployments, the 5G standard defines the identities to be used by the private network and devices as well as how devices should behave

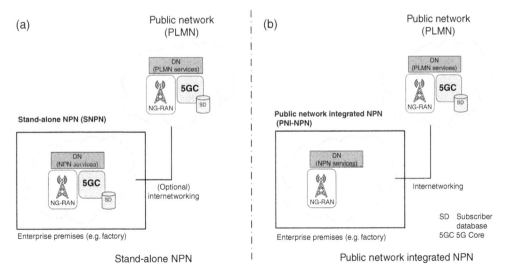

Figure 7.30 Network configurations for private networking.

when configured or not in SNPN access mode (e.g. network selection, access control). Moreover, even though a SNPN can operate autonomously, interworking with public networks may be of interest in some cases. For instance, interworking with a public network can be used to access the services provided in the public network (e.g. voice PSTN services) when the UE is connected to the SNPN as well as access to the SNPN services when the UE is connected to the public network. To support such cases, dual subscription is necessary and the interworking approach considered in the specifications relies on the same type of solutions defined for the interworking with non-3GPP access networks such as Wi-Fi networks.

In contrast, a PNI-NPN refers to a private network which is deployed in conjunction with a public network. In this case, various levels of integration are possible building upon technologies such as RAN sharing, network slicing, dedicated DNN selection, and Closed Access Groups. In the particular example illustrated in Figure 7.30, it is assumed that there is some NG-RAN equipment and service platforms located in the local premises so that PNI-NPN traffic remains confined locally. However, the control functions and subscribers' database are considered to be part of the public network, which may use network slicing to properly isolate the CP of the PNI-NPN users. Under such a configuration, PNI-NPN devices on the enterprise premises can also connect directly to the public network and get access to its services there, including roaming.

The suitability of these different deployment options highly depends on the use case. For example, SNPN deployed on-premises could be best suited for the manufacturing sector due to latency, control, and security requirements. On the other hand, sectors such as logistics and transportation where wide-area coverage is more important, a PNI-NPN could be the preferred option.

7.5 Applicability to Smart Grids

From a historical perspective, the use of cellular technologies through communication services provided by MNOs has progressively replaced the use of PMR technologies in the utility sector, based on the convenience of the radio coverage created "by default" by MNOs for their target customer base. Typically, MNOs' customer base has been the general public, and the coverage of average MNOs' public networks has been driven by their local (country) licenses, and intended to achieve deep population coverage, while often territory coverage has been sacrificed. Thus, ample territory areas may have been left out of MNOs footprints. While this is not a problem for most industries, it is a hurdle for those that are spread over all the territory and need to reach every single human-inhabited corner. Moreover, in some of these highly-spread infrastructure-intensive industries, many of the assets tend to be hidden from human sight (buried) and, in some of them, the availability of electric power is not the rule (e.g. water and gas utilities). This basic fact, along with other more sophisticated reasons (see Chapter 1), justifies the unequal adoption of public network services in utilities, specifically Smart Grid-related premises.

From a service perspective, the first cellular network service widely adopted by utilities was voice. Utilities have always relied on voice communications to support their operations. As described in Chapters 1 and 5, since the electric power system operations started to be supported by UCCs, there was a need to connect the UCC operator with the on-field crews to coordinate their interventions. This is a voice-assisted process still today, and while there is a trend to automate and drive these interventions through automated workflows that can help to increase efficiency and reduce mistakes in operations, it is probably unavoidable that some sort of voice-ready communication service will be needed. Utilities had predominantly used PMR technologies for voice communications, almost in all cases supported on self-owned and self-operated private networks.

However, the difficulty to compete with the simplicity, feature-rich, and wide-coverage in populated areas of public mobile networks resulted in two major trends to upgrade or complement the sort of PMR solutions initially used in the utility sector (e.g. MPT1327). A first trend was the adoption of more advanced PMR technologies (e.g. TETRA), with a technology substrate similar to that of 2G/GSM cellular networks but enhanced with multiple mission-critical voice communications features (e.g. group calls, direct-mode operation). This first trend was indeed two-fold: it had one branch, that adopted them in a dedicated, private network fashion (mostly industries with coverage requirements spanning limited extension territories), and another branch that used services provided by shared PMR network platforms run by specialized operators (e.g. UK Airwave [69] for emergency services and other mission-critical communications users). A second trend was the adoption of the regular commercial MNO provided services, complemented sometimes with satellite communications platforms for coverage extension and/or used as backup solution in case of terrestrial mobile networks unavailability.

While MNOs continued expanding their networks, the availability of a more consistent and competitive coverage started to stimulate the imagination of MNOs and utilities alike to start testing the first data services made available in public cellular networks. The data services provided by 1G were probably an anecdote in terms of use in any utility network. 2G networks were clearly different, although the perspective was not so much the same at their early stage when 2G shared 1G marketspace, and while 2G coverage was being created. Indeed, 2G CSD and SMS were the first services adopted in the 1990s, as soon as they started to be available. Their success and wide-spread adoption are such that they are still used today, being this among the factors that have prevented MNOs from phasing out completely their 2G networks despite the advent of 3G/4G/5G technologies.

2G CSD communications showed utilities a simple evolutionary way to transition some remote-control services in HV and specially MV grids. Prior to the emergence of CSD, SCADA-connected RTUs could run on services provided by private networks (typically optical fiber, microwaves, and PLC in PSs; and PMR in SSs and other elements of the MV grid) or wireline dial-up PSTN-based services. Most of them ran in the former, while some, especially in the less critical premises (some SSs), in the latter. Thus, SCADA systems started to prepare connectivity that transitioned from "always on," always open communication connection with the RTUs, to a connection that "needed to be established" before it was ready (CSD is operational once the connection -AT commands- has been established [70]). As it can be easily derived, from the usability perspective, UCC operators had to understand that CSD-based connectivity was to be used "in a more patient way." And, from the perspective of the connection from the RTU to the SCADA, some logic had to be defined either in the RTU or in communication modems, to trigger and be able to keep the connection when needed. However, be it for cost reasons or for the sake of testing these promising cheaper alternatives, some CSD-connected RTUs started to appear in SCADAs. 2G CSD services helped to avoid the connection to the wireline PSTN, which has always been challenging in substations for electrical reasons (see Chapter 1). 2G CSD service has also been popular and is still widely used, in Commercial and Industrial (C&I) metering systems, to connect meters that had to be remotely accessed mainly for billing purposes [71].

The use of SMS services was different. Understanding that SMS was an unreliable, data-constrained service, SMS adoption was based in the possibility to have RTUs with the capability of sending messages that were not critical to the operation of the system. SMS adoption in utility industry has probably not been any different from its use in other industries, and remote units capable of sending simple alarm messages when some conditions happened, have and are still in use. Complementary uses of SMS services to perform simple, non-real-time actions to remote units have also been adopted.

The upgrades of 2G networks with GPRS/EDGE capabilities were a big leap forward in the use of data services for utility operations. The connectivity provided did not need any utility systems-lead connection establishment procedure and was, from this perspective, always on. GPRS/EDGE services came along with IP technology. IP technology (or packet-based technologies in general) adoption in utilities has been an uneven and lengthy process across utilities and utility departments (each department classically assigned a different "service" – protection, control, metering, etc.). Telecommunications (or more often eclectically referred to as "Communications" in utilities) departments were for obvious reasons the first to use packet-based technologies and networks. GPRS/EDGE services started to appear in RTUs, IEDs, and C&I meters communications and were enabled in SCADAs and metering Head-End systems. The roll-out of GPRS/EDGE services was only limited by the cost of the solution (GPRS/EDGE modem cost and data traffic fee), the availability of the coverage, and the availability, and performance of the service itself. For this last aspect, the discussion related to the nature of public mobile networks needs to be present, along with the adjustment of MNOs' performance targets, as compared with the different utilities' services requirements (e.g. automation availability is different from metering; and different depending on Transmission and Distribution network, and within Distribution network for urban or rural MV, or for LV). GPRS/EDGE connectivity has been understood by utilities as a bandwidth constrained data transfer service, not usable for bandwidth needs above 20 kbps (as a common reference), and provided by MNOs only under rather limited service availability requirements (i.e. the service could be unavailable for long periods or its bandwidth importantly reduced due to network congestion). This prevented the adoption of these services in the more critical operations and areas of the grid.

3G technologies such as UMTS/HSPA came with higher data rate services. However, benefiting from these improved services was not straightforward for some end-users, since a significant effort on terminals/Customer Premises Equipment (CPE) replacement was necessary. Along with higher data rates, 3G also started to consider M2M services, while not specifically addressed to utility needs [72]. In utility practical terms, 3G in its successive releases:

- Offered higher bandwidths but was tied to terminal/CPE replacements, which was not always well-aligned with grid technology lifecycles and utility projects.
- Did not solve limitations on QoS and service availability requirements for critical Smart Grid services such as protection and control.
- Appeared mainly as a service for commercial Internet access.

The emergence of 4G systems focused on several aspects that started to be attractive for utilities. 4G provides higher reliability, higher availability, and lower latency services that can allow prioritized and guaranteed communications to adapt to Smart Grid requirements if a proper 4G network design and infrastructure is deployed. Such infrastructure requirements can be translated into appropriate service or network proposals stemming either from commercially available networks, or from join interest groups that may develop those services on specific-purpose networks for certain industry needs, or from private networks given that the proper access to spectrum is granted to utilities.

As a general rule, utilities have started to use 4G/LTE services as an evolutionary service process, once LTE radio modules have been introduced in the modems and routers utilities use for their existing services. LTE radio modules are typically integrated transparently in utility devices as long as their integration is not a burden in terms of development (re-design) or cost. Thus, there are some industry trends that, if the cost of the radio module is a significant percentage of the communication device cost, propose the use of lower-cost, though less-capable, LTE derivatives such as LTE-M and NB-IoT (see Chapter 8). However, the adoption of such alternatives has been rather limited for the time being.

Building upon the success of 4G/LTE, the latest 5G technologies entail now the promise of a new opportunity to tailor solutions to the expectations of Smart Grid services. Despite initial scepticism [73], 5G technologies come with multiple features that were never supported before in cellular technologies and deserve full consideration in the utility sector. Furthermore, with the addition of the appropriate mechanisms to offer service control (e.g. management of connectivity status and performance, not only referred to individual SIM cards but also including mechanisms to understand how the network status affects service availability and performance), 5G services may become attractive for utilities even if provided by a MNO, should the rest of the requirements be fulfilled. In particular, backward compatibility and lifetime terminal requirements are very important to utilities. Use cases for utilities need to consider lifetime durations typically above 15 years and even reaching 40 years for some assets. If the utility service infrastructure was to be refreshed sooner (e.g. following the most common three to four years cycles of the large commercial smartphone market), there might be unsurmountable barriers for MNOs' provided services adoption.

Last but not least, for 5G to reach a variety of applications and use cases with ambitious throughput, reliability, availability, and latency, a boost in infrastructure is needed, especially for the exploitation of the mm-wave bands. This infrastructure is a simple way to refer to the three components needed for a much denser deployment of base stations that happens around the availability of room (physical space), energy (electricity), and optical fiber. And here is where a connection with the utility world comes into perspective, not only as "one more vertical" to use 5G telecommunication services but also as an enabler of the effective and smooth deployment of mobile networks. The symbiotic relationship of energy and telecommunications world converges with the idea of pervasive infrastructure, instrumental to support our Society.

7.5.1 Smart Metering

The interest to remotely read meters has been present in the industry since long. However, there have been many obstacles to have meter reading implemented, not all related to technical or telecommunication aspects. On one hand, metering is not a unique homogeneous activity in a utility or across utilities. On the other hand, the benefits of the metering process are low if only applied to reading meters for billing purposes and do not address other aspects that can introduce further savings. In recent times, all these elements are put together in Cost-Benefit Analysis (CBA) assessments, to evaluate the convenience of implementing Smart Metering as one of the stages in the evolution toward Smart Grids. Other than that, the evolution of these systems, and the appetite to automate and digitalize metering processes, accompanied by the evolution of telecommunication alternatives to expedite it, can leverage existing telecommunication networks and technologies. Cellular technologies are some of those.

As explained in Chapter 5, Smart Metering can, at least, be classified into three broad groups: residential, C&I, and substation-related. The convenience and adoption of different telecommunication options depend on the group, the utility nature, and strategy, and the historic period in which each one needs to be addressed. Thus, although in some of these groups, certain telecommunication technologies are more prevalent due to its better adaptation to the requirements, exceptions can be found.

Some aspects are to be considered when evaluating a telecommunications technology for Smart Metering are:

- Integration of the telecommunication technology with the Smart Meter.
- Suitability of the telecommunications technology to the Smart Metering scenario.
- Volume of the Smart Meter market.
- Nature of the customer behind the meter.

The previous considerations are key to determine the best solution for each scenario. The reasons are connected to the nature of the meter and the heavy regulation around it, and where and how it needs to be installed (including physical premises and customers).

The meter is a metrologically controlled device. The meter needs to be controlled and certified by the appropriate government entity (national or local; always different for each country), so that there is a certainty over its accuracy. Every change in the meter must be verified by the appropriate authority, and the process involves time and cost. When a smart meter is considered, apart from the metrological component, there is the communications part. Smart meter vendors have developed designs that, without changing the metering part, can incorporate different telecommunication modules, ready to work with different technologies and networks.

We cannot forget the validation often needed in certain markets that use cellular technologies. In addition to the standard certification processes, some MNOs perform a validation process before the MNO can guarantee proper operation of the device in its network. There are three trends in smart meters designs, to address this context:

- Full integration of the telecommunication's component, i.e. inside the meter and electronically integrated with the rest of the elements. This approach is typical of PLC technologies and for massive markets (e.g. residential Smart Metering) where millions of low-cost smart meters are needed for pre-established time projects.
- Integration of the telecommunication's component in a Network Interface Card (NIC). The NIC is still internal to the meter but not electronically integrated. This approach allows incorporating different communications alternatives (e.g. 3GPP cards or any other standard or proprietary technology) without fully modifying the meters internal design. Different NICs can address different regional markets where volumes may not be massive.
- Externally attachable modems or routers, through standard serial interfaces (typically RS 232 or RS 485, and less often Ethernet). This design is not used in residential markets, but has the advantage that requires no meter re-certification.

C&I Smart Metering usually counts on internal NIC or externally provided telecommunication modems/routers. C&I smart meters are usually the responsibility of the C&I customer and often their property. C&I customers have the freedom to choose its own telecommunications alternative, compliant with certain requirements. For instance, acceptable connectivity options can be dial-up PSTN services (the customer offers a regular telephone line that is dialed by the utility) or public IP addresses offered through xDSL, FTTH, or cellular technologies. In both cases, these C&I customers' smart meters are accessed from utilities' Head-End systems. As it can be imagined, these C&I customers may not change their smart meters or connectivity technology for long periods. Thus, the utility needs to cope with many legacy technologies.

There are several aspects that deserve further attention:

- Dial-up connections are highly inconvenient, for efficiency and technical possibilities matters. On the one side, the connection phase and the data exchange with the appropriate command set

take longer than IP-based alternatives. Thus, there is a scalability problem at the Head-End system that can just be solved increasing the central resources. On the other side, it is a question of time that this connectivity type may be deprecated.
- Public IP addresses are not the efficient way of using these resources and tend not to be favored by TSPs. Further than that, modems and routers need to have the capability to remain connected, despite the fact that they are not to be activated by any other but a machine (the smart meter or the device on its own). If we consider that cellular IP connectivity started with the 2.5G, we can conclude that there are devices of over 20 years, with the limitations imposed on the state-of-the-art 20 years ago.
- C&I smart meters need to be directly accessed not only by the utility central systems but also by its owner and often by other entities that can coordinate or supervise utility activity. This aspect determines that either the smart meters, the modems or the routers may have other non-expected telecommunication possibilities, making them complex multi-purpose devices.

Residential Smart Metering is often addressed with fully integrated solutions. Cellular technologies have not been adopted in this context in a generalized way. Technology longevity aspects, full integration need (SIM card included), size and location constraints (reduced rooms, hidden-from-view placing, underground or metal-enclosed cabinets) among others, have favored PLC technologies in many world regions (e.g. Europe).

However, there are some markets where cellular technologies play a role, although combined with other technologies in the final access to the meter. The Smart Metering Program in Great Britain is one of those examples (covered with more detail in Chapter 8). Interestingly, the project started as a radio-only solution, assuming wireless connectivity for accessing the smart meters and the rest of the components independently of their location. The project has suffered delays and, more importantly, successive adaptations to identify the smart meters that could be realistically accessible. The new end date targets 85% of the 53 million electricity and gas meters by 2024 [74]. There are different reasons to exclude a certain percentage of smart meters, and the dependency on radio signal propagation at meter location is a very relevant one. The factors that negatively influence coverage refer to "meters located deep inside the premises or underground; property construction or property type; and metallic obstructions or enclosures likely to disrupt radio communications in the immediate environment" [75].

7.5.2 Distribution Grid Multiservice Access

Plenty of services in the Transmission and Distribution networks depend on connectivity within a facility, which may contain different services for grid elements within a short range. This is typically the case of PSs and SSs, where remote control needs can be grouped with other needs such as substation metering, protection management, surveillance, and all sorts of physical sensors and actuators connected to local controllers. In this situation, a single wideband or broadband connectivity network termination point at the facility can be leveraged to communicate and support remote access to different end points and functions of the Smart Grid within the facility (see Chapter 5).

Multiservice access connectivity is a readily available data service in cellular networks since M2M/MTC communications were enabled. This connectivity is already present and being used in different industries and activities, including utilities in some of the substations. This connectivity provides a unique connection service to different users or services, that while separated, are transported using a single CPE.

There are different options to configure private services across public networks, agnostic to the nature (i.e. wireline or wireless) of the underlying network. However, as the commercial implementation of cellular networks generally has a lower availability than other wireline alternatives, some solutions need to be found to increase service availability.

A commonly deployed solution is based on the use of dual-SIM card routers and multiple MNO subscriptions. With this approach, utilities can benefit from increased coverage (e.g. one of the networks can reach some areas not covered by the other and vice versa) and, in case of coverage overlapping, the connection to one of the networks can be used as a back-up of the other. Indeed, when coverage is available from the two networks, it is not common to use them simultaneously; on the contrary, the routers connect to one or the other depending on service conditions such as availability maximization, or MNOs traffic cost conditions. A similar but alternative solution to this dual-SIM card model is based on the Mobile Virtual Network Operator concept [76], i.e. the utility acting as a virtual operator in charge of its own SIM cards and with connection arrangements with different MNO's for network access. However, again, although some examples could be found, this is not the general rule.

On the technical side, Dynamic Multipoint VPN (DMVPN) based on Layer 3 IP connectivity provides field-proven connectivity mechanism for multiservice access across different distributed facilities, over public networks. DMVPN provides the capability to create a dynamic-mesh VPN without having to pre-configure (static) all possible tunnel end-point peers. Although there are other Layer 3, and Layer 2 options in the market, DMVPN is a solid one, well-proven and with a good track record of success, based on an ecosystem of different companies with valid implementations and products, and a vast number of stable field deployments [77].

DMVPN [78] relies on standard technologies such as multipoint Generic Routing Encapsulation (GRE) and Next Hop Resolution Protocol (NHRP). Multipoint GRE tunnels are used to establish a private network between the spokes and the central hub. These tunnels can use IPSec to protect the traffic when traversing the MNO network. NHRP creates a distributed mapping database of all the spoke tunnels to their public interface addresses (the one shared between the MNO and the utility, based on private IP addresses). The DMVPN hub is a router, and all the spokes are configured with tunnels to the hub. GRE and NHRP work dynamically creating the tunnel between them. Inside the tunnels, dynamic routing protocols can be used, as multicast and broadcast are supported. Thus, the multiservice network access gets implemented.

As it has already been mentioned, connectivity itself is not enough and minimum availability requirements have to be fulfilled as well. Availability may be hindered in any of two circumstances. First, service disruption can be caused by a degradation of the radio access connectivity, when received signal level drops below acceptable thresholds and an alternative cellular connection (e.g. in a different band or using other radio access technology) cannot be provided by the MNO. Second, the MNO network service may become unavailable due to a network failure or malfunctioning for whatever reasons (even if implemented with physical and logical protection mechanisms). In any of these circumstances, if none of the traffic routes within one MNO's network works, the router needs to look for alternative connectivity options within the back-up MNO's network.

To properly manage the dual interconnection capability and so increase availability, DMVPN supports a dual hub DMVPN layout. The concept consists of two separate DMVPN environments, one for each different MNO interconnection point connected to one DMVPN subnet. The spokes are connected to each of the DMVPN subnets. The dual configuration is managed with dynamic routing and the appropriate metrics to access each of the interconnection points, balancing or not their use, or simply using one of them as a back-up connection to be activated only if the primary is not available. In this case, if loss of connection when switching MNOs need to be avoided, dual router or radio modules should be used.

References

1 Ericsson (2020). *Ericsson Mobility Report* (November 2020) [Online]. https://www.ericsson.com/en/mobility-report (accessed 2 April 2021).

2 Scholefield, C. (1997). Evolving GSM data services. *Proceedings of ICUPC 97 – 6th International Conference on Universal Personal Communications*, San Diego, CA, USA, vol. 2, pp. 888–892. doi: 10.1109/ICUPC.1997.627291.

3 Romero, J.P., Sallent, O., Agusti, R., and Diaz–Guerra, M.A. (2005). Radio Resource Management Strategies in UMTS. Wiley.

4 ITU-R Recommendation M.1457 (2020). *Detailed Specifications of the Terrestrial Radio Interfaces of International Mobile Telecommunications-2000 (IMT-2000)* (October 2020).

5 ITU-R Recommendation M.2012-4 (2019). *Detailed Specifications of the Terrestrial Radio Interfaces of International Mobile Telecommunications-Advanced (IMT-Advanced)* (November 2019).

6 European Commission (2020). *Study on the Current And Prospective Use of the 900 MHz Band by GSM as a Technology of Reference, Considering Present and Future Union Policies.* https://ec.europa.eu/digital-single-market/en/news/study-current-and-prospective-use-900-mhz-band-gsm-technology-reference-considering-present-0 (accessed 27 April 2021).

7 Telefónica (2020). *Telefónica Switches on 5G and 75% of the Spanish Population Will Obtain a Signal This Year* (September 2020) [Online]. https://www.telefonica.com/documents/737979/145927704/pr-tef-5g.pdf/30e1abb9-f6fd-4579-a65f-31fabfc76d07?version=1.0 (accessed 5 February 2021).

8 3GPP (2020). *3GPP Meets IMT-2020* (November 2020) [Online]. https://www.3gpp.org/news-events/2143-3gpp-meets-imt-2020 (accessed 2 April 2021).

9 5G Americas (2020). *Mobile Communications Beyond 2020 – The Evolution of 5G Towards the Next G* (December 2020) [Online]. https://www.5gamericas.org/wp-content/uploads/2020/12/Future-Networks-2020-InDesign-PDF.pdf (accessed 2 April 2021).

10 ITU-R M.2083-0 (2015). *IMT Vision – Framework and Overall Objectives of the Future Development of IMT for 2020 and Beyond* (June 2015).

11 Report ITU-R M.2410-0 (2017). *Minimum Requirements Related to Technical Performance for IMT-2020 Radio Interface(s)* (November 2017).

12 3GPP TS 22.261 v17.3.0 (2020). *Service Requirements for the 5G System; Stage 1 (Release 17)* (July 2020).

13 3GPP TS 22.104 v17.3.0 (2020). *Service Requirements for Cyber-Physical Control Applications in Vertical Domains* (July 2020).

14 3GPP TS 22.186 v16.2.0 (2019). *Enhancement of 3GPP Support for V2X Scenarios; Stage 1 (Release 16)* (June 2019).

15 3GPP TS 22.289 v17.0.0 (2019). *Mobile Communication System for Railways* (December 2019).

16 3GPP TR 37.910 v16.1.0 (2019). *Study on Self-evaluation Towards IMT-2020 Submission (Release 16)* (September 2019).

17 Ferrús, R. and Sallent, O. (2015). Mobile Broadband Communications for Public Safety: The Road Ahead Through LTE Technology. Wiley.

18 ABIresearch (2020). *Private Cellular Networks to Generate over US$64 Billion in Equipment Revenues by 2030* (October 2020) [Online]. https://www.abiresearch.com/press/private-cellular-networks-generate-over-us64-billion-equipment-revenues-2030 (accessed 2 April 2021).

19 5GACIA (2019). *5G Alliance for Connected Industries and Automation (White Paper)* (July 2019) [Online]. https://www.5g-acia.org/publications/5g-non-public-networks-for-industrial-scenarios-white-paper (accessed 2 April 2021).

20 ITU M.2411-0 (2017). *Requirements, Evaluation Criteria and Submission Templates for the Development of IMT-2020* (November 2017).

21 ITU-R Radio Regulation. Edition of 2020 [Online]. https://www.itu.int/pub/R-REG-RR-2020 (accessed 2 April 2021).
22 Pujol, F., Manero, C., Carle, B., and Remis, S. (2020). *Quarterly Report 8, Up to June 2020, European 5G Observatory* (July 2020) [Online]. http://5gobservatory.eu/wp-content/uploads/2020/07/90013-5G-Observatory-Quarterly-report-8_1507.pdf (accessed 2 April 2021).
23 Radio Spectrum Policy Group (RSPG) (2019). *Strategic Spectrum Roadmap Towards 5G for Europe – RSPG Opinion on 5G Implementation Challenges (RSPG 3rd Opinion on 5G)*. RSPG19-007 FINAL (January 2019).
24 RadioResource Mission Critical Communications (2020). *Benelux Countries Demonstrate Significance of Spectrum for Private LTE Networks* (August 2020) [Online]. https://www.rrmediagroup.com/Features/FeaturesDetails/FID/1016 (accessed 2 April 2021).
25 Sohul, M.M., Yao, M., Yang, T., and Reed, J.H. (2015). Spectrum access system for the citizen broadband radio service. *IEEE Communications Magazine* 53 (7): 18–25.
26 OnGo Alliance. *OnGo for Power & Utilities* [Online]. https://www.cbrsalliance.org/industrial-iot/power-and-utilities (accessed 2 April 2021).
27 Matinmikko, M., Okkonen, H., Palola, M. et al. (2014). Spectrum sharing using licensed shared access: the concept and its workflow for LTE-advanced networks. *IEEE Wireless Communications* 21 (2): 72–79. https://doi.org/10.1109/MWC.2014.6812294.
28 Third Generation Partnership Project (3GPP) [Online]. www.3gpp.org (accessed 2 April 2021).
29 Ericsson (2020). *5G Evolution: 3GPP Release 16 and 17 Overview*. White paper (March 2020).
30 Nokia (2020). *5G Releases 16 and 17 in 3GPP*. White paper (April 2020).
31 Qualcomm (2020). *Propelling 5G Forward: A Closer Look at 3GPP Release 16*. White paper (July 2020).
32 Hewlett Packard (2020). *VRAN 2.0 on HPE Infrastructure* (May 2020) [Online]. https://h50146.www5.hpe.com/products/servers/document/pdf/edgeline/vran2.0.pdf (accessed 2 April 2021).
33 ETSI (2017). *Network Functions Virtualisation – White Paper on NFV Priorities for 5G*. White paper (February 2017) [Online]. https://portal.etsi.org/NFV/NFV_White_Paper_5G.pdf (accessed 2 April 2021).
34 Dahlman, E., Parkvall, S., and Sköld, J. (2013). 4G: LTE/LTE-Advanced for Mobile Broadband. Academic Press.
35 Cox, C. (2014). An Introduction to LTE, LTE-Advanced, SAE, VoLTE and 4G Mobile Communications, 2e. Wiley.
36 3GPP TS 38.401 v16.4.0 (2021). *NG-RAN Architecture Description* (January 2021).
37 Johnson, C. (2019). 5G New Radio in Bullets. Independently Published.
38 Dahlman, E. and Parkvall, S. (2018). 5G NR. The Next Generation Wireless Access Technology. Elsevier.
39 Bertenyi, B., Burbidge, R., Masini, G. et al. (2018). NG radio access network (NG-RAN). *Journal of ICT* 6 (1&2): 59–76.
40 Open RAN Alliance (2018). *O-RAN: Towards and Open and Smart RAN*. White paper (October 2018) [Online]. https://www.o-ran.org/resources (accessed 2 April 2021).
41 Small Cell Forum (2020). *5G FAPI: PHY API Specification*. SCF222 (March 2020) [Online]. https://scf.io/en/documents/222_5G_FAPI_PHY_API_Specification.php (accessed 2 April 2021).
42 Olsson, M., Sultana, S., Rommer, S. et al. (2009). SAE and the Evolved Packet Core. Academic Press.
43 Agustí, R., Bernardo, F., Casadevall, F. et al. (2010). LTE: Nuevas tendencias en comunicaciones móviles. Fundación Vodafone España https://proyectolte.files.wordpress.com/2012/09/lte-nuevas-tendencias.pdf (accessed 2 April 2021).

44 Mademann, F. (2018). The 5G system architecture. *Journal of ICT* 6 (1&2): 77–86.

45 Mayer, G. (2018). RESTful APIs for the 5G service based architecture. *Journal of ICT* 6 (1&2): 101–116.

46 3GPP TS 23.501 v16.7.0 (2020). *System Architecture for the 5G System*. (December 2020).

47 3GPP TS 29.500 v17.1.0 (2020). *Technical Realization of Service Based Architecture; Stage 3 (Release 17)* (December 2020).

48 Coker, O. and Azodolmolky, S. (2017). Software-Defined Networking with OpenFlow, 2e. Packt Publishing.

49 Camarillo, G. and Garcia-Martin, M.-A. (2008). The 3G IP Multimedia Subsystem (IMS): Merging the Internet and the Cellular Worlds, 3e. Wiley.

50 Huawei (2018). *Vo5G Technical White Paper* (July 2018) [Online]. https://www-file.huawei.com/-/media/corporate/pdf/white%20paper/2018/vo5g-technical-white-paper-en-v2.pdf (accessed 2 April 2021).

51 Ericsson (2018). *Communication Services over LTE, Wi-Fi and 5G*. White paper (April 2018) [Online]. https://www.ericsson.com/49d361/assets/local/reports-papers/white-papers/wp-voice-and-video-calling-over-lte.pdf (accessed 2 April 2021).

52 Nokia (2020). *Voice over 5G: The Options for Deployment*. White paper [Online]. https://onestore.nokia.com/asset/f/207122 (accessed 2 April 2021).

53 GSM Association (2020). *Official Document IR.92 – IMS Profile for Voice and SMS*, Version 15.0 (14 March 2020).

54 Chitturi, S., 3GPP SA6 Chair (2020). *[Blog] 5G Application Standards: Enabling B2B Services* (December 2020) [Online]. https://research.samsung.com/news/-Blog-5G-Application-Standards-Enabling-B2B-Services (accessed 2 April 2021).

55 3GPP (2020). *5G Application Standards* (September 2020) [Online]. https://www.3gpp.org/news-events/partners-news/2137-sa6_webinar (accessed 2 April 2021).

56 3GPP S6-191415 (2019). *5G Vertical User Workshop Presentation on SA6 Activities* (July 2019) [Online]. https://www.3gpp.org/ftp/tsg_sa/WG6_MissionCritical/TSGS6_032_Roma/docs/S6-191415.zip (accessed 2 April 2021).

57 FirstNet [Online]. www.firstnet.com (accessed 2 April 2021).

58 3GPP TS 23.502 v16.7.0 (2020). *Procedures for the 5G System* (December 2020).

59 3GPP TS 36.101 v17.0.0 (2021). *User Equipment (UE) Radio Transmission and Reception (Release 12)* (January 2021).

60 3GPP TS 38.101 v17.0.0 (2021). *NR; User Equipment (UE) Radio Transmission and Reception; Part 1: Range 1 Standalone* (January 2021).

61 ETSI (2018). *MEC in 5G Networks*, 1e. ETSI White Paper No. 28 (June 2018).

62 Ferrús, R., Sallent, O., Pérez-Romero, J., and Agustí, R. (2018). On 5G radio access network slicing: radio interface protocol features and configuration. *IEEE Communications Magazine* 56 (5): 184–192. https://doi.org/10.1109/MCOM.2017.1700268.

63 Kinney, S. (2020). *Bringing 5G NR into Unlicensed Spectrum: The Private Network Opportunity*. RCRWirelessNews Report (December 2020).

64 GSA Report (2020). *Unlicensed Spectrum: August 2020 – LTE & 5G Executive Summary* (August 2020).

65 MulteFire Alliance [Online]. www.multefire.org (accessed 2 April 2021).

66 Chandramouli, D. (2020). *5G for Industry 4.0* (May 2020) [Online]. https://www.3gpp.org/news-events/2122-tsn_v_lan (accessed 2 April 2021).

67 Prakash, R. (Qualcomm) (2020). *Transforming Enterprise and Industry with 5G Private Networks* (October 2020).

68 3GPP TR 23.734 v16.2.0 (2019). *Study on Enhancement of 5G System (5GS) for Vertical and Local Area Network (LAN) Services (Release 16)* (June 2019).

69 Airwave. *Emergency Services Network* [Online]. https://www.airwavesolutions.co.uk/the-service/emergency-services-network (accessed 2 April 2021).

70 Hussein, O. (2019). *Circuit Switched Data (CSD) Application Note* (January 2019) [Online]. https://www.siretta.com/2019/01/circuit-switched-data-csd-setting-application-note (accessed 2 April 2021).

71 Red Eléctrica de España (2012). *Sistema de información de medidas electrónicas*. Version 1.0 (July 2012) [Online]. https://www.ree.es/sites/default/files/01_ACTIVIDADES/Documentos/Documentacion-Simel/SIMEL_Guia_Configuracion_Modem_GSM_v1.0.pdf (accessed 2 April 2021).

72 Antón-Haro, C. and Dohler, M. (eds.) (2015). Machine-to-Machine (M2M) Communications. Architecture, Performance and Applications. Woodhead Publishing Limited.

73 Utilities Technology Council (2019). *Cutting Through the Hype: 5G and Its Potential Impacts on Electric Utilities* (March 2019) [Online]. https://utc.org/wp-content/uploads/2019/03/Cutting_through_the_Hype_Utilities_5G-2.pdf (accessed 2 April 2021).

74 Hinson, S. (2019). *Energy Smart Meters*. Briefing paper. Number 8119 (7 October 2019). House of Commons Library.

75 Data Communications Company (DCC) (2020). *DCC Statement of Service Exemptions* (June 2020) [Online]. https://www.smartdcc.co.uk/media/3952/dcc_statement_of_service_exemptions_2020_v10-final.pdf (accessed 2 April 2021).

76 Camarán, C. and De Miguel, D. (Valoris) (2008). Mobile Virtual Network Operator (MVNO) basics:What is behind this mobile business trend. *Telecom Practice* (October 2008).

77 Sendin, A., Sanchez-Fornie, M.A., Berganza, I. et al. (2016). Telecommunication Networks for the Smart Grid. Artech House.

78 Cisco Systems (2006). *Dynamic Multipoint IPsec VPNs (Using Multipoint GRE/NHRP to Scale IPsec VPNs)* [Online]. http://www.cisco.com/c/en/us/support/docs/security-vpn/ipsec-negotiation-ike-protocols/41940-dmvpn.pdf (accessed 2 April 2021).

8

Wireless IoT Technologies

8.1 Introduction

The Internet of Things (IoT) embodies the vision of connecting virtually anything with everything. Wireless technologies play a fundamental role in the realization of this vision, allowing IoT devices to communicate without the need of cables.

Multiple sets of wireless IoT technologies coexist today because of the broad range and diversity of communications requirements associated with current and emerging IoT applications.

Which IoT connectivity solution fits best for a given application depends on multiple aspects [1]. Selection criteria may include whether the solution is indoor or outdoor, the mobility of the IoT devices (e.g. static or moving terminals), costs, availability of a specific technology, availability of spectrum, required power, battery life, data volumes and frequency, the needed bandwidth, the needed performance, proven stability, maturity, security, future evolution, and developments in the specific technologies, ecosystem (developers, partners, integrators), and far more. Also, the development of a given application may require the combination of more than one wireless IoT technology.

This chapter focuses on the most relevant wireless IoT technologies for the Smart Grid.

8.2 Mainstream Wireless IoT Technologies for the Smart Grid

Three dimensions are commonly used to compare the scope of wireless IoT technologies: communications range, achievable data rate, and type of spectrum used (licensed or license-exempt).

The short-range, low-data rate segment shows the highest technological diversity. Solutions here cover from mass-market consumer IoT applications (e.g. healthcare/fitness wearables, smart home appliances) to industrial applications (e.g. factory automation, critical infrastructure monitoring, asset tracking, SCADA systems). Among the most popular technologies in this segment, Bluetooth Low Energy (BLE), Zigbee, Z-Wave, and Thread [2] can be found. A key commonality among them is the use of license exempt spectrum, mainly sub-1 and 2.4 GHz ISM bands. Another common trait in some of these technologies is an implementation of the lower layers of the IoT protocol stack based on, or derived from, the IEEE 802.15.4 standard, as it is the case of ZigBee and Thread technologies. Indeed, an adaptation of the IEEE 802.15.4 standard is also at the core of the Wireless Smart Ubiquitous Network (Wi-SUN) technology, specifically targeting the needs of the smart utility sector.

Smart Grid Telecommunications: Fundamentals and Technologies in the 5G Era, First Edition. Alberto Sendin, Javier Matanza, and Ramon Ferrús.
© 2021 John Wiley & Sons, Inc. Published 2021 by John Wiley & Sons, Inc.

In the high-data rate but short-to-medium coverage range segment, the prevalent technology as of today is Wi-Fi. This technology, based on the IEEE 802.11 standards, makes use also of license-exempt spectrum, currently in the 2.4 and 5 GHz bands, and further extending into the 6 GHz band in some countries. From the early versions of the standard (IEEE 802.11b) to the latest one (IEEE 802.11ac, also known as Wi-Fi 6), data rates have increased from few Mbps to several Gbps, allowing the technology to properly cope with bandwidth demanding local area connectivity in ranges typically from tens of meters to a few hundred meters. Wi-Fi technology also counts with extensions focused on long range, low power IoT applications such as IEEE 802.11ah, marketed as Wi-Fi HaLow [3].

Within the wide area coverage segment, unlicensed spectrum-based LPWAN technologies have emerged more recently as a very low cost alternative technology choice to cellular networks. Among the most prominent solutions in this segment are Sigfox and LoRaWAN, both operating in sub-1 GHz unlicensed spectrum [4]. Coverage ranges of these technologies may span some few kilometers, with data rates in the order of a few kbps or below, and devices only transmitting low amounts of data (e.g. up to 140 messages of 12 bytes a day in Sigfox).

Finally, cellular technologies provide nowadays multiple features that allow them to address from the "high-end" IoT applications segment, requiring low-latency, high-reliability, and high-data rates communications requirements – which can be built upon the URLLC/Industrial IoT services and features described in Chapter 7, to the "low-end" IoT applications, in need of long-range, low power low data rate requirements – which can be built upon mMTC services and features such as EC-GSM-IoT, LTE-M, and NB-IoT, known collectively as *Cellular IoT* [5]. In contrast to the other mentioned wireless IoT technologies, cellular technologies have traditionally been deployed over licensed spectrum. However, the latest specifications also come with some support for operation in unlicensed spectrum.

In this context, several control and management applications in the Smart Grid may benefit from the applicability of wireless IoT technologies. Table 8.1 outlines the main characteristics of the set of wireless IoT technologies that are of special relevance for the Smart Grid. These technologies are described in the following sections.

8.3 IEEE 802.15.4-based Technologies: Zigbee and Wi-SUN

8.3.1 Scope and Standardization

IEEE 802.15.4 is a technical standard for Low-Rate Wireless Personal Area Networks (LR-WPANs). A LR-WPAN is conceived as [7]: "a simple, low-cost communication network that allows wireless connectivity in applications with limited power and relaxed throughput requirements. The main objectives of an LR-WPAN are ease of installation, reliable data transfer, extremely low cost, and a reasonable battery life, while maintaining a simple and flexible protocol." IEEE 802.15.4 specifies the PHY and MAC layers for LR-WPANs, covering a wide variety of PHY layer configurations and operating frequency bands. The IEEE 802.15.4 standard constitutes the basis for Zigbee and Wi-SUN technologies, each of which further extends the IEEE 802.15.4 standard by specifying the networking and application layers sitting above the PHY and MAC layers.

8.3.1.1 IEEE 802.15.4 Standard

The IEEE 802.15.4 standard is maintained by the IEEE 802.15 WG [8]. The first version of the IEEE 802.15.4 standard was released in 2003 and, since then, it has evolved and extended its applicability domains.

Table 8.1 Main wireless IoT technologies for smart grid applications.

	Zigbee	Wi-SUN FAN	Sigfox	LoRa/LoRaWAN	NB-IoT	LTE-M
Applicability domain	LAN, NAN	FAN	WAN, FAN	WAN, FAN	WAN, FAN	WAN, FAN
Baseline standards	IEEE 802.15.4	IEEE 802.15.4, IETF, ANSI	Proprietary	Proprietary	3GPP specifications	3GPP specifications
Channel bandwidth (kHz)	600–5000	100–600	125	50–125	200	1400
Spectrum	Unlicensed	Unlicensed	Unlicensed	Unlicensed	Licensed Some support for unlicensed	Licensed Some support for unlicensed
Frequency bands	Sub-1 and 2.4 GHz	Sub-1 GHz	Sub-1 GHz	Sub-1 GHz	LTE operating bands	LTE operating bands
Transmission scheme	Multiple	FSK	UNB GFSK/BPSK	LoRa CSS	OFDM based	OFDM based
Data rate (kbps)	250	50–300	0.6	0.3–50	200	1000
Latency (multiples of milliseconds)	10×	10×	1000×	1000×	100×	10×
Application payload	Not limited	Not limited	~10 bytes	~200 bytes	Not limited	Not limited
Network topology	Star, Mesh	Star, Mesh	Star	Star	Star	Star
Radio link range or MCL (Note 1)	~10–100 m	~500 m (2–3 km Line of Sight -LOS-)	MCL: ~150/170 dB	MCL: ~150/170 dB	MCL: 164 dB	MCL: 159 dB

Note 1: Maximum Coupling Loss (MCL) is a link budget metric computed as the difference between transmit power and receiver sensitivity [6]. Values given here are only indicative.

The initial standard, IEEE Std. 802.15.4-2003, defined only two PHY layers and a simple MAC layer based on the CSMA/CA protocol. The latest revision of the standard, IEEE Std. 802.15.4-2020, published in July 2020 [7], comes with more than 20 PHY layer options in different frequency bands and a significant variety of MAC enhancements incorporated after several amendments and revisions of the standard in 2006, 2011, and 2015. Some of the features added in this process have resulted from the consideration of special application domains, such as Low-Energy Critical Infrastructure Monitoring (LECIM) and Smart Utility Networks (SUNs) [9].

8.3.1.2 Zigbee

Layered above the IEEE 802.15.4 PHY and MAC layers, Zigbee defines the upper layer functionality (e.g. routing, security, applications) to deliver a full-stack solution for very low-cost, very low-power-consumption, two-way, wireless communications. Zigbee is nowadays a very popular implementation of the IEEE 802.15.4 standard, with a wide adoption in the smart home market and different applicability levels in other spaces such as building automation, industrial lighting, Smart Metering, etc.

Zigbee specifications are developed and promoted by the Zigbee Alliance[1] [10], which conducts product certification and interoperability testing in order to ensure that products from different vendors that use the Zigbee stack are compatible with one another.

The first version of the Zigbee specifications was completed in 2003. Since then, it has been revised multiple times, with the latest revision, commonly referred to as Zigbee 3.0, published in 2017 [11]. In this process, different versions and adaptations of the Zigbee stack have been produced. The most complete one is known as ZigBee PRO, developed as an enhancement of the original ZigBee protocol by providing a number of extra features that are particularly useful for supporting networks with a large number of nodes (e.g. hundreds or even thousands of nodes).

In addition to the networking features, an important aspect developed as part of the Zigbee specifications are the so-called *application profiles* (or simply *profiles*). A profile defines the roles and functions of devices in a Zigbee network. For example, there is a Smart Energy Profile (SEP) that defines various *device types* such as energy service portal, load controller, thermostat, in-home display, and so on. And then, for each device type, it specifies the required functionality. For example, a load controller must respond to a defined command to turn a load on or off.

The Zigbee Alliance promotes different flavors of the Zigbee technology in order to adapt it to different market segments and needs. Among them, related to utility applications:

- *Zigbee*, also known as Zigbee 3.0. It is the core solution, targeting applicability in a wide range of segments. ZigBee 3.0 employs the ZigBee PRO protocol and is designed to facilitate the deployment of general Zigbee networks that are not market-specific (i.e. devices from different market sectors may connect to the same wireless network).
- *Zigbee Smart Energy*. This is a Zigbee solution that specifically addresses the smart energy segment, intending to provide interoperable products that monitor, control, inform, and automate the delivery and use of energy and water [12]. Compared to Zigbee 3.0, security requirements are stricter in the Zigbee Smart Energy (ZSE) solution.
- *JupiterMesh*, though with limited public information available [13]. JupiterMesh addresses the smart energy segment, with a focus on delivering a NAN solution. Like Wi-SUN (see Section 8.3.1.3), JupiterMesh is based on the IEEE 802.15.4g and 802.15.4e amendments of the IEEE 802.15.4 standard.

---

1 The Zigbee Alliance is now the Connectivity Standards Alliance (CSA), as can be read in https://csa-iot.org/.

8.3.1.3 Wi-SUN

The Wireless Smart Ubiquitous Network (Wi-SUN) is a wireless communication technology intended to provide a communications infrastructure for very large-scale outdoor networks and industrial devices (e.g. smart meters and street lights) for use by utilities, city developers, and other service providers. In addition to IEEE 802.15.4 for the PHY and MAC layers, Wi-SUN is based on various IEEE, IETF, ANSI/TIA, and ETSI standards to support the upper layers and security framework of the Wi-SUN solution (in contrast to Zigbee, whose upper layers have been entirely specified by the Zigbee Alliance).

The Wi-SUN Alliance [14] is the industrial forum established for the specification and promotion of the Wi-SUN technology. The Wi-SUN Alliance started operations in Japan in 2012 and is focused on identifying functionality required for different application areas to develop technical *profiles*, consisting of several open standards to meet the functional requirements for the applications. Along with the technology specifications, the Wi-SUN Alliance also develops test plans to verify the compliance of products with the profile and interoperability with other conformant products. The Wi-SUN Alliance claims that the US and Japanese markets are ramping up in the adoption of Wi-SUN solutions.

Following are the relevant profiles for electricity utilities; Wi-SUN has also defined Resource Limited Monitoring and Management (RLMM) and Japan Utility Telemetering Association (JUTA) profiles for other markets [15]:

- *HAN profile*. It is a profile for HANs, also referred to as ECHONET profile. It is a specification to establish a Japanese official application profile for electricity energy management, named as ECHONET [16]. It is used for the communication between a smart electricity meter and HEMS. The Wi-SUN HAN/ECHONET profile is based on IEEE 802.15.4, 6LowPAN, IPv6, UDP, and Protocol for carrying Authentication for Network Access (PANA), as PHY, MAC, convergence layer, network layer, transport layer, and authentication protocol, respectively. The HAN/ECHONET profile also includes two hops relay communication in datalink layer and supports communication between HEMS and radio devices set in the home electrical appliances.
- *FAN profile*. It is a profile for FANs that connects field devices into utility operations [17]. The field devices include AMI and DA components such as meters, power line sensors, capacitor banks, DERs, etc. The FAN profile is also based on IEEE 802.15.4 and IPv6 connectivity and allows for both Layer 3 and Layer 2 packet routing and forwarding. The FAN profile is the one further described in this section.

8.3.2 Network and Protocol Stack Architecture

Baseline network components and topologies supported by Zigbee and Wi-SUN are generally taken from the IEEE 802.15.4 standard. These baseline aspects are described below under Section 8.3.2.1, which is then followed by Sections 8.3.2.2 and 8.3.2.3 that further elaborate on the specifics of, respectively, the Zigbee and Wi-SUN protocol stack architectures.

8.3.2.1 Network Components and Topologies

The basic component of an IEEE 802.15.4 network is called *device*. A device represents any equipment with a radio module that implements the MAC and PHY layers of the IEEE 802.15.4 standard. Two or more devices communicating on the same physical radio channel are said to constitute a Personal Area Network (PAN). There is one device in the PAN that takes the role of the PAN *coordinator*, which is in charge of the overall control of the PAN (e.g. selection of the radio channel, decisions on admission of devices into the PAN, etc.). In PANs involving multihop communications and clusters of devices communicating in different channels (see Section 8.3.2.1.3), in addition to the PAN coordinator, there could be other devices taking the role of coordinators, which in this case refers to having control of the channel used to interact with the in-range devices.

Devices are classified in two types: Full-Function Devices (FFDs) and Reduced-Function Devices (RFDs). An FFD is a device that is capable of serving as a PAN coordinator or coordinator. In contrast, an RFD is a device that is not capable of serving as a PAN coordinator or coordinator. An RFD is intended for applications that are extremely simple, such as a light switch or a passive infrared sensor; it does not have the need to send large amounts of data and only associates with a single FFD at a time. Consequently, the RFD can be implemented using minimal resources and memory capacity.

8.3.2.1.1 Network Topologies

Depending on the application requirements, an IEEE 802.15.4 network operates under two baseline topologies: the star topology or the peer-to-peer topology. Illustrations of both are shown in Figure 8.1.

In the star topology, the communication is established always between devices and the PAN coordinator. The PAN coordinator is typically line-powered, while the devices are more likely to be battery-powered. Applications that benefit from a star topology include home automation, personal computer peripherals, games, and personal health care.

In turn, the peer-to-peer topology differs from the star topology in that any device is able to communicate with any other device as long as they are in range of one another. A peer-to-peer topology also has a PAN coordinator, which can be nominated as such, for instance, by virtue of being the first device to trigger the network formation. Applications such as industrial control and monitoring, wireless sensor networks, asset and inventory tracking, intelligent agriculture, and security would benefit from such a network topology.

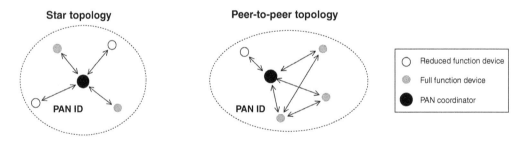

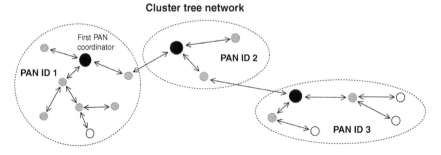

Figure 8.1 IEEE 802.15.4 network topologies.

The peer-to-peer topology is the basis for implementing more complex network formations, such as mesh or tree network topologies. Indeed, mesh/tree network topologies with multihop communications are key to extend the range of the IEEE 802.15 network further than the range of the radios, which is a limitation in the case of the star topology. Moreover, a mesh network topology may support dynamic route determination as well as redundant routing paths. Such features enable the realization of *self-healing* systems, able to cope with disruptions in a given path (radio link) and/or failures in a routing node. In any case, the routing capabilities to enable such multihop communications are not part of the IEEE 802.15.4 standard and shall be implemented in the upper layers.

8.3.2.1.2 Addressing

IEEE 802.15.4 supports both extended (64-bit) and short (16-bit) addressing for information routing. An extended address is assigned to every device radio module that complies with the IEEE 802.15.4 specification. This is a globally unique number made up of an Organizationally Unique Identifier (OUI) plus 40 bits assigned by the manufacturer of the radio module. OUIs are obtained from IEEE to ensure global uniqueness.

In addition, a device can be assigned a short address during the association process to the PAN, which is unique in that network. Thus, a device will use either the extended address or the short address for communication within the PAN.

In addition to device addresses, each independent PAN selects a unique network identifier, known as the PAN ID in Figure 8.1. This PAN ID allows communication between devices within a network using short addresses and enables transmissions between devices across independent networks. The PAN coordinator assigns the PAN ID when it creates the network. A device can try and join any network or it can limit itself to a network with a particular PAN ID. Importantly, the PAN ID shall be different among all networks that may be currently operating within the radio communications range.

8.3.2.1.3 Network Formation with Mesh Topologies

While the network formation has to be orchestrated with the involvement of the higher layers, the IEEE 802.15.4 standard provides a number of mechanisms and procedures for starting and maintaining PANs, including beacon frame generation, scanning, synchronization, device discovery, PAN ID conflict resolution, association and disassociation.

To explain the principles of how network formation works in IEEE 802.15.4-based systems, we will consider a complex network topology such as the so-called *cluster tree network*, as in Figure 8.1. This is a special case of the peer-to-peer topology in which most devices are FFDs. To create such a network, a device referred to as the first PAN coordinator will trigger the formation of the first cluster by choosing an unused PAN ID (e.g. PAN ID 1) and will start broadcasting beacon frames to neighboring devices. A candidate neighboring device receiving a beacon frame is able to join the network directly at the PAN coordinator. If the device is admitted, the PAN coordinator adds the new device as a child device in its neighbor list. Likewise, the newly joined device adds the PAN coordinator as its parent in its neighbor list and, depending on its capabilities, could also begin transmitting periodic beacons so that other candidate devices found in its range would be able to join the network at that device. The simplest form of a cluster tree network is a single cluster network, but larger networks are possible by forming a mesh of multiple neighboring clusters, as in this example. To do so, the first PAN coordinator will instruct another device to become the PAN coordinator of a new cluster adjacent to the first one (e.g. PAN ID 2) and, similar to the first cluster formation, other devices could gradually connect to the new cluster. The advantage of such a

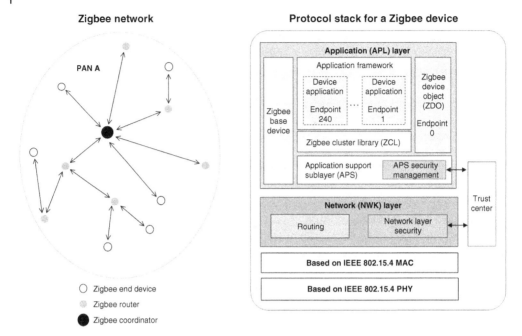

Figure 8.2 Zigbee network architecture and protocol stack of a Zigbee device.

multicluster structure is increased coverage area, while the disadvantage is a higher message latency due to the multihop transfer when source and destination devices are not in direct range.

8.3.2.2 Zigbee Network Architecture and Protocol Stack

A Zigbee network supports star, tree, and mesh topologies [11]. Three different types of logical devices are specified (see Figure 8.2): Zigbee *coordinator*, *router*, and *end device*. The Zigbee coordinator is the PAN coordinator, responsible for initiating and maintaining the devices on the network. Zigbee routers are FFD devices able to move data and control messages through the network and Zigbee end devices can be either FFD or RFD devices just acting as leaf nodes, with no routing capabilities. Zigbee specifications describe only intra-PAN communications.

The protocol stack architecture of a Zigbee device is also illustrated in Figure 8.2. The PHY and MAC layers are based on the IEEE 802.15.4 standard, as previously mentioned. Above them, the Zigbee specification defines the Network (NWK) layer and the Application Layer (APL), both developed by the Zigbee Alliance. The NWK layer handles network addressing and routing, with support for mesh networking. The NWK also provides security for the network, ensuring both authenticity and confidentiality (128-bit AES encryption is supported). The APL is the framework for the development of Zigbee applications. More details on the features and capabilities of these layers are addressed later in Section 8.3.3.3.

8.3.2.3 Wi-SUN FAN Network Architecture and Protocol Stack

A Wi-SUN FAN is a self-forming and self-healing system with a flexible architecture to build mesh or star topologies [17]. A Wi-SUN FAN may contain one or more PANs, as in Figure 8.3. Within a PAN, devices assume one of three operational roles:

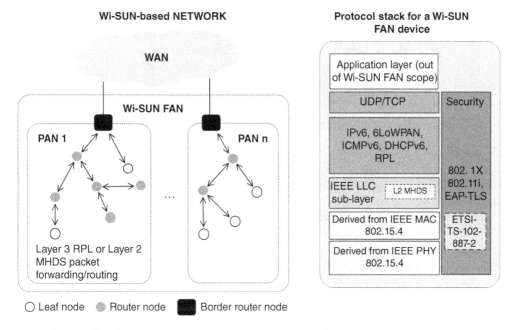

Figure 8.3 Wi-SUN FAN architecture and protocol stack of a Wi-SUN FAN device.

- Border Router, providing WAN connectivity to the PAN. The Border Router maintains source routing tables for all its nodes, provides node authentication and key management services, and disseminates information (e.g. broadcast schedules).
- Router nodes, providing upward and downward packet forwarding within a PAN.
- Leaf nodes provide minimum capabilities: discovering and joining a PAN, send/receive IPv6 packets, etc.

The protocol stack of a generic Wi-SUN FAN node is also illustrated in Figure 8.3. The PHY and MAC layers are derived from the IEEE 802.15.4 standard.

At the network layer, FAN nodes support IPv6 (RFC2460) with IPv6 over Low power Wireless Personal Area Networks (6LoWPAN, RFC6282) adaptations for header compression and fragmentation. ICMPv6 (RFC4443) is used for control plane information exchange along with DHCPv6 (RFC3315) for automated address management. Both unicast and multicast forwarding are supported. Two methods are available for packet routing: Layer 3 Routing Protocol for Low-Power and Lossy Networks (RPL), specified in RFC6550, and Layer 2 Multihop Delivery Service (MHDS), standardized in ANSI/TIA-4957.210 [18]. Layer 3 RPL is the mandatory method while Layer 2 MHDS is optional. In a PAN, only one routing method shall be used (i.e. L3 and L2 routing methods are not operated simultaneously). On top of the IPv6 connectivity, both UDP and TCP services are supported and at application layer, which is outside the scope of the Wi-SUN FAN specification, different protocols can be used, such as DLMS/COSEM, ANSI C12.22, DNP3, IEC 60870-5-104, ModBus TCP, and CoAP-based management protocols [19].

With respect to security mechanisms, Wi-SUN FAN supports Layer 2 network access control, mutual authentication, and the establishment of secured pairwise links between a FAN node and its PAN Border Router based on an adaptation of the IEEE 802.1X over IEEE 802.15.4 protocol

(IEEE 802.15.9 [20]). Access control uses EAP-TLS (RFC5216) and supports digital certificates. IEEE 802.11i Group Key Management and Node Pairwise (N2NP) Authentication (defined in ETSI TS 102.887-2 [21]) are also supported, the latter as an optional feature.

More details on the features and capabilities of PHY, MAC, and network layers of Wi-SUN FAN are addressed later in Section 8.3.3.4.

8.3.3 Main Capabilities and Features

8.3.3.1 IEEE 802.15.4 Physical Layer

IEEE Std. 802.15.4-2020 [7] includes over 20 PHY layer variants to consider different applications and regulatory environments. Supported modulations (see Table 8.2) range from simple Frequency Shift Keying (FSK) to Direct Sequence Spread Spectrum (DSSS), Chirp Spread Spectrum (CSS), Ultra-Wide-Band (UWB), and OFDM-based modulations. Operating frequency ranges cover from frequencies as low as 169 MHz up to the frequencies in the 2.4 GHz band. Operation within the 3–10 GHz range is also defined for UWB modulations. Supported data rates are quite dependent on the PHY layer option, with lower, medium, and highest data rates in the order of ~5 kbps, 50–300 kbps, and 2–3 Mbps, respectively (excepting UWB, which can deliver up to 27 Mbps).

Table 8.2 IEEE 802.15.4 PHY layer options.

Application space	PHY layer variants
General	• O-QPSK PHY (combines DSSS and O-QPSK modulation) • BPSK PHY (combines DSSS and BPSK modulation) • CSS PHY (uses D-QPSK modulation) • High Rate Pulse (HRP) UWB PHY • GFSK PHY • MSK PHY • Low Rate Pulse (HRP) UWB PHY • TASK PHY (employs ternary sequence spreading followed by ASK modulation) • Rate Switch (RS) – GFSK PHY
Smart Utility Network (SUN)	• SUN FSK PHY • SUN OFDM PHY • SUN O-QPSK PHY
Low-Energy Critical Infrastructure Monitoring (LECIM)	• LECIM DSSS PHY • LECIM FSK PHY
Television White Space (TVWS)	• TVWS-FSK PHY • TVWS-OFDM PHY • TVWS-NB-OFDM PHY
Rail Communications and Control (RCC)	• RCC Land Mobile Radio (LMR) PHY (uses one of GMSK, 4FSK, QPSK, π/4 DQPSK, or DSSS employing DPSK) • RCC DSSS BPSK PHY
China Medical Band (CMB)	• CMB O-QPSK PHY (combines DSSS and O-QPSK modulation) • CMB GFSK PHY

In addition to data transmission and reception, the IEEE 802.15.4 physical layer is also responsible for:

- Activation and deactivation of the radio transceiver.
- Energy Detection (ED) within the current channel.
- Link Quality Indication (LQI) for received packets.
- Clear Channel Assessment (CCA) function to support CSMA/CA in the MAC layer. Different CCA modes are supported (e.g. CCA Mode 1 shall report a busy medium upon detecting any energy above the ED threshold; CCA Mode 2 shall report a busy medium only upon the detection of a signal compliant with the 802.15.4 standard with the same modulation and spreading characteristics).
- Channel frequency selection.

8.3.3.2 IEEE 802.15.4 MAC Layer

The MAC layer defines how multiple IEEE 802.15.4 devices operating in the same PAN share the radio channel. Key features of the MAC sublayer are beacon management, channel access, guaranteed time slots management, frame validation, acknowledged frame delivery, association, and disassociation. In addition, the MAC sublayer provides hooks for implementing application-appropriate security mechanisms.

The IEEE 802.15.4 MAC layer supports different operation modes [7, 15]. In the most basic one, referred to as non-beacon-enabled PAN, devices compete for the medium at all times using an unslotted CSMA/CA access method. Instead, in beacon-enabled PANs, the PAN coordinator transmits beacons at specific intervals that define a so-called superframe structure used to organize transmissions. The superframe (see Figure 8.4) can be split into an active portion and an inactive portion. The coordinator may enter a low-power (sleep) mode during the inactive portion.

The active portion is divided into equally sized slots that are distributed into a Contention Access Period (CAP) and, optionally, a Contention-Free Period (CFP). In the CAP, devices compete with each other for channel access using a slotted CSMA/CA mechanism. In contrast, the CFP provides so-called Guaranteed Time Slots (GTS) that can be allocated to specific devices for low latency applications or applications requiring specific data bandwidth. The PAN coordinator may allocate

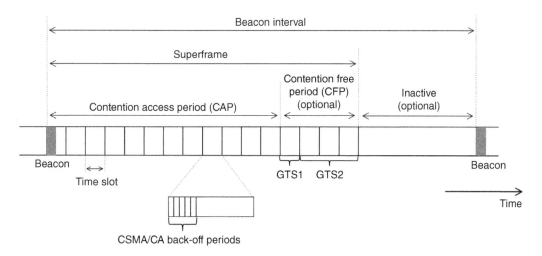

Figure 8.4 IEEE 802.15.4 MAC superframe structure.

up to seven of these GTSs, and a GTS may occupy more than one slot period. However, a sufficient portion of the CAP shall remain for contention-based access of other networked devices or new devices wishing to join the network. All contention-based transactions shall be complete before the CFP begins. Each device transmitting in a GTS shall also ensure that its transaction is complete before the time of the next GTS or the end of the CFP.

The IEEE 802.15.4e amendment brought multiple enhancements into the MAC sublayer, adding new MAC options such as Deterministic Synchronous Multichannel Extension (DSME) and Time Slotted Channel Hopping (TSCH) [22]. DSME targets to better support applications that require high levels of reliability or deterministic latency, while TSCH was created to improve operation in a harsh industrial environment. For instance, DSME defines a multisuperframe structure, which is a cycle of repeated superframes, and within the CFP in a superframe the DSME is capable of allocating GTSs across multiple channels. In this way, DSME can check the link quality and use channel adaptation to switch a GTS (assigned to a specific device) to a different channel in consecutive GTSs to overcome interference. On the other hand, in TSCH, superframes are replaced with a slotframe structure that repeats cyclically. A slotframe is formed by a sequence of time slots, inside which transmissions can occur with or without contention. However, unlike the superframe, slotframes and a device's assigned time slot(s) within the slotframe can be initially communicated by beacon or configured by a higher layer as the device joins the network. In this way, because all devices share common time and channel information, devices hop over the entire channel space to minimize the negative effects of multipath fading and interference and do so in a slotted way to avoid collisions, minimizing the need for retransmissions.

8.3.3.3 Zigbee Specifics
8.3.3.3.1 PHY and MAC Layers

The ZigBee specification is based on a subset of the PHY layer options included in the IEEE 802.15.4 standard. In particular, a Zigbee device shall support at least one of the following options [11]: O-QPSK PHY at 2.4 GHz frequency band or the BPSK PHY at both 868 and 915 MHz bands or the FSK PHY located at 863–876 MHz and 915–921 MHz. Each of the frequency bands incorporates its own set of channels.

Due to its global availability, the 2.4 GHz ISM band is the one that counts with a higher adoption in most of Zigbee products. Data rates of 250 kbps per channel can be delivered with O-QPSK PHY in this band, with up to 16 channels available (channel spacing is of 5 MHz). According to the IEEE 802.15.4 standard, the receiver sensitivity for O-QPSK PHY should be −85 dBm or better. Therefore, the communication range of an 802.15.4 device in the 2.4 GHz band, assuming e.g. a 15 dBm transmit power, can be in the order of 100–200 m outdoors and 20–50 m indoors [23].

At the MAC layer, Zigbee adopts the mechanisms specified under IEEE 802.15.4 for forming and joining a network, a CSMA/CA mechanism for devices to listen for a clear channel, as well as a link layer to handle retries and acknowledgment of messages for reliable communications between adjacent devices. However, more sophisticated options such as GTS, DSME and TSCH defined in IEEE 802.15.4 standard are not used in Zigbee.

8.3.3.3.2 Network Layer

When a coordinator attempts to establish a Zigbee network, it does an energy scan to find the best RF channel for its new network. When a channel has been chosen, the coordinator assigns the logical network identifier (PAN ID), which will be applied to all devices that join the network. A Zigbee device can join the network either directly or through association. To join directly, the system designer must somehow add beforehand the device's address into the device neighbor tables. The direct joining device will issue a so-called orphan scan, and the device with the matching extended address (in its neighbor table) will respond, allowing the device to join. To join by association, a device sends out a beacon request on a channel, repeating the beacon request on other channels until it finds an acceptable network to join.

8.3.3.3.3 Application Layer

The APL layer is a framework for the development of Zigbee applications. The APL layer (see Figure 8.2) consists of the Application Framework (AF), Application Support Sub-layer (APS), the Zigbee Device Objects (ZDO), and the manufacturer-defined applications that give the device its specific functionality.

The AF allows for the interaction with the applications supported in the Zigbee device through service access points. A single Zigbee device may run several applications. For example, an environmental sensor may contain separate applications to measure temperature, humidity, and atmospheric pressure. Each application communicates via a so-called *endpoint* in the AF. Up to 240 endpoints are supported.

The APS is responsible for message handling between the NWK layer and the relevant application. For example, when a message arrives via the NWK layer (e.g. to illuminate a LED), the APS relays this instruction to the responsible application using the *endpoint* information in the message. The APS also takes care of maintaining the binding tables, which allows nodes to be paired in such a way that a certain type of output data from one node is automatically routed to the paired node, and of application-level security, allowing communications between Zigbee devices to be encrypted/decrypted. Other functions of the APS include group address definition and management, packets SAR, and reliable data transport.

The ZDO is a special application, common to all Zigbee devices, which is necessary to manage the core Zigbee processes such as node initialization to allow applications to be run, device discovery and service discovery, security services, etc. as well as communications with remote nodes, including network creation/joining/leaving processes and binding requests and responses.

The ZCL is the component that provides the standard Zigbee *clusters* used by device applications that run on the endpoints. A cluster is defined by a set of attributes (parameters that relate to the functionality) and a set of commands (that can typically be used to request operations on the cluster attributes). The clusters collected in the ZCL cover the functionalities that are most likely to be used. However, an application can also be developed using private extensions for specific features from a manufacturer.

To ensure the interoperability of Zigbee devices from different manufacturers, the Zigbee Alliance has defined a set of standard device types. A device type defines the functionality of a device, which is itself defined by the clusters included in the device type, where each cluster corresponds to a specific functional aspect of the device. Hence, a Zigbee device includes an instance of a device type, which is implemented by the application that runs on an endpoint. A device type usually supports both mandatory and optional clusters, so a device can be customized in terms of the optional clusters used. A physical Zigbee device can actually support more than one device type via multiple device application instances associated with different end points. In addition to whatever functionalities a Zigbee node may support, all Zigbee 3.0 nodes must implement the *Zigbee Base Device*, which is not associated with an endpoint, and is in charge of handling fundamental operations such as commissioning.

8.3.3.4 Wi-SUN FAN Specifics
8.3.3.4.1 PHY and MAC Layers

Wi-SUN PHY layer specification [24] is derived from the subset of IEEE Std. 802.15.4-2015 SUN FSK PHY [25]. 2-FSK modulation schemes, with channel spacing range from 100 to 600 kHz, are defined to provide raw data rates from 50 to 300 kbps, with FEC as an optional feature.

Table 8.3 Wi-SUN PHY layer operating modes.

PHY operating modes	Symbol rate (ksymbols/s)	Modulation index
Operating mode #1a	50	0.5
Operating mode #1b	50	1.0
Operating mode #2a	100	0.5
Operating mode #2b	100	1.0
Operating mode #3	150	0.5
Operating mode #4a	200	0.5
Operating mode #4b	200	1.0
Operating mode #5	300	0.5

The specified Wi-SUN PHY layer operating modes are given in Table 8.3. Wi-SUN FAN profile devices shall support Operating Mode #1a or Operating Mode #1b for at least one of the frequency bands from 863 to 928 MHz. For each supported band, a Wi-SUN FAN profile device shall also support Operating Mode #3 if the band supports channel spacing of at least 400 kHz and shall support Operating Mode #2a where spectrum is limited to a maximum 200 kHz channel spacing. Minimum receiver sensitivity for a SUN FSK receiver shall be better than −97 dBm with FEC and 50 kbps. According to [26], effective communication distances of around 650 m (and up to 2 km for LOS) in a Japanese city were achieved with a wide area Wi-SUN deployment using powers of 20 mW for low-power radios and 250 mW in devices serving as base stations with directional antennas.

The Wi-SUN FAN MAC sublayer is constructed using data structures defined in IEEE Std. 802.15.4-2015 [25]. The MAC sublayer operates in non-beacon enabled mode (no periodic beacon frames are defined) and hence no CFP exists. All channel access operates using unslotted CSMA/CA with CCA Mode 1.

The MAC sublayer also supports neighbor synchronized channel hopping for both unicast and broadcast frame transmissions. The total number of PHY channels available depends on the selected regional band and channel spacing (for instance, band 863-870 available in EU is arranged in 69 channels with 100 kHz channel separation for Operation Mode #1a). A "channel function" defines a method used to determine, from the list of available PHY channels, the specific channel upon which a node is operating at a given time. Each node advertises its channel schedules with information necessary for a neighbor node to determine on which channel a node will be operating at any given time. A node is required to listen on its advertised unicast channel schedule when a node is not transmitting or listening for broadcast messages. A fixed channel mode of operation is supported for situations in which channel hopping is not desired.

8.3.3.4.2 Network Layer
Along with the MAC layer mechanisms, Wi-SUN relies on Extensible Authentication Protocol over LAN (EAPOL) and IEEE 802.1X security standards for network formation. When a node is first powered up and has no information regarding available neighbors or PANs (nor channel hopping schedules, etc.), it transmits PAN Advertisement Solicit (PAS) frames and waits for PAN Advertisement (PA) frames. A received PA frame is accepted if information carried in the frame matches the factory preset information (e.g. the network name and routing method). If a node receives several PA frames, it must select the node which advertises the lowest PAN cost [27] as its

EAPOL target node. After selecting an EAPOL target node, the joining node must perform the IEEE 802.1X/802.11i security flow via the selected EAPOL target node, to authenticate itself to the network and obtain the Group Transient Keys (GTKs) from the Border Router. If the authentication succeeds, the node sets its PAN ID to be that of the EAPOL target node and transmits PAN Configuration Solicit (PCS) frames. These are responded with a PAN configuration (PC) frame so that the node can select the source node of the received PC frame as its initial source of broadcast transmissions. At this point, the node is also properly configured with its channel-hopping schedule and active group keys.

Once a FAN node has joined the network, it shall be able to auto configure a link-local IPv6 address as described in RFC4862. For Layer 3 routing, Wi-SUN FAN adopts the nonstoring mode of the RPL protocol. RPL is a distance-vector protocol based on so-called Destination-Oriented Directed-Acyclic Graphs (DODAG), which have only one route from each leaf node to the root through which all the traffic from the leaf node will be routed to. Initially, each FAN node sends a DODAG Information Object (DIO) advertising itself as the root. DIO is propagated on the network and the whole DODAG is gradually built. RPL requires the construction and maintenance of DODAGs. A DODAG rooted at the Border Router is called a "grounded DODAG." According to the nonstoring mode of RPL, downward packets are routed using source routing from the root. For communication between two peers, the packet first goes up to the Border Router and is then sent to the destination node via source routing.

If Layer 2 routing is used, MHDS provides similar functionality to RPL with both announcement and maintenance of routes. In a PAN of the Wi-SUN FAN, MHDS nodes provide forwarding for packets from a so-called MHDS Service, which is the equivalent to the RPL Border Router, to downstream neighbor nodes and for packets from downstream neighbors toward the MHDS Service. As in RPL, the downstream transmissions are source routed and the upstream transmissions are hop-by-hop. In MHDS, upstream transmissions may also be source routed.

8.4 Unlicensed Spectrum-based LPWAN: LoRaWAN and Sigfox

8.4.1 Scope and Standardization

The market for unlicensed spectrum-based LPWAN solutions has been rapidly developing since early 2015 [4]. Specifically, LPWAN technologies in unlicensed spectrum primarily target IoT use cases that are delay tolerant, do not need high data rates, and typically require very low power consumption and cost on the device side. To achieve favorable power budgets, LPWAN technologies use sub-GHz bands which have advantages in terms of attenuation and are less congested compared to other unlicensed bands such as 2.4 GHz where most of the popular wireless technologies like Wi-Fi, Bluetooth, and ZigBee operate. Modulation techniques like ultra-narrow band and spread spectrum are utilized to achieve MCLs in the order of 150–170 dB. Most prominent unlicensed spectrum-based LPWAN technologies are SigFox and LoRaWAN [28, 29].

Sigfox is a proprietary technology developed by a French firm bearing the same name [30]. The Sigfox technology is exploited in a commercial model under which the Sigfox company itself, or in partnership with other network operators, are the providers of Sigfox connectivity services. Devices require a compliant radio transceiver to connect to SigFox's network and the technology is freely available for chip/device vendors. Sigfox-branded networks are available in several countries, with the Sigfox company directly operating the networks in France and the US markets. Sigfox has a "one-contract one-network" model allowing devices to connect in any country.

Like Sigfox, LoRaWAN is also based on a proprietary radio technology, known as LoRa (Long Range), developed and commercialized by the chip company Semtech Corporation [31]. However, unlike SigFox, LoRa technology is licensed by device manufacturers and public or private LoRaWAN networks can be deployed with no restrictions in how organizations use them or charge for access to their customers (i.e. Semtech builds LoRa technology into its chipsets and these chipsets are then built into end-devices that can be deployed in different networks). In this respect, Semtech promoted the creation of the LoRa Alliance [32], with the goal of building a LoRaWAN ecosystem based on an open, end-to-end networking solution, called the LoRaWAN protocol, used along with the LoRa chipsets. A certification and compliance program has been created by the LoRa Alliance to ensure interoperability and facilitate network and IoT platform integration.

8.4.2 LoRaWAN

8.4.2.1 Network Architecture and Protocol Stack

A LoRaWAN network [33] follows a star-of-stars topology (Figure 8.5) in which End-Devices (ED) exchange messages with Gateways (GWs), and these GWs relay the ED messages from/to a central Network Server (NS).

The communication between EDs and GWs is based on the LoRa PHY protocol. An ED only communicates with the GWs and not directly with other EDs. A LoRaWAN network does not require the EDs to be associated with a specific GW for data exchange. A GW acts basically like a relay entity that forward the data received in the radio interface from EDs to the NS using encapsulation methods, after inserting additional details regarding channel quality. The channel quality information is necessary because more than one GW can receive a message sent by an ED, so that

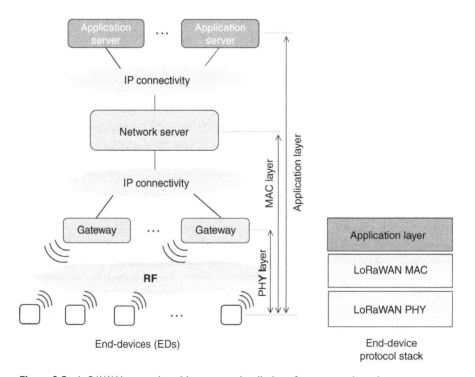

Figure 8.5 LoRaWAN network architecture and radio interface protocol stack.

the NS determines the most suitable GW to reach the ED from the network side. Thus, no handover mechanism is needed when EDs moves across areas covered by different GWs. A GW is connected to a NS using standard IP networks.

Because of the simplicity of the interaction between EDs and GWs, the intelligence and complexity of the network is pushed to the NS, which serves as the brain of the LoRaWAN network. Multiple GWs can be connected to a central NS. The NS carries out different functions such as handling ED associations, filtering redundant received packets, performing security checks, scheduling acknowledgments back to EDs through the most suitable GW and performing adaptive data rate mechanisms.

The NS acts as a bridge among GWs and the Application Servers (ASs), where an AS represents the system or platform that hosts the IoT application layer with the specific functionality (e.g. asset tracking, water and gas smart metering, etc.). ASs can be owned by third parties other than the LoRaWAN network operator. Multiple ASs can be connected to a single NS. Interfaces and message flow between NS and ASs are specified by the LoRa Alliance [34].

8.4.2.2 Protocol Frame Structure

The radio protocol stack of a LoRaWAN device consists of only three layers, as illustrated in Figure 8.5: PHY, MAC, and Application. Figure 8.6 shows the frame structure corresponding to these three layers [33, 35].

The PHY layer frame defines a preamble with a programmable duration to synchronize the receiver with the transmitter. Optionally, the preamble is followed by a physical header that contains the payload length in bytes, the FEC code rate of the payload and the header CRC. If these parameters are predefined in advance between EDs and GWs, the header can be removed.

At the MAC layer, the MAC header defines the protocol version and message type, i.e. whether it is a data or a management frame (e.g. join request, join accept messages), whether it is transmitted in uplink or downlink, and whether it shall be acknowledged or not. A Message Integrity Code

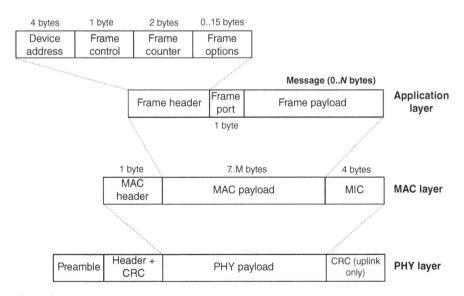

Figure 8.6 LoRaWAN protocol frame structure.

(MIC) is added to protect the entire MAC header and payload. The MIC value is used to prevent the forgery of messages and authenticate the ED. LoRaWAN relies on symmetric cryptography using session keys derived from the ED's root keys.

At application layer, the frame header contains, among others, the device address, which in LoRaWAN is 32 bit long and formed by a combination of a network identifier and a temporary identity assigned dynamically to the ED when joining the network. The frame payload can be encrypted with an application session key using the AES 128 algorithm. The maximum length of the frame payload is region specific and can be up to around 200 Bytes [36].

8.4.2.3 Physical Layer

The LoRaWAN physical layer uses a special Chirp Spread Spectrum (CSS) technique, which spreads a narrowband input signal over a wider channel bandwidth. The resulting signal has noise like properties, making it harder to detect or jam.

The transmitter makes the chirp signals vary their frequency over time without changing their phase between adjacent symbols. As long as this frequency change is slow enough so to put higher energy per chirp symbol, distant receivers can decode a severely attenuated signal several dBs below the noise floor. LoRaWAN supports multiple Spreading Factors (SFs; between 7 and 12) to decide the trade-off between range and data rate. Higher SFs delivers long range at an expense of lower data rates and vice versa. LoRaWAN also combines FEC with the spread spectrum technique to further increase the receiver sensitivity. The data rate ranges from 300 bps to 37.5 kbps depending on SF and channel bandwidth. Further, multiple transmissions using different SFs can be received simultaneously by a GW. LoRaWAN employs an Adaptive Data Rate (ADR) mechanism in which each ED's SFs is adjusted to select the highest practical data rate while maintaining an acceptable SNR.

The physical layer operates in the 433, 868, or 915 MHz ISM frequency bands. Specific details are provided in the Regional Parameters technical documents of the LoRa Alliance [37]. In Europe, only the 433 and 868 MHz bands can be used. In the 868 MHz band, there are three 125 kHz channels that are mandatory to be implemented in every ED. There are another five 125 kHz channels in the 867 MHz sub-band that can be optionally used for LoRa communication. In Europe due to transmission regulations, each transmission in any of the 868 and 867 MHz sub-bands should comply with a 1% radio duty cycle or implement a LBT or adaptive frequency agility mechanism. When the duty cycle regulation is followed, it means that, if the radio transmitted for 1 second, it cannot transmit for the next 99 seconds. The default transmit power by a device is the maximum transmission power allowed for the device considering device capabilities and regional regulatory constraints. Lower power levels are also supported.

8.4.2.4 MAC Layer

An ED may transmit on any channel available at any time, using any available data rate, as long as the following rules are respected:

- The ED changes channel in a pseudo-random fashion for every transmission. The resulting frequency diversity makes the system more robust to interferences.
- The ED respects the maximum transmit duty cycle relative to the sub-band used and local regulations.
- The ED respects the maximum transmit duration (or dwell time) relative to the sub-band used and local regulations.

Message transmission in the uplink follows a simple unslotted ALOHA mechanism. In downlink, LoRaWAN end devices only receive downlink messages when a *receive window* is open, which results in three categories of devices (Classes A, B, and C) depending on how this occurs:

- Class A devices only open two temporary receive windows after sending a message uplink. Downlink messaging is only used for acknowledgement purposes.
- Class B devices share the acknowledgement functionality of class A devices but also open receive windows at scheduled times known to the NS.
- Class C devices have reception windows constantly open except for when a message is transmitted.

8.4.3 Sigfox

8.4.3.1 Network Architecture and Protocol Stack

Like LoRaWAN, Sigfox follows a star-of-stars topology [38, 39]. Sigfox defines its architecture (see Figure 8.7) with two layers, namely, the Network Equipment and the Sigfox Support Systems (SSS).

The Network Equipment layer consists of all the base stations that are in charge of transmitting/receiving messages from/to Sigfox devices. The connection between base stations and the SSS is carried out over IP connectivity, typically using the public Internet with connections secured with Virtual Private Network (VPN) technology. In this respect, base stations store credentials to securely communicate with the SSS.

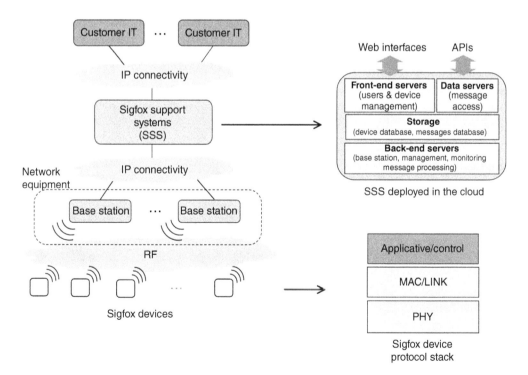

Figure 8.7 Sigfox network architecture and radio interface protocol stack.

No pairing is required between Sigfox devices and base stations to transmit or receive data. When a Sigfox device has data to be sent, the device broadcasts a radio message that could be potentially detected by several base stations, which convey them to the SSS.

The SSS is hosted in a cloud environment by SigFox and consists of different servers and services carrying out most network functionality. Specifically, Back-End Servers communicate directly with base stations and are responsible for monitoring and managing the base stations and processing incoming/outcoming messages (e.g. when they receive multiple copies of the same message, the back-end servers determine which to keep). Storage keeps in databases all messages retrieved from devices as well as metadata regarding each device. Front-End Servers communicate with the device metadata database in Storage and present a web interface for subscribers to manage devices and configure user or group permissions. Data Servers communicate with the message database in Storage, to offer them to subscribers through APIs. Sigfox allows both receiving and sending messages to devices, although the system is optimized for uplink communications, i.e. device-originated traffic.

8.4.3.2 Protocol Frame Structure

Like LoRaWAN, the protocol stack used in the Sigfox radio interface (see Figure 8.7) consists of only three layers [40]: PHY, MAC/LINK and Applicative/Control layer. The frame structures associated with these layers are shown in Figure 8.8.

The PHY is in charge of frequency selection, modulation, error detection, and, only for the downlink, error correction. The maximum frame size at PHY layer is 29 bytes for both uplink and downlink.

The MAC/LINK layer handles the shared access to the radio channel, defines the uplink, and downlink procedures for message transmission and includes several security mechanisms such as integrity check (CRC), authentication, and replay attach protection with message counters. To compute the message authentication token, each Sigfox device is provisioned with a unique symmetrical authentication key during manufacturing.

The Applicative/Control layer deals with user and control messages and security for payload encryption. The content of the uplink message may be applicative data or control data (e.g. keep

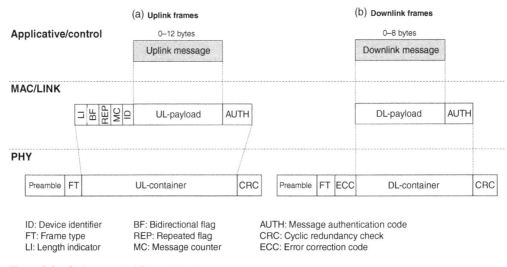

Figure 8.8 Sigfox protocol frame structure.

alive messages). The format of an applicative message content is freely defined by the application. Maximum size is 12 bytes. The content of downlink message is a fixed-length field of 8 bytes. It carries applicative data prepared by user's distant application server in response to an uplink message. Format of the content is user dependent. By default, data are not encrypted over the air. However, if privacy is required, payload encryption is supported.

8.4.3.3 Physical Layer

Sigfox operates in unlicensed frequency spectrum bands below 1 GHz [40]. As these frequency bands are ruled by local regulations that may vary from country to country, Sigfox has defined Radio Configurations (RC) that define radio parameter values that comply with local regulations for a set of countries (see Table 8.4). Radio parameters values corresponding to a RC are called a regional profile. For example, access to the unlicensed spectrum is ruled by duty cycle limits and transmit power limits in European countries, by dwell time limits and Frequency Hopping (FH) constraints in North American countries, and by Listen Before Talk (LBT) and transmit power limits in Japan.

Within each RC, one so-called uplink macro-channel and one downlink macro-channel are defined. A macro-channel represents the frequency interval within an unlicensed spectrum where a Sigfox device may communicate. These channels are 192 kHz wide. When local regulations require FH techniques, Sigfox transmissions shall implement at least 6 contiguous micro-channels of 25 kHz in each macro-channel.

Within the 192 or 25 kHz wide channels, Sigfox uplink transmissions are only 100 or 600 Hz wide depending on the regional profiles (e.g. 100 Hz is used in Europe and 600 Hz in the US). Uplink modulation is differential BPSK (D-BPSK) so that messages are transferred with a data rate of 100 or 600 bps. In contrast, downlink transmissions always use radio bursts of 600 bps and modulation is GFSK with a frequency deviation of 800 Hz.

Table 8.4 Sigfox Radio Configurations.

	RC1	RC2	RC3	RC4	RC5	RC6	RC7
Uplink center frequency (MHz)	868.130	902.200	923.200	920.800	923.300	865.200	868.800
Downlink center frequency (MHz)	869.525	905.200	922.200	922.300	922.300	866.300	869.100
Uplink data rate (bps)	100	600	100	600	100	100	100
Downlink data rate (bps)	600	600	600	600	600	600	600
Sigfox recommended EIRP (dBm)	16	24	16	24	14	16	16
Specifics	Duty cycle 1%[a]	FH[b]	LBT[c]	FH[b]	LBT[c]		Duty cycle 1%[a]

[a] Duty cycle is 1% of the time per hour (36 seconds). For an 8–12 bytes payload, this means 6 messages per hour, 140 messages per day.
[b] Frequency hopping: The device broadcasts each message 3 times on 3 different frequencies. Maximum On Time 400 ms per channel. No new emission before 20 seconds.
[c] Listen Before Talk: Devices must verify that the Sigfox-operated 192 kHz channel is free of any signal stronger than −80 dBm before transmitting.

With respect to transmit power, EIRP values given in Table 8.4 represent target values used to evaluate coverage of Sigfox systems in each country. A Sigfox device that transmits at those values is said to be "Class 0." When local regulation allows, end points may transmit at higher transmit power, but the drawback is an increased power consumption. A Sigfox device may also transmit at lower transmit power, but the drawback is a reduced quality of service in uplink. Of note is that, depending on local regulations, uplink and downlink transmit power can be different. For instance, the maximum uplink transmit power is 25 mW in Europe (158 mW in the USA), whereas the maximum downlink transmit power is 500 mW in Europe (4 W in the United States) [41].

8.4.3.4 MAC Layer

Sigfox transmissions are asynchronous and always device-initiated, which allows the device to stay in sleep state by default and minimize its energy consumption. Two types of message exchanges are supported: Uplink only procedure (U-procedure), with no onward downlink message; and bidirectional procedure (B-procedure).

In either case, the device starts by transmitting the uplink message, which can be repeated two times (i.e. three transmissions in total) on random frequencies at different time intervals. Uplink messages may be received by multiple nearby base station, adding spatial diversity. This is called 3D or triple diversity in Sigfox (i.e. time, frequency and spatial diversity). In the case of the U-Procedure, there is no response from the network, i.e. a U-procedure transaction is unconfirmed.

In the B-procedure, a downlink message may be transferred after the uplink message, within a given delay between the first frame transmitted (e.g. 20 seconds for RC1). A maximum reception time is also defined (e.g. 25 seconds for RC1). If a downlink message is received successfully in the device, it sends an uplink confirmation message.

8.5 Cellular IoT: LTE-M and NB-IoT

8.5.1 Scope and Standardization

As already explained in Chapter 7, cellular technologies have evolved to provide massive Machine-Type Communications (mMTC) services. Two main technologies are part of the latest 3GPP Release 16 specifications for mMTC: *Long-Term Evolution for Machine-Type Communications* (LTE-M) and *Narrowband Internet of Things* (NB-IoT).

LTE-M is a backward compatible extension of LTE. Work in LTE-M started in 2011, under 3GPP Release 12. At that time, GSM/GPRS was the main cellular technology used for wide-area IoT because of its maturity and, its low modem cost compared with 3G and 4G devices [5]. In this context, LTE-M design focused on the realization of an LTE-based solution with device complexity and cost comparable with existing GSM/GPRS devices [6]. Cost reduction techniques introduced in LTE-M include half-duplex operation, use of a single-receive antenna, operation in reduced bandwidth, and reduced maximum transmit power classes. In subsequent releases, LTE-M has been further improved with, among others, coverage enhancing features allowing up to 20 dB link budget margin beyond normal LTE coverage.

In contrast, NB-IoT is a new radio-access technology, not backward compatible with LTE but designed to be deployable on existing LTE networks. 3GPP NB-IoT work started in 2014 and first normative specifications came with Release 13 in 2016. NB-IoT design target was to find a cellular solution able to compete with technologies such as LoRaWAN and Sigfox in the LPWAN segment. In this respect, more challenging performance requirements than those achievable with LTE-M

Table 8.5 Comparison of the scope of LTE-M and NB-IoT technologies.

	LTE-M	NB-IoT
Channel bandwidth	1.4 MHz, 5 MHz	200 kHz
Number of reception chains	1	1
Duplexing mode	Full or half duplex FDD, TDD	Half duplex FDD
Terminal mobility	No limitations	Limited mobility. Handover not supported
Data rates	Up to few Mbps	Up to 250 kbps
Voice support	VoLTE	No support
Transmit power	23, 20, 14 dBm	23, 20, 14 dBm
Deployment flexibility	Compatible with LTE cells	Need separate NB-IoT cells
Device categories	Cat-M1 (Rel-13), Cat-M2 (Rel-14)	Cat-NB1 (Rel-13), Cat-NB2 (Rel-14)
Maximum coupling Loss	159 dB	164 dB

and GSM/GPRS were set up for NB-IoT, especially in terms of coverage range, number of supported devices per cell, and battery lifetime. In parallel to the NB-IoT work, the so-called *Extended Coverage Global System for Mobile Communications Internet of Things* (EC-GSM-IoT) enhancements were also brought into GSM/GPRS specifications, matching the same performance pursued with NB-IoT [42].

LTE-M and NB-IoT technologies address different market segments. LTE-M allows for IoT applications with data rate requirements of a few Mbps, latencies of tens of milliseconds, and full device mobility. LTE-M also supports voice communications. In contrast, NB-IoT targets IoT applications with data rate of a few kbps or even lower, message delivery times of seconds, and more limited mobility (e.g. handover is not supported). A comparison of the scope of both technologies is given in Table 8.5.

8.5.2 Network and Protocol Stack Architecture

Building upon the baseline LTE network architecture (see Chapter 7), some enhancements and optimizations have been introduced to support of LTE-M and NB-IoT services [43]. This enhanced architecture, depicted in Figure 8.9, is commonly referred to as *network architecture for MTC (Machine Type Communications)* or simply as *Cellular IoT* (CIoT) network architecture.

8.5.2.1 New Network Attach Method and Connectivity Options

A new network attach option, known as *Attach without PDN connection*, is supported. It allows a UE to remain registered to the network without the need to establish any PDN connection (before this, a PDN connection was always established along with the network registration in LTE). This option facilitates handling huge numbers of devices that may keep a connection inactive for very long period of times and seldom transmit data over it.

Four different options are supported for data connectivity:

- *IP over User Plane*, which is the conventional option supported in LTE to send user IP traffic via the S-GW and P-GW elements.

- *IP over Control Plane*, which allows sending IP traffic over NAS protocols between the UE and the MME.
- *Non-IP over User Plane*, for the support of non-IP data delivery (NIDD) over the S-GW and P-GW elements, allowing a device to exchange non-IP data encapsulated in a transparent container without using an IP stack.
- *Non-IP over Control Plane*, for the support of NIDD over the NAS protocols.

8.5.2.2 New Network Entities

On the E-UTRAN side, there is no difference with the conventional LTE architecture. The base station functionality is the eNB, which now is able to provide, in addition to radio access to conventional LTE devices:

- Access to LTE-M devices over the same LTE cells used by conventional LTE devices (see Section 8.5.3.1), and/or,
- Access to NB-IoT devices through the configuration of NB-IoT cells on separate NB-IoT carriers (see Section 8.5.3.2).

On the EPC side, the most noticeable differences are the introduction of the so-called Service Capability Exposure Function (SCEF) and MTC Interworking Function (MTC-IWF).

The SCEF provides a means to securely expose the services and capabilities of the CIoT network to external IoT platforms through network APIs (T8 interface in Figure 8.9). One key feature provided by the SCEF is the support for NIDD. In this respect, the non-IP data are transferred from/to devices across the MME, which routes the data to/from the SCEF so that it can be offered to the external servers by means of the network APIs. In addition to NIDD, other services and capabilities offered by SCEF via T8 include monitoring events (e.g. UE reachability, UE location), QoS management, network parameter configuration, group message delivery, support for high-latency communications in the case of devices with long power saving mode cycles, etc.

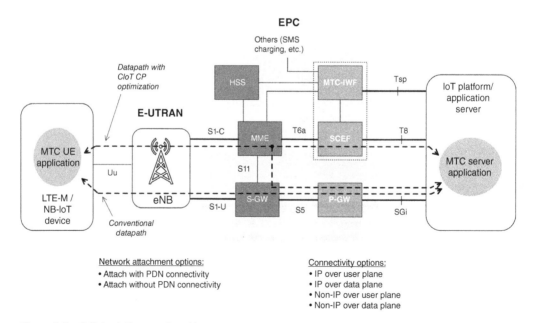

Figure 8.9 Cellular IoT network architecture.

The MTC-IWF is intended to relay or translate signaling protocols (Tsp interface in Figure 8.9) used to invoke specific functionality in the network. One key functionality provided by MTC-IWF is device triggering. This is required when an IP address for the UE is not available or reachable from the external IoT platform, and there is the need to trigger the UE to perform application-specific actions, which may include initiating communication with the network. One common mechanism to implement device triggering is SMS-based triggering, so that, internally, the MTC-IWF has access to the proper interfaces to send SMS to the CIoT devices. In certain deployments, the MTC-IWF may be colocated with the SCEF, in which case MTC-IWF functionality is exposed to the external network via the T8 interface.

8.5.2.3 Control Plane and Data Plane Optimizations

In addition to the new network entities, two optimizations with end-to-end impact have been introduced on top of the basic LTE functionality. These are *User Plane (UP) CIoT EPS optimization* and *Control Plane (CP) CIoT EPS optimization*. While both optimizations were initially conceived for NB-IoT devices, they are currently not limited to NB-IoT and could be also used to handle LTE-M as well as conventional LTE devices.

The *CP CIoT EPS optimization*, which is a mandatory feature for NB-IoT, provides an efficient solution for the transmission of infrequent and small data packets, including non-IP traffic and SMS. With this option, user data are transmitted between devices and the MME over the CP, encapsulated in NAS protocols, avoiding the establishment of Data Radio Bearers (DRBs) in the radio interface. This results in fewer signaling messages and reduced air time, which ultimately impact on the power consumption of these devices. Two options are supported to route the data between the MME and the external application servers: data transfer via the conventional S-GW/P-GW data plane (supporting both IP and non-IP user traffic); and data transfer through the SCEF. The path for UP traffic routing under *CP CIoT EPS optimization* is illustrated with the upper dotted arrow in Figure 8.9.

In contrast, the *UP CIoT EPS optimization* is an optional feature that adds some enhancements to the conventional data transfer mechanisms over the UP path (i.e. DRBs over the radio interface and traffic exchange over the S1-U, S5, and SGi interfaces). Indeed, the use of DRBs and conventional UP is more suitable for connection-oriented sessions where sequences of messages need to be exchanged between the server and end-device. For instance, one of these enhancements is the possibility for an RRC connection to be suspended and resumed (in LTE, an RRC connection is usually released after 10–20 seconds of inactivity and a new RRC context must be established when activity restarts). This feature reduces the signaling overhead and latency incurred in the CP activation.

8.5.3 Main Capabilities and Features

8.5.3.1 LTE-M Radio Access

LTE-M backward compatibility with LTE means that transmissions of LTE-M devices can be properly arranged within the conventional time–frequency resource grid used in LTE cells, and coexist with the transmissions of other LTE devices. LTE-M devices relies on the same physical signals and channels as LTE devices for many procedures such as cell search, cell synchronization, system information acquisition, and random access.

LTE-M supports the same system bandwidths at the network side as LTE (not at the device side, as described later), that is, from 1.4 to 20 MHz. Hence, if an operator has a large spectrum allocation for LTE, then there is also a large bandwidth available for LTE-M traffic because the resources on a LTE carrier can be fully dynamically shared between LTE and LTE-M devices. Indeed, as LTE-M traffic may be more delay-tolerant in general, it is possible to schedule LTE-M traffic during periods when the ordinary LTE users are less active, thereby minimizing the performance impact from the LTE-M traffic on the LTE traffic.

On this basis, LTE-M introduces a number of modifications to (1) reduce the complexity and cost of LTE-M devices, (2) improve their coverage range, (3) improve their energy consumption and so extend their battery lifetime, and (4) better handle a massive large number of LTE-M devices within the coverage of an connected to a LTE cell.

8.5.3.1.1 Reduced Device Complexity

Device complexity reduction, which goes along with device cost reduction, is primarily achieved in LTE-M by:

- Enabling half-duplex FDD (HD-FDD) operation.
- Reducing the maximum transmit power that shall be supported by the device.
- Enabling the operation through only a portion of the LTE cell bandwidth.
- Reducing the peak rate and transmission modes that shall be supported by the terminals.

In contrast to full-duplex FDD, a HD-FDD device switches back and forth between reception and transmission in the uplink and downlink carriers. Two HD-FDD operation modes are supported: *HD-FDD operation type A* and *HD-FDD operation type B*. In the former, which is not specific to LTE-M, the switching between reception and transmission can be done in consecutive subframes (e.g. one subframe used for downlink and the very next one for uplink). Implementation of this fast switching commonly requires using two separate local oscillators for downlink and uplink carrier frequency generation. In contrast, HD-FDD operation type B, which is the one introduced as part of the LTE-M adaptions, a subframe is used as a guard time at every switch, giving the device more time to retune its carrier frequency with no need of a second oscillator.

In terms of transmit power, power classes with maximum transmit power of 20 and 14 dBm have been specified, complementing the most common power class of 23 dBm used in LTE devices. Lower power classes allow simplifying the integration of the Power Amplifier (PA) on the LTE-M chipsets for low complexity System on Chip (SoC) designs as well as the use of simpler and more compact battery types. However, uplink coverage for these devices is reduced.

Another cost reduction technique is the operation in reduced bandwidth, in particular using only 1.4 MHz even if the LTE cell is configured with a larger bandwidth (e.g. 20 MHz). This requires new arrangements for downlink control channels in the resource grid given that conventional control channels in LTE (i.e. PDCCH, PCFICH, and PHICH) are spread over the full channel bandwidth and thus cannot be used. LTE-M defines the concept of *narrowband,* which consists of a block of six contiguous PRBs. Figure 8.10 illustrates this concept for a 15 MHz channel, where the full LTE cell bandwidth (75 PRBs) is divided into 12 narrowbands (leaving 3 PRBs, one at the centre and at each edge, that are not part of any narrowband). An LTE-M device may operate using only one of these narrowbands. This solution, while not being a problem for the data channels (i.e. PDSCH), because the scheduler itself may enforce this allocation, requires that control channels are also placed within a *narrowband*. This has resulted in the definition of a new control called MTC PDCCH (MPDCCH) mapped within the data region of a *narrowband* (see Figure 8.10) so that collisions between the MPDCCH and the full-bandwidth LTE control channels are avoided.

LTE-M also introduces restrictions on the mandatory modulations and multiantenna transmission schemes to facilitate low-cost receiver implementations with relaxed requirements on processing power. These restrictions are reflected in reduced data peak rates. In particular, LTE-M

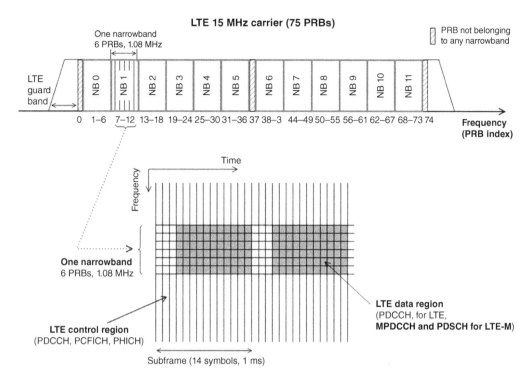

Figure 8.10 Illustration of LTE-M *narrowbands*.

devices support modulations up to 16QAM compared to ordinary LTE devices which support at least up to 64QAM. Moreover, LTE-M are only required to support a single receive antenna, instead of at least two established for LTE terminals.

8.5.3.1.2 Coverage Enhancement Modes

LTE-M introduces Coverage Enhancement (CE) modes based on the use of repetitions. In particular:

- CE mode A, supporting a maximum of 32 repetitions.
- CE mode B, supporting up to 2048 repetitions.

Figure 8.11 shows an example of how repetition-based modes work. In this example, the MPDCCH carrying the Downlink Control Information (DCI) is repeated in four subframes, followed by the sending of data in the PDSCH with a repetition number of eight subframes. As in LTE, the DCI includes the PRB allocation, MCS, and information needed for supporting HARQ. In contrast to LTE, LTE-M allows DCI and the scheduled data transmission not to occur in the same subframe in order to further facilitate a low-complexity device implementation (this feature is called *cross-subframe scheduling*). Once the device has decoded the data, HARQ feedback may also use repetitions, which in this example considers four repetitions and a 2 ms uplink frequency hopping interval in the PUCCH. Repetition-based techniques decrease the radio utilization efficiency and increase of the overall transmission time.

The use of the CE modes is not limited to LTE-M devices. An ordinary LTE device can indicate support for CE mode A, or A and B. Furthermore, the device can also indicate support for a maximum data channel bandwidth (1.4, 5, 20 MHz) in CE mode. The configuration of the CE mode

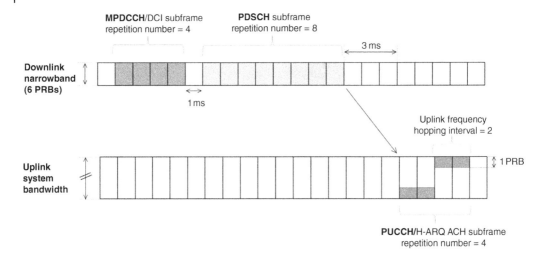

Figure 8.11 LTE-M downlink scheduling with repetitions.

could provide coverage improvements to ordinary LTE devices in poor coverage conditions and, even for devices in good coverage, a configuration in CE mode with a relatively small bandwidth can be beneficial to save power.

8.5.3.1.3 Improved Device Energy Consumption and Battery Lifetime

In IoT applications that mainly require infrequent transmission of short packets, device battery lifetime depends at a large extent on how the device behaves when the device is not transmitting or receiving data. The reason is that a device in idle mode still needs to monitor paging and perform mobility measurements for cell reselection. Although the energy consumption for these tasks is much lower compared with data transmission and reception in connected mode, further energy saving can be achieved increasing the periodicity between paging occasions or not requiring the device to monitor paging at all during a given period of time. In this respect, two key features are supported, namely, Power Save Mode (PSM) and extended Discontinuous Reception (eDRX). These features are not exclusive to LTE-M, and are also supported in NB-IoT and LTE devices [44, 45].

PSM allows a device to shut down its radio for a predetermined time interval without getting de-registered from the network. While it has always been possible for a device's application to turn its radio module off to conserve battery power, without PSM the device is required to reattach to the network every time that the radio module is turned back on. This reattach procedure consumes a small amount of energy, but the cumulative energy consumption of reattaches can become significant over the lifetime of a device. In contrast, with PSM, if the device awakes before the expiration of a time interval agreed with the network, a reattach procedure is not required. This time interval can go from seconds up to a year [5]. After the expiration of this interval, the device has to perform a periodic Tracking Area Update (TAU) procedure.

Figure 8.12 depicts the operation of a device configured for PSM in the case that there is no application data traffic and the device is only forced to perform the periodic TAUs. After completion of the TAU procedure, the device starts an Active Timer and remains reachable for Mobile Terminating (MT) traffic by monitoring the paging channel until the Active Timer expires. Then, the device enters into a power saving state and becomes unreachable for MT until the next TAU procedure.

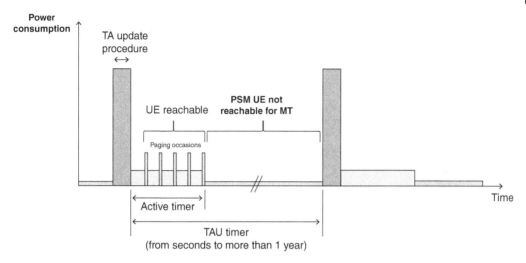

Figure 8.12 PSM cycle and TAU periods.

eDRX is an enhancement of the basic LTE DRX feature (see Chapter 7) that allows extending the time interval during which a device in idle mode is not required to listen to paging messages or "occasions." In particular, an eDRX cycle corresponds to a sequence of DRX listening/sleep cycles, called Paging Time Window (PTW), followed by a long sleep period, which may last up to ~44 minutes for LTE-M and ~3 hours for NB-IoT devices [5]. eDRX can be used standalone or in combination with PSM.

8.5.3.1.4 Handling Large Number of Devices

Already since the first releases, LTE specifications have included capabilities intended to handle a large number of devices while protecting both the E-UTRAN and EPC from potential congestion. Some of these mechanisms, such as Access Class Barring (ACB), were adapted from GSM networks, while other mechanisms, such as Extended Access Barring (EAB), have been developed to aid with congestion and overload control in scenarios where a high number of devices can be connected to the eNB for delay-tolerant, low-bit-rate data services [46].

ACB is a basic mechanism that associates each device with 1 out of 10 *access classes* based on the last digit of their IMSI. In addition, five special access classes (e.g. "emergency services," "security services," "public utilities," "PLMN staff", and "For PLMN use") are defined and potentially assigned to devices. Thus, ACB allows the network to bar devices of different *access classes* by broadcasting in BCCH system information messages a bitmap of the barred access classes.

EAB is a more selective mechanism to restrict network access to devices that are explicitly configured for EAB as *low access priority* devices. When EAB is activated in the network, access to the network is not allowed for a UE configured for low access priority, unless the UE is accessing the network with a special access class (access classes 11–15). The triggering of the EAB mechanism can be initiated when MMEs request to restrict the load, or if requested by the network management system. A further EAB enhancement was the support of dual priority access so that a device can hold dual-priority applications for, e.g. *normal* (default) priority access, which could be used to send infrequent service alerts/alarms; and *low priority/delay tolerant* access, which can be used for the vast majority of their connection establishments.

In addition to access restrictions, low-access priority indicators have also been included in the RRC and NAS signaling messages so that the network can use them to support congestion control of delay-tolerant MTC devices. For instance, in case the RRC connection request message signals that the access was made by a delay-tolerant device, the eNB has the option to reject the connection and force the device, via an *RRC Connection Reject* message, to wait for the duration of a timer before making a new attempt.

8.5.3.2 NB-IoT Radio Access

NB-IoT has been defined as a new radio access technology that leverages extensively the LTE design (i.e. OFDM numerologies, time structure, etc.) even though it is not backward compatible with LTE (i.e. NB-IoT and LTE traffic is not mixed in the same cell) [47, 48]. A NB-IoT cell includes its own physical signals and channels for cell detection, synchronization, access, and so on. A NB-IoT cell uses 180 kHz in downlink and 180 kHz in uplink. This channel bandwidth choice enables a number of deployment options that provide flexibility and facilitate the coexistence of NB-IoT with legacy LTE and GSM/GPRS deployments.

8.5.3.2.1 Further Reduced Device Complexity

NB-IoT is also designed to fulfil the performance requirements of IoT such as significant coverage extension, low device complexity, long battery lifetime, and supporting a massive number of IoT devices. In this respect, NB-IoT incorporates the same sort of features already introduced in LTE-M, such as repetition-based transmissions, HD-FDD, reduced modulation orders, single receiver and transmitter chains and 23, 20, and 14 dBm power classes. In addition, NB-IoT enables low-complexity UE implementation through:

- Significantly reduced Transport Block (TB) sizes.
- Support only single HARQ process.
- No need for a Turbo decoder, which requires iterative receiver processing, at the UE. Instead, Tail-Biting Convolution Code (TBCC) is used in downlink.
- No connected mode mobility measurement is required. A UE only needs to perform mobility measurement in idle mode.
- Low sampling rate due to lower UE bandwidth.
- Relaxed oscillator accuracy in the device.

8.5.3.2.2 Deployment Options

NB-IoT can be deployed in three different operation modes (see Figure 8.13):

- Stand-alone, as a dedicated carrier. In stand-alone operation, a NB-IoT cell can use any available spectrum exceeding 180 kHz. NB-IoT can be used as a replacement of GSM carriers, as it fits well within a GSM channel (200 kHz).
- In-band within the occupied bandwidth of a wideband LTE carrier. In in-band operation, a NB-IoT cell can be deployed using the resources corresponding to a single PRB inside an LTE carrier. Within this resources, NB-IoT signals are allocated so that LTE signals and control channels are not overlapped, as illustrated in the lower part of Figure 8.13. This allows keeping the orthogonality between the LTE cell and the superposed NB-IoT cell. Moreover, as typically the LTE cell and the in-band NB-IoT cell are handled by the same eNB, the total transmission power is shared between LTE and NB-IoT and Power Spectral Density (PSD) boosting techniques can be used on the NB-IoT PRB to improve the coverage of the NB-IoT service.
- Within the guard-band of an existing LTE carrier. In guard-band operation, the NB-IoT cell would utilize the PRBs left within the guard-band of the LTE carrier. In this case, additional guard band for the adjacent LTE carrier or a filter with a faster roll off (i.e. transition bandwidth) may be required.

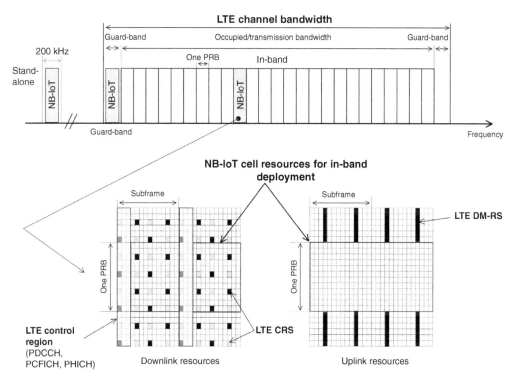

Figure 8.13 NB-IoT operation modes and illustration of in-band resource allocation.

NB-IoT supports multicarrier operation with one NB-IoT carrier, known as the anchor carrier, providing all the signals and control channels for UE initial synchronization and access, and one or more additional carriers, known as secondary carriers, used to handle NB-IoT traffic. Similar to existing LTE UEs, an NB-IoT UE is only required to search for an (anchor) carrier on a 100 kHz raster.

8.5.3.2.3 Physical Channels Structure
In downlink, subcarrier spacing (15 kHz) and as well as slot, subframe, and frame durations (0.5, 1, and 10 ms, respectively) used in a NB-IoT cell are identical to those in LTE. Downlink transmissions use the 12 subcarriers of the PRB. In uplink, NB-IoT adds support for so-called multitone and single-tone transmissions. Multitone transmission allows the use of either 6 or 3 out of the 12 PRB subcarriers, keeping the same numerology and temporal structures. Single-tone transmission allows transmitting on a single subcarrier. Two numerologies are supported in this case: 15 and 3.75 kHz. With 3.75 kHz single-tone, the slot duration is set to 2 ms to accommodate the larger OFDM symbol duration (in contrast to the 0.5 ms slot duration in the 15 kHz numerology).

Figure 8.14 illustrates the main traits of the NB-IoT radio access design, showing one example of a potential mapping of the physical signals and physical channels specified for NB-IoT operation.

In the downlink, Narrowband Primary Synchronization Signal (NPSS), Narrowband Secondary Synchronization Signal (NSSS), and Narrowband Physical Broadcast Channel (NPBCH) are used by an NB-IoT device to perform cell search, which includes time and frequency synchronization, cell identity detection, and system information acquisition. NPSS is transmitted in subframe #5 (SF #5) in every frame. NSSS has 20 ms periodicity and is transmitted in SF #9. NPBCH carries the MIB and is transmitted in SF #0 in every frame. This allocation is the same for each of the deployment

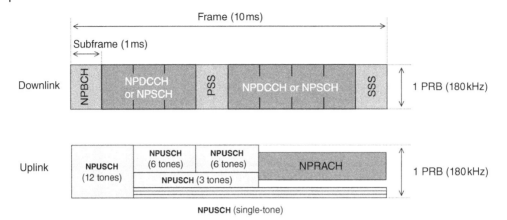

Figure 8.14 Illustrative NB-IoT physical channels structure for downlink and uplink.

modes (stand-alone, in-band, or guard-band) so that when a UE is first turned on and searches for an NB-IoT carrier, there is no need for the UE to know the type of deployment. The operation mode is learned by the UE, once system information has been acquired.

The rest of downlink subframes are used for Narrowband Physical Downlink Control Channel (NPDCCH) and Narrowband Physical Downlink Shared Channel (NPDSCH), which unlike LTE, are multiplexed now only in the time domain. NPDCCH carries scheduling information for both downlink and uplink data channels, HARQ feedback, paging indications, and random access response (RAR) scheduling information. NPDSCH carries data from upper layers, paging messages, system information, and RAR messages.

In the uplink, NB-IoT only includes two channels: Narrowband Physical Uplink Shared Channel (NPUSCH) and Narrowband Physical Random Access Channel (NPRACH).

NPUSCH is used to carry data and control information from higher layers. Additionally, as there is no uplink control channel in NB-IoT, NPUSCH also carries HARQ feedback for NPDSCH. NPUSCH can occupy less than 1 PRB via the support of single-tone or multitone transmissions. Such sub-PRB bandwidth transmission options are preferable under poor coverage conditions. Moreover, NPUSCH single-tone transmission in NB-IoT can use modulations with PAPR close to 0 dB to best serve devices at the edge of coverage (i.e. $\pi/2$-BPSK or $\pi/4$-QPSK with phase continuity between symbols).

Similar to the PRACH in LTE, the NPRACH in NB-IoT is used by the device to support the random procedure and start an RRC connection. However, in contrast to LTE, the NPRACH is also used to indicate scheduling requests given that there is no uplink control channel. NPRACH preambles are based on single-tone frequency-hopping waveforms. One NPRACH preamble consists of four symbol groups, with each symbol group comprising one CP and five symbols. The CP length is 66.67 μs (Format 0) for cell radius up to 10 km and 266.7 μs (Format 1) for cell radius up to 40 km.

8.5.3.2.4 Operational Aspects

A NB-IoT cell can operate with up to three different coverage enhancement (CE) levels, referred to as *normal, robust* and *extreme*, which allow serving UEs with dissimilar path loss ranges. The applicable CE level for a given UE is determined based on its received signal power. Each CE level is associated with a certain setting of some transmission parameters, including the transmit power, the subset of subcarriers, the number of repetitions of NPRACH preambles, and the maximum number of transmission attempts. For example, a NPRACH preamble can be configured to be repeated up to 64 times under the *extreme* CE level.

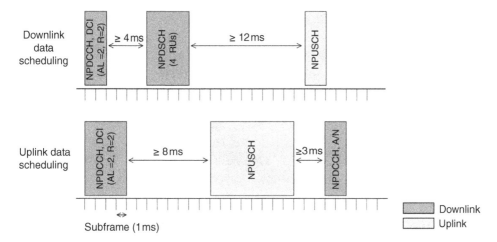

Figure 8.15 Downlink and uplink NB-IoT scheduling and HARQ operation.

NB-IoT allows the use of only one HARQ process and supports longer UE decoding times for both NPDCCH and NPDSCH. Asynchronous, adaptive HARQ procedure is adopted to support scheduling flexibility. Figure 8.15 illustrates how data scheduling works in downlink and uplink.

In both cases, a scheduling command is always conveyed first through the NPDCCH. NPDCCH may use Aggregation Levels (AL) 1 or 2 for transmitting the DCI. With AL-1, two DCIs are multiplexed in one subframe; otherwise, one subframe only carries one DCI (i.e. AL-2), giving rise to a lower coding rate and improved coverage. Further coverage enhancement can be achieved through repetition. Each repetition occupies one subframe.

For downlink scheduling, the time offset between NPDCCH and the associated NPDSCH is indicated in the DCI, which is at least 4 ms (in comparison, LTE PDCCH schedules PDSCH in the same subframe). After receiving NPDSCH, the UE needs to send back the HARQ acknowledgement. The resources of NPUSCH carrying HARQ acknowledgement are also indicated in DCI. The time offset between the end of NPDSCH and the start of the associated HARQ acknowledgement is at least 12 ms. This offset is longer than that between NPDCCH and NPDSCH because the TB carried in NPDSCH might be much longer than the DCI, which is only 23 bits long.

For uplink scheduling, the DCI also specifies which subcarriers a UE is allocated. The time offset between the end of NPDCCH and the beginning of the associated NPUSCH is at least 8 ms. After completing the NPUSCH transmission, the UE monitors NPDCCH to learn whether NPUSCH is received correctly by the base station, or a retransmission is needed.

8.5.3.3 Operation in Unlicensed Spectrum

As discussed in Chapter 7, the MulteFire Alliance develops specifications to adapt 3GPP cellular-based technology for standalone operation in unlicensed or shared spectrum. In the MulteFire Release 1.1 specification [49], new optimizations were introduced especially for IoT, such as support for NB-IoT and LTE-M (referred to as eMTC in the MulteFire specifications) in unlicensed spectrum. These adaptations are referred to as NB-IoT-U and eMTC-U, respectively.

With respect to eMTC-U, the specification covers operation into the 2.4 GHz band, which is globally available. While layers above the physical layer are practically the same as LTE-M, eMTC-U employs Listen-Before-Talk (LBT) and Frequency Hopping (FH) technology to coexist and avoid harmful interference with other technologies that use the 2.4 GHz band (e.g. Wi-Fi, Bluetooth,

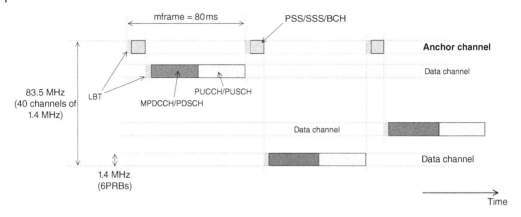

Figure 8.16 Time-frequency structure for eMTC-U operation.

IEEE 802.15.4). In particular, the eMTC-U eNB performs LBT before transmitting in each of the channels that visits according to a hopping sequence (the UE does not use any channel access mechanism to acquire the medium since its transmissions are scheduled by the eNB).

Figure 8.16 illustrates the time–frequency structure used by eMTC-U. The channel bandwidth is 1.4 MHz (like LTE-M) so that the 82.5 MHz of bandwidth available in the 2.4 GHz band are arranged into 40 channels. In the time domain, the system defines a new frame structure with a length of 80 ms and arranged in two time periods: an anchor channel dwell of 5 ms duration, and a data channel dwell of 75 ms duration. The anchor channel dwell is always transmitted on the same frequency channel. The reason is that the anchor channel is used to allow UEs to find the cell and determine the FH sequence of the data channel. Three anchor channel frequencies are defined in the standard, and the eNB selects one of them to perform transmissions on it. As a UE performs initial acquisition, it may cycle through the defined anchor channels, attempting cell acquisition, until it detects the cell. Once the cell is detected, the UE gets synchronized with the hopping sequence of the corresponding data channels. The data channel dwell may consist of several alternate uplink and downlink transmissions. Each of the dwells in anchor and data channels is preceded by LBT, and the data dwell always starts with a downlink transmission [50].

With respect to NB-IoT-U, the solution adapted by the Multefire Alliance has been particularly optimized for the 902–928 MHz band, governed by FCC regulations, and for the 865–870 MHz band, governed by EU regulations. The underlying channel bandwidth (180 kHz at receiver), modulation schemes (OFDM on downlink, SC-FDMA on uplink), coverage enhancement techniques, and FEC schemes (Convolutional Coding on downlink, Turbo Coding on uplink) are identical between 3GPP NB-IoT and NB-IoT-U [51]. The key changes between 3GPP NB-IoT and NB-IoT-U relate to the lower physical layer, where the synchronization and broadcast channels (NPSS, NSSS and NPBCH) have been modified to satisfy the license-exempt regulations while achieving the target performance for NB-IoT-U. Higher layer procedures are also almost unchanged compared with 3GPP NB-IoT.

8.5.3.4 LTE-M and NB-IoT Roadmap in 5G

Since their introduction in Release 12 and 13, new optional features have been added to LTE-M and NB-IoT in successive releases until Release 16. Among them [44, 45]:

- Support for multicast/broadcast transmissions for group messaging.
- Connected mode mobility for NB-IoT.

- Paging and random access support in nonanchor carriers, with configurations of up to 15 nonanchor carriers.
- Local RRM policy information storage for UE differentiation, enabling the network to handle device-specific IoT information to improve scheduling according to, e.g. its battery lifetime or power supply, mobility, and traffic pattern.
- Support for TDD for NB-IoT.
- Wake-Up Signals (WUS) to save UE power. When a UE is in DRX or eDRX, it must regularly check for paging occasions. At most of these occasions, no message arrives for the UE. This feature allows the eNB to send the UE a WUS to instruct the UE that it must monitor NPDCCH for paging.
- Early data transmission (EDT), allowing an idle mode UE to transmit a short amount of data in during the random access procedure, reducing the latency, and signaling overhead for short messages.

NB-IoT and LTE-M actually constitute the only solution that latest 5G specifications provide for the mMTC use cases. Performance studies conducted within 3GPP [52] concluded that NB-IoT and LTE-M technologies already fulfilled the requirements established within the ITU-2020 program for mMTC services. Accordingly, 3GPP decided not to specify any mMTC solution in 3GPP Release 15 and 16 for the new NR interface.

Release 17 is developing new features that are relevant for the continued expansion of the cellular IoT technologies:

- *NR light*, also referred to as Reduced Capabilities (REDCAP) UE. This pursues the specification of a category of NR-based IoT devices that fits between high-end, full-featured NR devices and low-end IoT LTE-M and NB-IoT devices.
- Support of Non-Terrestrial Networks (NTN) such as satellites and high-altitude platforms. This pursues the adaptation of the NR interface as well as NB-IoT and LTE-M interfaces for operation in satellite networks that could complement terrestrial deployments.

8.6 IoT Application and Management Layer Protocols

Several industrial alliances and standardization organizations (e.g. OMA SpecWorks [53], OASIS [54], IETF) are promoting the use of IP-based frameworks and open protocols to connect and managing smart objects and devices [55, 56]. Figure 8.17 provides an overview of the envisioned framework, which has at its center the common IP connectivity protocols, such as IPv4,

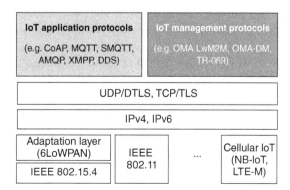

Figure 8.17 IoT application and management protocols.

IPv6, TCP, UDP, and the security protocols Transport layer Security (TLS)/Secure Socket Layer (SSL) and Datagram Transport Layer Security (DTLS).

At the application layer, different protocols have been developed for resource-constrained devices. These include: MQ Telemetry Transport (MQTT), Secure MQTT (SMQTT), Advanced Message Queuing Protocol (AMQP), Constrained Application Protocol (CoAP), Extensible Messaging and Presence Protocol (XMPP), and Data Distribution Service (DDS). The selection of the most appropriate protocol is highly dependent on the specific use case. For example, if the application requires REST functionality with HTTP-based transport, then CoAP could be a valid option. On the other hand, if the application is overhead and power sensitive, a publish/subscribe messaging model such as the one supported by MQTT could be more convenient. Among the mentioned application layer protocols, CoAP and MQTT are among the most widely used.

On the management side, there are also several protocols suitable for the management of heterogeneous, resource-constrained devices. Some of them are: Technical Report 069 (TR-069), developed within Broadband Forum for remote management of M2M devices using HTTP; Open Mobile Alliance Device Management (OMA-DM), overseen by OMA SpecWorks and designed for remote provisioning, updating and managing faulty issues of M2M devices; and Lightweight M2M (LwM2M), another OMA SpecWorks protocol specifically designed for IoT device management.

8.6.1 CoAP

CoAP is a request/response application layer protocol designed by the IETF (RFC 7252) to target resource-constrained devices. It follows a client-server architecture based on a subset of the HTTP methods (e.g. HTTP commands GET, POST, PUT, and DELETE). Both synchronous and asynchronous request/response exchanges are supported.

CoAP runs over UDP to keep the overall implementation lightweight. This allows CoAP to support unicast as well as multicast transactions.

While CoAP does not include any built-in security features, it may be used in conjunction with the DTLS to secure the CoAP transactions. DTLS runs on top of UDP and is the analogous of TLS for the TCP. DTLS provides authentication, data integrity, confidentiality, automatic key management, and cryptographic algorithms.

8.6.2 MQTT

MQ Telemetry Transport (MQTT) is a lightweight application protocol that follows a publish/subscribe model. The protocol started as a proprietary protocol developed by IBM (as part of a product called IBM MQ) and was later approved as an OASIS standard [57].

In a publish/subscribe model, there are *publishers*, which are devices or applications that send the messages, *subscribers*, which are devices or applications that receive the messages they are interested in, and *brokers*, which are devices or applications that pass the messages from the publishers to the subscribers. The use of the publish/subscribe model allows one-to-many message distribution, and application decoupling.

MQTT runs over TCP. However, publish/subscribe protocols typically need fewer computational resources than request/response since clients do not have to request updates. MQTT supports three message delivery options: "at most once," where a message is sent once and no acknowledgement is required; "at least once," where a message is sent until an acknowledgement is received (though duplicates can occur); and "exactly once," where a four-way handshake mechanism is used to ensure the message is delivered exactly one time. To ensure security, MQTT brokers may rely on TLS/SSL protocols.

8.6.3 OMA LwM2M

Lightweight M2M (LwM2M) is an open protocol for managing resource-constrained devices on a variety of networks. The specifications are overseen by OMA SpecWorks [58]. The LwM2M protocol follows a REST architectural model (e.g. stateless transactions) and builds on the CoAP protocol for management data transfer. This makes the protocol suitable for constrained-resources and battery-powered devices, including 3GPP CIoT and LoRaWAN devices.

LwM2M supports multiple management capabilities [59] such as device bootstrapping, device configuration, firmware updates, diagnostics, connection management (e.g. managing basic parameters needed for the technologies to function e.g. network security credentials etc.), device control (e.g. wake-up, device rebooting), security control (e.g. lock and wipe the device), and data reporting.

8.7 Applicability to Smart Grids

Wireless IoT technologies are applicable to Smart Grids. Although there are many examples of this applicability, the following sections focus on the most representative ones.

8.7.1 Great Britain Smart Metering System

The Great Britain (GB) Smart Metering program started preparations in 2009 and planned to roll out an estimated quantity of 53 million smart electricity and gas meters to domestic and nondomestic properties in GB by 2020. However, today, the roll-out is still underway with a total of 22.2 million smart and advanced meters operating in GB in homes and small businesses [60]. This is an evidence of the difficulties of this kind of Smart Grid-related programs, and specifically, within the ambitions particular of the electricity system in UK.

Electricity meters in GB are neither operated nor managed by the Distribution utilities, but are under the responsibility of retailers. This in unusual compared to rest of approaches around Europe and partially explains the differences with other solutions for Smart Metering in Europe. Differences can be found in the customer-oriented approach, but also in the selection of technical, and specifically, telecommunications solutions.

The Smart Metering program in GB is a customer-centric program where the roll-out is led by the suppliers with a framework of services provided by different companies and a hybrid network with several technical components that are not homogeneous across the different service areas. These services are provided by a so-called Data Communications Company (DCC) [61] managing a shared Smart Metering national infrastructure with different service providers. Energy suppliers but also network operators (using the data of the smart meters to better manage the grid), and other third parties use these services and cover their cost. The Smart Energy Code (SEC [62]) is the multiparty agreement with the rights and obligations of the relevant parties involved in the end-to-end management of Smart Metering in GB (Figure 8.18).

The anchor telecommunication's device in the customer's place (home, building) is the Communications Hub (Comms Hub or CH). The telecommunications network is designed with two areas, namely the Smart Metering Wide Area Network (SMWAN) and the energy consumers domain or Smart Metering Home Area Network (SMHAN). The CH is the gateway between the two network segments. The application layer protocol of the system is DLMS/COSEM, and

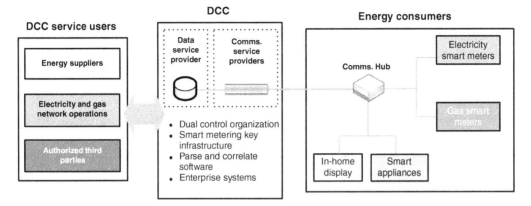

Figure 8.18 Great Britain Smart Metering system general view. *Source*: Based on Ref. [63].

CGI [64] acts as the Data Service Provider (DSP), managing the messages exchange among the system and the smart meters.

There are two telecommunication network solutions to reach these CHs in the SMWAN, particular of the North, and Center and South zones. The telecommunication services are provided by Arqiva in the North and by Telefonica in the Center and South (Arqiva and Telefonica are referred to as Communications Service Providers -CSPs-). Arqiva's telecommunications network is based in a UHF (412–414 MHz for uplink, and 422–424 MHz for downlink) point-to-area proprietary network. Telefonica's telecommunications network is based in 2G (GSM/GPRS/EDGE) in 900/1800 MHz bands, and 3G (HSPA) in 900/2100 MHz bands. Telefonica complements its solution with a mesh radio by former Connode (today CyanConnode [65]) to reach cellars and some locations of specific building types were cellular networks may have poor coverage. This complementary mesh radio solution is proprietary and based on IEEE 802.15.4g, and according to [66] uses the 869.4–869.65 MHz band.

The CH as the central piece of the telecommunications network that connects to the SMWAN and, as such, is presented in different variants [67], as its telecommunications module facing the WAN is different. However, the connectivity toward the smart meters, the In-home Display (IHD), and any other smart appliance is provided over IEEE 802.15.4-based Zigbee. Thus, the HAN is a Zigbee radio network working in 2.4 GHz, where all connected smart meters communicate with the system, and the rest of the devices (so-called Consumer Access Devices [CADs]) paired to their HAN will be able to access consumption and tariff information directly from the smart meter (electricity information every 10 seconds and gas information every 30 minutes – the gas meter is battery operated [68]).

The electricity smart meters are defined in the Smart Metering Equipment Technical Specifications 2 (SMETS2) [69]. A SMETS2 meter must be also be compliant with the so-called Great Britain Companion Specification (GBCS) [70]; the GBCS describes the detailed requirements for communications between Smart Metering devices in consumers' premises, and between them and the DCC. This specification is in fact the reference to which individual system components should refer to, and this applies to the CHs as well; the technical definition of the CH can be found in [71].

The specifications above refer to the Zigbee SEP as the central piece of the Zigbee Smart Energy (ZSE) profile specification identified in the GBCS. The ZSE standard [72] defines all the aspects of Zigbee needed in CADs and CHs. It includes device descriptions and standard practices for smart energy applications in residential or light commercial environments, with installation scenarios

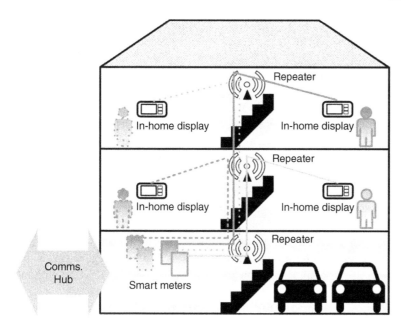

Figure 8.19 Example of Zigbee configuration to reach multidwelling units. *Source*: Modified from Ref. [72].

ranging from a single home to an entire apartment complex. It includes metering (credit and prepayment modes), pricing (based on scheduling), messaging and demand response and load control. The GBCS requires, for all Smart Metering equipment, the implementation of functionality equivalent to a subset of the Zigbee standards, including all mandatory components required to achieve ZSE certification; it also identifies the parts of the Zigbee standards that are not used or used in a different way.

As a result of the use of Zigbee, HAN RF propagation (Figure 8.19) is a key concern since the beginning of the project. Early trials suggested that 2.4 GHz radio would achieve approximately 70% coverage of homes in GB without using additional equipment. Thus, the 868 MHz option (solutions in the 862–876 MHz and 915–921 MHz ranges) started to be pursued with the intention to reach a coverage of 96.5% of premises with the so-called Dual Band CHs [73].

As a final note, Zigbee is not expected to cover all building types and installation scenarios. For that purpose, alternative HAN communication solutions have been explored, and the AltHANCo [74] has been created. This company will develop telecommunication infrastructures in buildings, to guarantee that smart meters can communicate with the different CADs (including IHD's) inside customers' homes.

8.7.2 Unlicensed Spectrum-based LPWAN Technologies for Smart Metering

The connection of IoT and the Smart Grid is a blurry space. On the one side, utilities have diverse telecommunication needs connected both to their traditional services and those involved in their Smart Grid evolution. On the other, utilities are also an industry with needs similar to others, that may take advantage of commercially existing proposals, to develop opportunities within their non-critical grid-related services (e.g. for their retail companies to improve their service offer to third parties; for their asset management departments with non-operations related assets, etc.). In this context, vendors and service providers finding markets to develop new businesses, try to capture utility needs and devote resources to develop proofs of concepts and eventually pilot projects to demonstrate feasibility. As IoT market is still in its infancy, it is common to find references to such limited field tests in literature.

The term "utility" is not helping to clarify the applications developed for electricity utilities either. "Utility," as we have seen, refers to a wide range of power systems-related companies, but it also refers to water and gas industries, to name the most outstanding ones. When IoT literature refers to Smart Metering for utilities, it cannot be taken for granted that it is referring to electricity utilities, which is the scope of this book. The evolution into smartness of water and gas industries is different to power systems', for a variety of reasons. Some of the reasons are related to the maturity of the industry in terms of control, reliability, and criticality of the service; but some others, relate to the simple fact of the availability of electricity in the points where the smartness is needed. Following the example of the smart meter, it is not to be taken for granted that power supply exists where a water or a gas meter is located and consequently, telecommunication devices that need to provide service to the smart meters, need to operate on batteries [75]. This aspect is deeply affecting not only the telecommunications technology to be used, but the operation regime (i.e. if the connection can be constantly operational, when it needs to work for over 10–15 years without being replaced [76]).

The most common utility use case claimed by IoT market is Smart Metering. Although there are some references to the Smart Grid (see e.g. the case where power lines need to be monitored, and the devices to do that can only operate on the power induced from the power line [77]), literature focuses on Smart Metering, and although there are cases associated to electricity utilities, most of them are with water and gas industries. When we move into real massive deployment decisions, the focus of IoT on Smart Metering for water and gas industries becomes clearer [28, 78].

The evolution of electricity utilities Smart Metering uses cases can be traced parallel to the evolution of 3GPP-related technologies. IoT is in this case an evolution of the initial M2M (Machine to Machine) concept. Electricity Smart Metering experiences started to be popular with GPRS-connected devices [79]. These smart meters leveraged the pervasiveness of mobile networks to connect to them. At the same time, they developed mechanisms to control power consumption [80] and the stress imposed on the mobile networks (high number of connected devices in a system not prepared to host them). These solutions could not easily be applied over water and gas industries due to the need of the smart meters to run on battery, and although there are also experiences in this domain, GPRS never became the rule but the exception [81].

Continuing with GPRS as the example of a 3GPP technology for this kind of solutions, the time-lapse between its origin and the advent of its NB-IoT and LTE-M successors is long. During this period of time, a diverse range of technologies outside the realm of traditional TSPs (both to deploy private networks and to be run by alternative TSPs) have emerged and made the market grow, and some of them have been widely adopted. Even some traditional TSPs have adopted non-3GPP technologies in some markets [82], to continue growing and capturing a market that was emerging. These alternative-to-3GPP technologies are now in the market and in the field and have filled the delay gap [83].

With this situation, it became apparent to utility industry in general (companies and regulators) that IoT LWPAN technology started to be mature to gradually evolve from meter operations based on human local intervention, to remotely connected systems. Thus, pricing considerations came into play [84], and markets and technologies started to be developed. While in some industries (water and gas), very low-consumption radio technologies started to be considered the only solution, in electricity industry competing alternatives (e.g. short-range radio and PLC) could be considered. It is also true that smart meter vendors offer all sorts of telecommunication alternatives within their smart meters, and it is unusual to find that the focus is put on a single technology. However, this need to keep a broad range of technology solutions in their portfolio imposes tensions in their evolution capabilities and commercial margins.

Once any utility decides that a LPWAN solution is suiting its service needs, there is no clear decision tree to select the most convenient technology alternative. As we have seen in this chapter, there are multiple technologies and service model (private network or commercial service) options to the extent that each interest group around the different technologies has produced their own reports and white papers to show their market position [85–88]. Despite the technical considerations, that are to be analyzed carefully to identify the constraints of each technology option (influencing aspects such as coverage range, scalability -users, load-, interference management, energy efficiency, etc. [28]), there are other considerations that should be borne in mind [89] to analyze the rest of performance and compliance with requirements aspects (cost, integration, evolution, etc. [28]):

- Scope of the project, in terms of data to be exchanged (operation regime, including firmware upgrade needs), area to be covered, and number of devices to be connected.
- Amount of data to be exchanged, with daily, weekly, and yearly patterns.
- Power supply autonomy and needed battery lifetime of end-devices.
- Availability of private infrastructure.
- Availability of private spectrum, or spectrum in service providers' networks.
- Coverage of LPWAN network, related to the location of the end-devices to be reached.
- Need of a single technology, or of a hybrid network to cope with different areas and needs.
- Total cost of ownership in the lifespan of the assets and expected operation, considering technology refresh cycles.
- Vendor and service providers' ecosystem and their engagement with the technology alliances supporting the technologies.

Notwithstanding the consideration above, IoT market is much wider that the utility use cases, and it is safe to state that smart cities, asset tracking, logistics, environmental and wildlife monitoring, agriculture, etc., are today the highest priorities from the different IoT vendors, service providers, and stakeholders in general.

References

1 I-SCOOP. *Comprehensive and Visual Classification of Wireless IoT Protocols and Network Technologies* [Online]. https://www.i-scoop.eu/internet-of-things-guide/wireless-iot-protocols-technologies (accessed 7 February 2021).
2 Chew, D. (2019). The Wireless Internet of Things: A Guide to the Lower Layers. Wiley-IEEE Standards Association.
3 Wi-Fi Alliance (2020). *Wi-Fi HaLow™: Wi-Fi® for IoT Applications* (May 2020).
4 Raza, U., Kulkarni, P., and Sooriyabandara, M. (2017). Low power wide area networks: an overview. *IEEE Communications Surveys & Tutorials* 19 (2): 855–873. https://doi.org/10.1109/COMST.2017.2652320.
5 Liberg, O., Sundberg, M., Wang, Y.-P.E. et al. (2018). Cellular Internet of Things Technologies, Standards, and Performance. Academic Press, Elsevier Ltd.
6 3GPP TR 36.888 v12.0.0 (2013). *Study on Provision of Low-Cost Machine-Type Communications (MTC) User Equipments (UEs) Based on LTE (Release 12)* (June 2013).

7 IEEE Std 802.15.4-2020 (2020). *IEEE Standard for Low-Rate Wireless Networks* (Revision of IEEE Std 802.15.4-2015), pp. 1–800 (23 July 2020). doi: https://doi.org/10.1109/IEEESTD.2020.9144691.
8 IEEE 802.15. *Working Group for Wireless Specialty Networks (WSN)* [Online]. https://www.ieee802.org/15 (accessed 1 April 2021).
9 Ramonet, A.G. and Noguchi, T. (2020). IEEE 802.15.4 now and then: evolution of the LR-WPAN standard. *22nd International Conference on Advanced Communication Technology (ICACT)*, Phoenix Park, PyeongChang, Korea (South).
10 Zigbee Alliance [Online]. www.zigbeealliance.org (accessed 1 April 2021).
11 Zigbee Alliance (2017). *Zigbee Specification*. ZigBee Document 05-3474-22 (April 2017).
12 Zigbee Alliance (2014). *Zigbee Smart Energy Standard*. Zigbee Document 07-5356-19 (December 2014).
13 Zigbee Alliance. *JupiterMesh* [Online]. https://zigbeealliance.org/solution/jupitermesh (accessed 7 February 2021).
14 Wi-SUN Alliance [Online]. www.wi-sun.org (accessed 1 April 2021).
15 Harada, H., Mizutani, K., Fujiwara, J. et al. (2017). IEEE 802.15.4g based Wi-SUN communication systems. *IEICE Transactions on Communications* E100.B (7): 1032–1043.
16 ECHONET [Online]. https://echonet.jp/english (accessed 1 April 2021).
17 Wi-SUN Alliance (2018). *Technical Profile Specification Field Area Network*.
18 ANSI/TIA-4957.210 (2013). *Layer 2 Standard Specification for the Smart Utility Network* (January 2013).
19 Heile, B. (2016). *Wi-SUN Alliance Field Area Network (FAN) Overview*, IETF 97 (November 2016).
20 IEEE 802.15.9-2016 (2016). *IEEE Recommended Practice for Transport of Key Management Protocol (KMP) Datagrams* (August 2016).
21 ETSI TS 102.887-2 (2013). *Smart Metering Wireless Access Protocol; Part 2: Data Link Layer (MAC Sub-layer)* (September 2013).
22 Kurunathan, H., Severino, R., Koubaa, A., and Tovar, E. (2018). IEEE 802.15.4e in a nutshell: survey and performance evaluation. *IEEE Communications Surveys Tutorials* 20 (3): 1989–2010.
23 Muñoz, J., Riou, E., Vilajosana, X. et al. (2018). Overview of IEEE802.15.4g OFDM and its applicability to smart building applications. *Wireless Days (WD)*, Dubai, United Arab Emirates (April 2018), pp. 123–130, doi: 10.1109/WD.2018.8361707.
24 Wi-SUN Alliance (2018). *PHYWG Wi-SUN PHY Specification 1V07_RC2*.
25 IEEE Std 802.15.4-2015 (2016). *IEEE Standard for Low-Rate Wireless Networks* (Revision of IEEE Std 802.15.4-2011), pp. 1–709 (22 April 2016). doi: 10.1109/IEEESTD.2016.7460875.
26 Mochizuki, K., Obata, K., Mizutani, K., and Harada, H. (2016). Development and field experiment of wide area Wi-SUN system based on IEEE 802.15.4g. *2016 IEEE 3rd World Forum on Internet of Things (WF-IoT)*, Reston, VA, pp. 76–81. doi: 10.1109/WF-IoT.2016.7845425.
27 IETF (2017). *Wi-SUN FAN Overview: draft-heile-lpwan-wisun-overview-00*.
28 Buurman, B., Kamruzzaman, J., Karmakar, G., and Islam, S. (2020). Low-power wide-area networks: design goals, architecture, suitability to use cases and research challenges. *IEEE Access* 8: 17179–17220. https://doi.org/10.1109/ACCESS.2020.2968057.
29 IETF RFC 8376 (2018). *Low-Power Wide Area Network (LPWAN) Overview* (May 2018).
30 Sigfox [Online]. www.sigfox.com (accessed 6 November 2020).
31 Semtech [Online]. www.semtech.com (accessed 6 November 2020).
32 LoRa Alliance [Online]. www.lora-alliance.org (accessed 6 November 2020).
33 LoRa Alliance (2017). *LoRaWAN 1.1 Specification* (November 2017) [Online]. https://lora-alliance.org/wp-content/uploads/2020/11/lorawantm_specification_-v1.1.pdf (accessed 1 April 2021).
34 LoRa Alliance (2020). *LoRaWAN Backend Interfaces Technical Specification (TS002-1.1.0)* (October 2020).

35 Haxhibeqiri, J., De Poorter, E., Moerman, I., and Hoebeke, J. (2018). A survey of LoRaWAN for IoT: from technology to application. *Sensors* 18 (11): 3995.
36 Adelantado, F., Vilajosana, X., Tuset-Peiro, P. et al. (2017). Understanding the limits of LoRaWAN. *IEEE Communications Magazine* 55 (9): 34–40.
37 LoRa Alliance (2020). *RP002-1.0.2 LoRaWAN Regional Parameters* (October 2020).
38 Sigfox (2017). *Technical Overview*, whitepaper [Online]. www.sigfox.com (accessed 1 April 2021).
39 Lavric, A., Petrariu, A.I., and Popa, V. (2019). Long range SigFox communication protocol scalability analysis under large-scale, high-density conditions. *IEEE Access* 7: 35816–35825. https://doi.org/10.1109/ACCESS.2019.2903157.
40 Sigfox (2020). *Sigfox Connected Objects: Radio Specifications*, Version 1.5 (February 2020).
41 Gomez, C., Veras, J.C., Vidal, R. et al. (2019). A Sigfox energy consumption model. *Sensors* 19 (3): 681.
42 3GPP TR 45.820 v13.0.0 (2016). *Cellular System Support for Ultra-Low Complexity and Low Throughput Internet of Things* (June 2016).
43 3GPP TS 23.682 v16.6.0 (2020). *Architecture Enhancements to Facilitate Communications with Packet Data Networks and Applications (Release 16)* (March 2020).
44 GSM Association (2018). *LTE-M Deployment Guide to Basic Feature Set Requirements*. Official Document CLP.29, Version 2.0 (5 April 2018).
45 GSM Association (2019). *NB-IoT Deployment Guide to Basic Feature set Requirements* (June 2019).
46 3GPP SP-100863 (2010). *Update to Network Improvements for Machine Type Communication*, 3GPP TSG SA Meeting #50.
47 Wang, Y.-P.E., Lin, X., Adhikary, A. et al. (2017). A primer on 3GPP narrowband internet of things. *IEEE Communications Magazine* 55 (3): 117–123. https://doi.org/10.1109/MCOM.2017.1600510CM.
48 Kanj, M., Savaux, V., and Le Guen, M. (2020). A tutorial on NB-IoT physical layer design. *IEEE Communications Surveys & Tutorials* 22 (4): 2408–2446. https://doi.org/10.1109/COMST.2020.3022751.
49 MulteFire Alliance (2018). *MulteFire Release 1.1 Technical Overview*. White Paper (December 2018).
50 MulteFire Alliance (2019). *MulteFire Release 1.1 eMTC-U*. White Paper.
51 MulteFire Alliance (2019). *NB-IoT-U: The Next Generation Dedicated IoT Network for Enterprise Customers*. White Paper (February).
52 3GPP TR 37.910 v16.1.0 (2019). *Study on Self-evaluation Towards IMT-2020 Submission (Release 16)* (September 2019).
53 OMASpecWorks [Online]. www.omaspecworks.org (accessed 1 April 2021).
54 OASIS [Online]. www.oasis-open.org (accessed 1 April 2021).
55 Karagiannis, V., Chatzimisios, P., Vazquez-Gallego, F., and Alonso-Zarate, J. (2015). A survey on application layer protocols for the Internet of Things. *Transaction on IoT and Cloud Computing* 3 (1): 11–17.
56 Salman, T. and Jain, R. (2019). A survey of protocols and standards for Internet of Things. *Computer Science, Business ArXiv* [Online]. https://arxiv.org/ftp/arxiv/papers/1903/1903.11549.pdf (accessed 1 April 2021).
57 OASIS. MQTT Version 5.0 [Online]. https://docs.oasis-open.org/mqtt/mqtt/v5.0/mqtt-v5.0.html (accessed 1 April 2021).
58 OMA SpecWorks LwM2M [Online]. https://omaspecworks.org/what-is-oma-specworks/iot/lightweight-m2m-lwm2m (accessed 1 April 2021).
59 OMA SpecWorks (2019). *Lightweight Machine to Machine v1.1*. White Paper [Online]. https://github.com/OpenMobileAlliance/OMA_LwM2M_for_Developers/wiki (accessed 1 April 2021).

60 Department for Business, Energy & Industrial Strategy (2020). *Smart Meter Statistics in Great Britain: Quarterly Report to End September 2020* [Online]. https://assets.publishing.service.gov.uk/government/uploads/system/uploads/attachment_data/file/937577/Q3_2020_Smart_Meters_Statistics_Report_FINAL.pdf (accessed 1 April 2021).

61 Data Communications Company [Online]. www.smartdcc.co.uk (accessed 1 April 2021).

62 The Smart Energy Code [Online]. https://smartenergycodecompany.co.uk/the-smart-energy-code-2 (accessed 1 April 2021).

63 Data Communications Company (DCC). *Annual Report 2020* [Online]. https://www.smartdcc.co.uk/media/4098/20204_dcc_report_and_accounts_9.pdf (accessed 1 April 2021).

64 CGI Smart Metering [Online]. https://www.cgi.com/uk/en-gb/utilities/smart-metering (accessed 1 April 2021).

65 CyanConnode. *Telefónica and Connode Set to Provide UK Smart Meter Communications Service* [Online]. https://cyanconnode.com/telefonica-connode-set-provide-uk-smart-meter-communications-service (accessed 1 April 2021).

66 Aegis Systems Limited (2014). *M2M Application Characteristics and Their Implications for Spectrum* (May 2014) [Online]. https://www.ofcom.org.uk/__data/assets/pdf_file/0040/68989/m2m_finalreportapril2014.pdf (accessed 1 April 2021).

67 Data Communications Company (DCC). *Communications Hubs* [Online]. https://www.smartdcc.co.uk/document-centre/communications-hubs (accessed 1 April 2021).

68 EAMA. *Consumer Access Devices* [Online]. http://www.beama.org.uk/asset/EBA9BBB9-756E-48A0-B6C38EBFBC3EEEAB (accessed 1 April 2021).

69 *Smart Metering Equipment Technical Specifications 2 (SMETS2)* (November 2020) [Online]. https://smartenergycodecompany.co.uk/download/29586 (accessed 1 April 2021).

70 *Smart Metering Implementation Programme Great Britain Companion Specification (GBCS)* (November 2020) [Online]. https://smartenergycodecompany.co.uk/download/29602 (accessed 1 April 2021).

71 *Smart Metering Implementation Programme Communications Hub Technical Specifications (CHTS)* (January 2020) [Online]. https://smartenergycodecompany.co.uk/download/29589 (accessed 1 April 2021).

72 Zigbee Alliance. *Downloads Documents Referenced in GBCS* [Online]. https://zigbeealliance.org/downloads-documents-referenced-in-gbcs (accessed 1 April 2021).

73 Data Communications Company (DCC). *Dual Band Communications Hubs* [Online]. https://www.smartdcc.co.uk/smart-future/dual-band-communications-hubs (accessed 1 April 2021).

74 AltHANCo [Online]. www.althanco.com (accessed 1 April 2021).

75 Reuning, H. and Joosse, M. (2013). Investigations into the lifetime of gas meter batteries in the Netherlands. *Metering International* (2) [Online]. https://tadiranbatteries.de/pdf/articles/Reuning-Joosse.pdf (accessed 1 April 2021).

76 Dittrich, T. *Battery Concepts for Smart Utility Meters – The Requirements and Proving Their Suitability* [Online]. https://tadiranbatteries.de/pdf/applications/battery-concepts-for-smart-utility-meters.pdf (accessed 1 April 2021).

77 Akpolat, A., Nese, S., and Dursun, E. (2018). Towards to smart grid: dynamic line rating, pp. 96–100 [Online]. doi: https://doi.org/10.1109/SGCF.2018.8408950. https://www.researchgate.net/profile/Alper_Akpolat/publication/326361438_Towards_to_smart_grid_Dynamic_line_rating/links/5fc4c486458515b7978a15c9/Towards-to-smart-grid-Dynamic-line-rating.pdf (accessed 1 April 2021).

78 Analysis Mason (2020). *Operators Must Build Credibility with Utility Companies to Win Smart Water and Gas Metering Contracts* [Online]. https://www.analysysmason.com/contentassets/

cf9ce89a6b1e403c9f56a28d1fce0c4c/analysys_mason_smart_water_gas_metering_comment_jul2020_rdme0.pdf (accessed 1 April 2021).

79 Vodafone. *Case Study: GPRS Mobile Network for Smart Metering* [Online]. https://www.iotone.com/casestudy/gprs-mobile-network-for-smart-metering/c465 (accessed 1 April 2021).

80 Vodafone. *Case Study: SMS-Helping Enexis Power Ahead with Smart Meter Rollout* [Online]. https://www.iotone.com/casestudy/helping-enexis-power-ahead-with-smart-meter-rollout/c464 (accessed 1 April 2021).

81 GSMA. *Utilities* [Online]. https://www.gsma.com/iot/mobile-iot-sectors-utilities (accessed 1 April 2021).

82 Weldon, K. *Orange Business Services Global Industrial IoT Services* [Online]. https://www.orange-business.com/sites/default/files/globaldata_global-industrial-iot_jul18.pdf (accessed 1 April 2021).

83 I-SCOOP. *LPWAN: Delayed Deployment of NB-IoT and LTE-M Benefits LoRa and Sigfox* [Online]. https://www.i-scoop.eu/internet-of-things-guide/lpwan/lpwan-2026-nbiot-ltem-delays-benefit-lora-sigfox (accessed 1 April 2021).

84 Smart Energy International (2019). *Affordable Connectivity Driving Smart Water and Gas* [Online]. https://www.smart-energy.com/magazine-article/affordable-connectivity-driving-smart-water-and-gas (accessed 1 April 2021).

85 Wi-SUN Alliance (2019). *Comparing IoT Networks at a Glance* [Online]. https://www.wi-sun.org/wp-content/uploads/WiSUN-Alliance-Comparing-IoT-Networks-2019-Nov-A4.pdf (accessed 1 April 2021).

86 ABI Research (2019). *LORAWAN® AND NB-IOT: Competitors or Complementary?* [Online]. https://lora-alliance.org/sites/default/files/2019-06/cr-lora-102_lorawanr_and_nb-iot.pdf (accessed 1 April 2021).

87 GSMA (2020). *LPWA Networks Comparison Paper* (August 2020) [Online]. https://www.gsma.com/iot/resources/lpwa-networks-comparison-paper (accessed 1 April 2021).

88 Link Labs (2018). *NB-IoT vs. LoRa vs. Sigfox* (25 June 2018) [Online]. https://www.link-labs.com/blog/nb-iot-vs-lora-vs-sigfox (accessed 1 April 2021).

89 I-SCOOP. *LPWA Network Technologies and Standards: LPWAN Wireless IoT Guide* [Online]. https://www.i-scoop.eu/internet-of-things-guide/lpwan (accessed 1 April 2021).

Index

a

Access and mobility management function (AMF)   247, 248, 253, 254, 257–260, 280
Access class barring (ACB)   321
Access control   135, 173, 220, 237, 250, 282, 301, 302
Access network   46, 179, 180, 184, 215, 216, 220, 254, 255, 282
Access point name (APN)   239, 240
Accuracy   31, 146, 158, 162, 175, 286, 322
Acknowledgement (ACK)   191, 203, 207, 265, 268, 274, 311, 325, 328
Add–drop multiplexer (ADM)   121, 124
Address resolution protocol (ARP)   278
Advanced distribution automation (ADA)   25
Advanced encryption standard (AES)   204, 300, 310
Advanced metering infrastructure   159, 175, 221
Advanced meter reading (AMR)   73, 139, 158, 159, 192, 256
Aggregate   46, 120, 230, 278
Aggregation   46, 47, 51, 150, 195, 266, 280, 325
Aging   66, 73, 185
African Telecommunications Union (ATU)   68, 77, 187
Alliance for Telecommunications Industry Solutions (ATIS)   29, 119, 120
Allocation   28, 77, 187, 188, 215, 220, 228, 231, 233, 234, 241, 244, 254, 264, 278, 280, 317–319, 323
Aluminum   17, 18, 65, 66, 71
American National Standards Institute (ANSI)   28, 38, 61, 119, 120, 158–161, 297, 301, 334
Amplifier   62, 115, 123, 124, 262, 318
Amplitude modulation (AM)   90, 91, 194, 213
Amplitude shift keying (ASK)   91, 104, 203, 302
Analog data   44
Antenna   67, 109, 113–116, 197, 225, 228, 238, 240–242, 246, 265–267, 276, 306, 314, 319
Application layer   57, 58, 142, 143, 150, 154, 161, 167, 248, 278, 294, 300, 301, 305, 309, 310, 328, 329, 335
Application programming interface (API)   254, 256, 290
Application protocol control information (APCI)   142
Application protocol data unit (APDU)   142
Application server   246, 251, 254, 309, 313, 317
Application service data unit (ASDU)   142
Arab Spectrum Management Group (ASMG)   69, 77
Asia Pacific Telecommunity (APT)   68, 77
Asset   4, 9, 11, 14, 19, 21–23, 25, 27, 29, 30, 34, 35, 60, 63, 70, 77, 137–140, 149, 165, 169, 170, 172, 173, 176, 216, 218, 282, 285, 291, 293, 298, 309, 331, 333, 336
Association of Radio Industries and Business (Japan) (ARIB)   29, 192, 206, 227
Asymmetric digital subscriber line (ADSL)   217
Asynchronous transfer mode (ATM)   135, 180, 183
Asynchronous transfer mode passive optical network (APON)   180
Attenuation   61, 62, 66, 91, 93, 94, 112, 114, 185, 193, 196, 211, 214, 215, 307
Authentication   161, 187, 204, 214, 237, 250, 251, 253, 254, 257, 259, 297, 301, 302, 306, 307, 312, 328
Authentication server function (AUSF)   253, 254, 257
Automatic metering infrastructure (AMI)   139, 159–161, 192, 193, 222, 297
Automatic meter reading   73, 139
Automatic repeat request (ARQ)   104, 203, 219, 244, 265, 268, 270, 320
Automation   20, 25, 32, 34, 55, 73–75, 139, 140, 147, 151, 155, 174, 176, 177, 193, 217, 218, 221, 222, 231, 255, 284, 289, 293, 296, 298
Autonomous   5, 25, 129, 149
Availability   13, 14, 22, 24, 25, 27, 30, 31, 34, 35, 59, 71, 73, 117, 120, 121, 133, 149, 153, 157, 159, 163, 164, 171, 172, 193, 217–219, 231, 234, 278, 282–285, 288, 293, 304, 332, 333
Average service availability index (ASAI)   153
Average system interruption duration index (ASIDI)   153
Average system interruption frequency index (ASIFI)   153
Awareness   162, 164

b

Backbone   6, 39, 46, 77, 177, 193–195
Backhaul   46, 75, 196, 238, 240, 247

Smart Grid Telecommunications: Fundamentals and Technologies in the 5G Era, First Edition. Alberto Sendin, Javier Matanza, and Ramon Ferrús.
© 2021 John Wiley & Sons, Inc. Published 2021 by John Wiley & Sons, Inc.

Backward error correction (BEC)   104
Band   61, 62, 68–71, 74, 75, 88, 94, 97, 99, 100, 112, 113, 116, 117, 134, 140, 183, 184, 188, 191, 192, 194, 198, 204, 206, 209, 210, 212, 214, 218, 228, 231–235, 241, 242, 247, 255, 262, 264, 266, 268, 271, 272, 276, 280, 281, 285, 288, 289, 293, 294, 302, 304, 306, 307, 310, 313, 318, 319, 322–326, 330, 331, 336
Bandwidth   30, 55, 60–62, 67, 70, 71, 79, 85, 88, 89, 97, 103, 108, 118, 121, 122, 128, 129, 141, 142, 157, 171, 173, 186–189, 191, 193–195, 204, 209, 227, 230, 264, 266, 267, 269–275, 284, 293, 294, 303, 310, 314, 315, 317–320, 322, 324, 326
Base node (BN)   200, 202, 203, 220
Base station   35, 46, 188, 225, 230, 237, 238, 240, 242, 246, 268, 269, 276, 277, 285, 306, 311, 312, 314, 316, 325
Battery   3, 24, 30, 33, 35, 99, 144, 150, 218, 230, 270, 271, 293, 294, 298, 315, 318, 320, 322, 327, 329, 330, 332, 333, 336
Battery EV (BEV)   150
Beamforming   113, 114, 248, 265, 276
Beamwidth   113
Bidirectional   45, 69, 149, 171, 179, 312, 314
Billing   11, 31, 158–160, 250, 283, 285
Binary phase shift keying (BPSK)   93, 207, 213, 302, 304, 313, 324
Bit error rate (BER)   97–99, 128, 146, 201, 208
bits per second (bps)   73, 88, 122, 192, 193, 310, 313
Blackout   172
Black start   70, 77, 149, 162
Bluetooth low energy (BLE)   293
Border gateway protocol (BGP)   131, 135
Bridge   125, 126, 135, 158, 211, 309
Broadband   33, 38, 60, 61, 63, 69, 71, 74, 75, 77, 193, 194, 197, 211, 216, 217, 221, 222, 227–229, 234, 255, 256, 280, 287, 289, 290, 328
Broadband over power line (BPL)   33, 75, 193–195, 216–219, 222
Broadband passive optical network (BPON)   180
Broadband PLC   71, 74, 75, 193, 211, 217
Broadcast   45, 116, 125, 133, 183, 192, 241, 257, 267, 270, 288, 301, 306, 307, 323, 326
Broadcast channel (BCH)   270
Broadcast control channel (BCCH)   270, 321
Broadcasting   45, 133, 194, 299, 321
Building energy management systems (BEMSs)   167, 176
Bus   14, 48, 49, 72, 108, 154, 157, 164, 175, 252,

C

Cable   6, 14–18, 25, 30, 31, 33, 35–38, 41, 42, 44, 48, 60, 61, 63–67, 71, 72, 108, 118, 183, 191, 193, 197, 214, 218, 293
Capacitor   14, 148, 197, 297
Capacity   9, 11, 13, 14, 17, 24, 44, 55, 62, 66, 70, 88, 119, 121, 122, 125, 128, 148–150, 169, 180, 183, 187, 188, 230–233, 238, 253, 265, 276, 280, 298
Capital expenditure (CAPEX)   238
Carrier   35, 59, 62, 89–91, 93, 97, 99, 100, 102, 103, 107–109, 118, 133, 171, 192–194, 196, 197, 199, 200, 205–207, 209–215, 220, 221, 230, 235, 240, 241, 243, 244, 247, 262, 266, 270, 272–275, 281, 316, 318, 322–324, 327
Carrier aggregation   230, 235, 266, 272, 281
Carrier sense multiple access/collision avoidance (CSMA/CA)   108, 110, 200, 211, 213, 215, 296, 303, 304, 306
Carrier sense multiple access/collision detection (CSMA/CD)   108, 132, 191
Carrier Ethernet   118, 133
Cell re-selection   242, 244
Cellular   29, 38, 55, 69, 70, 77, 113, 172, 217, 225, 226, 228, 229, 233, 235, 237, 238, 241, 280, 282, 283, 285–289, 291, 294, 314–316, 325, 327, 330, 333, 335
Cellular Internet of Things (CIoT)   229, 315–317, 329
Cellular network   225, 237, 241, 282, 283, 287–289, 294, 330
Century   2, 5, 10, 21, 36, 69, 73, 158, 168, 169, 192, 215
Channel coding   103, 106, 107, 197, 198, 206, 207, 246, 271, 273
Channel estimation   196, 246, 267, 268
Channel quality indicator (CQI)   264, 266
Channel state information (CSI)   264, 266, 268
Charging   28, 150, 151, 250, 251, 256
Checksum   53, 57, 133
Chirp spread spectrum (CSS)   264, 302, 310
Circuit   6, 7, 14, 18, 41, 44–46, 50–53, 56, 61, 66, 143–149, 151, 180, 197, 225, 226, 237, 240, 256, 292
Circuit breaker   144, 147, 151
Circuit-switched   44, 45, 50–52, 225, 226, 237, 240, 256, 292
Circuit switched data (CSD)   226, 283, 292
Circuit switching   45, 50, 51
Citizens broadband radio service (CBRS)   234, 235
Clear channel assessment (CCA)   303, 306
Clear to send (CTS)   110
Cloud   55, 94, 251, 312, 335
Cluster   298, 299, 305
Coarse wavelength division multiplexing (CWDM)   122
Coaxial   60, 61, 71, 118, 132, 179, 214, 216
Code division multiple access (CDMA)   108, 227
Coding   73, 81, 88, 103–107, 116, 197, 200, 207, 210, 213, 215, 227, 264–266, 271, 325, 326
Coding rate   105, 200, 210, 215, 325
Coding scheme   207, 227, 264
Coexistence   68, 183, 184, 190, 211, 272, 280, 322
Collision   108–110, 125, 132, 186–188, 191, 200, 304, 318
Comisión Interamericana de Telecomunicaciones (Inter-American Telecommunication Commission) (CITEL)   68, 77
Comité International Spécial des Perturbations Radioélectriques (International Special Committee on Radio Interference) (CISPR)   33
Command   69, 139, 140, 144–147, 151, 156, 162, 252, 264, 283, 286, 296, 305, 325, 328

Commercial 4, 8, 10, 22, 23, 26, 29, 30, 35, 38, 48, 69, 70, 77, 105, 110, 119, 139, 158–160, 166, 167, 170–172, 180, 183, 192, 216, 217, 225, 227–229, 235, 283–285, 288, 307, 330, 333
Commercial and industrial (C&I) 139, 283–287
Common API framework (CAPIF) 254, 256
Communications hub (CH) 36, 329, 330
Companion specification for energy metering (COSEM) 160, 161, 175, 219, 301, 329
Compatibility 32, 44, 64, 66, 191, 192, 196, 204, 205, 226, 247, 272, 275, 285, 317
Component carriers (CCs) 207, 273
Conductor 6, 12, 14, 16–18, 43, 60, 61, 63–67, 71, 113, 152, 169, 170, 197
Congestion 53, 128, 169, 243, 260, 278, 284, 321
Connected mode 242–244, 257, 270, 320, 322, 326
Connectionless 53
Connection-oriented 53, 133, 135, 143, 161, 246, 317
Connectivity options 286, 288, 315, 316
Conseil International des Grands Reseaux Electriques (International Council on Large Electric Systems) (CIGRE) 29, 38, 174, 223
Constrained application protocol (CoAP) 301, 328, 329
Consumption point 4, 5, 8, 15, 16, 23, 25, 137, 138
Contention access period (CAP) 303, 304
Contention-free period (CFP) 200, 202, 213, 303, 304, 306
Contention-free slot (CFS) 207, 208
Continuous-time Fourier transform 83, 88
Control center 13, 141, 143, 155, 173, 174
Control channel 187, 242–244, 247, 248, 264, 267, 268, 270, 274, 276, 318, 322–324
Control plane 59, 134, 135, 230, 244, 251, 301, 316, 317
Control resource set (CORESET) 275
Convergence layer 187–189, 194, 198, 297
Converter 6, 81, 141, 149
Convolutional encoder 106, 200, 210, 213
Cooling 167, 185
Copper 17, 18, 33, 60, 61, 71, 76, 99, 118, 132, 179, 183, 216
Copper cable 60, 179
Copper pair 33, 60, 61, 71, 99, 118, 132, 216
Core 2, 9, 17, 18, 22, 25, 29, 36, 46, 51, 58, 61, 65, 66, 71, 111, 112, 117, 118, 138–140, 168, 179, 183, 184, 192, 195, 196, 225–228, 235–237, 239, 248, 250, 252, 253, 255–257, 274, 275, 290, 293, 296, 305
Core network (CN) 46, 183, 184, 192, 225, 227, 235, 237–239, 250, 252, 253, 255, 256, 278, 280
Corona 17, 64, 66
Corporate 171, 238, 278, 291
Correction 44, 56, 104, 106, 123, 148, 207, 312
Cost–benefit analysis (CBA) 159, 285
Coverage enhancement (CE) 319, 320, 324
Coupling 168, 169, 194, 197, 218, 221, 295, 315
Coverage 14, 29, 30, 35, 51, 69, 70, 72, 113, 171, 172, 186, 192, 196, 211, 215, 217, 225, 229, 231–233, 238, 240–242, 244, 276, 282–284, 287, 288, 294, 300, 314, 315, 318–320, 322, 324–326, 330, 331, 333
Coverage area 113, 241, 300

Coverage level 30
Coverage range 72, 217, 294, 315, 318, 333
Credit 331
Critical peak pricing (CPP) 166
Critical peak rebates (CPR) 166
Current 3, 5, 12–14, 16, 17, 25, 36, 51, 66, 128, 137, 143–145, 148, 149, 159, 161, 169, 170, 176, 217, 226, 255, 257, 289, 293, 303
Current flow 149
Customer(s) 4, 6, 8, 9, 15, 16, 18, 20, 23, 24, 26, 139, 147, 150–153, 158, 161, 164–168, 194, 196, 222, 252, 278, 286, 308, 331, 335
Customer average interruption duration index (CAIDI) 153
Customer average interruption frequency index (CAIFI) 153
Customer engagement 23, 158, 159
Customer experiencing long interruption durations (CELID) 153
Customer experiencing multiple interruptions (n) (CEMIn) 153
Customer experiencing multiple sustained interruption (n) and momentary interruption events (CEMSMIn) 153
Customer management 137, 139, 147
Customer premise equipment (CPE) 184, 217, 284, 287
Customer relationship management (CRM) 139
Customer total average interruption duration index (CTAIDI) 153
Cybersecurity 35, 164, 171, 173
Cyclic Prefix 102, 198, 212–214, 261, 262, 264, 272
Cyclic Prefix OFDM (CP-OFDM) 261, 262, 272, 273
Cyclic redundancy check (CRC) 77, 174, 176, 204, 221, 309, 312

d

Data acquisition 32, 138, 139, 141, 142
Database 37, 42, 125, 131, 141, 171, 251, 281, 282, 288, 312
Data center 48, 55, 238, 277
Data communications company (DCC) 292, 329, 330, 336
Data concentrator 34, 160, 163, 196
Data encryption 57
Data exchange 107, 141, 143, 154, 161, 164, 260, 286, 308
Data flow 51, 179
Data-link layer (DLL) 56, 58, 124, 150, 200, 209–211, 334
Data network (DN) 239, 246, 253, 259, 277
Data network name (DNN) 239, 259, 282
Data radio bearers (DRBs) 243, 244, 246, 248, 250, 269, 270
Data rate 31, 36, 44, 53, 60–62, 74, 75, 79, 88, 89, 110, 119–122, 132, 181, 183, 187, 191–193, 200, 201, 204, 205, 207, 211, 226, 227, 230–232, 235, 264–266, 271, 278, 284, 293, 294, 302, 304, 305, 309, 310, 313, 315

Data transfer   53, 163, 170, 243, 251, 252, 255, 284, 294, 317, 329,
Data transmission   45, 46, 53, 102, 107, 110, 116, 122, 141, 192, 196, 197, 206, 211, 212, 235, 244, 264, 268, 271, 274, 303, 319, 320, 327
Delay   53, 112, 128, 129, 131, 140, 145–147, 164, 165, 187, 188, 212, 213, 248, 272, 274, 278, 307, 314, 318, 321, 322, 332
Demand   3, 6, 8, 12, 23, 24, 37, 38, 55, 70, 77, 78, 148, 150, 159, 164, 165, 173, 176, 208, 221, 255, 331
Demand response (DR)   5, 6, 12, 24, 37, 139, 165–168, 331
Demand-side management (DSM)   12, 24, 139, 159, 164–167
Demodulation   91, 97, 264, 268
DeModulation reference signal (DM-RS)   268, 273
Dense wavelength division multiplexing (DWDM)   62, 118, 121, 122
Department of Energy (USA) (DOE)   36
Dependability   146
Destination   43, 45, 49–51, 53, 56–59, 110, 117, 123, 125–128, 130–134, 187, 191, 208, 239, 300, 307
Deterministic synchronous multichannel extension (DSME)   304
Device complexity   314, 318, 322
Device language message specification (DLMS)   160, 161, 175, 219, 301, 329
Dielectric   15, 33, 61, 63, 64, 66, 111
Differential protection   145–147
Diffraction   114
Digital data   43, 44, 99, 118
Digital Fourier transform spread OFDM (DFT-S-OFDM)   261, 273
Digitalization   45, 158
Digital modulation   44, 62, 91, 93, 94, 197, 199, 200, 206, 207, 209, 210, 213
Digital modulator   102, 197
Digital telecommunication   44, 45, 60, 89, 146
Digital transmission   81, 97, 221
Direct load control   166
Direct sequence spread spectrum (DSSS)   302
Discharge   17, 33, 64
Disconnect   144, 147, 151, 168
Discontinuous reception (DRX)   270, 271, 321, 327
Discovery   130, 133, 191, 299, 305
Discrete Fourier transform (DFT)   83, 85, 88, 100, 102, 261, 273
Distance protection   145–147
Distributed energy resources (DER)   1, 4–6, 12, 19, 24, 25, 36, 37, 137, 139, 148–150, 155, 156, 159, 167, 168, 175, 223
Distributed generation (DG)   4–6, 12, 20, 24, 25, 33, 36, 148, 162, 168
Distributed network protocol (DNP3)   142, 149, 301
Distributed restart   149, 175
Distributed storage (DS)   12, 150
Distribution automation   25, 32, 38, 147, 152, 174
Distribution FACTS (D-FACTS)   148
Distribution grid   4, 5, 10, 11, 18, 25, 29, 35, 74, 139, 147, 148, 151, 158, 161, 168, 170, 195, 216, 221, 287
Distribution management system (DMS)   138, 139, 149, 168

Distribution network operator (DNO)   7
Distribution system operator (DSO)   16, 37, 151, 168, 218
Disturbance   72, 73, 162, 164
Domain master (DM)   123, 211, 215, 268, 273, 328
Double sideband (DSB)   91
Downlink   230, 241, 244, 251, 254, 259–261, 264, 266–271, 273, 274, 277, 309, 311–314, 318–320, 322–326, 330
Downlink control information (DCI)   267, 319, 325
Downstream   180, 181, 183–191, 307
Dual connectivity   235, 248, 255, 266, 272, 281
Ducted   16, 17
Duplex   45, 70, 71, 107, 132, 315
Duplexing   107, 273, 315
Dynamic bandwidth assignment (DBA)   188
Dynamic line rating (DLR)   169, 170, 177, 336
Dynamic multipoint VPN (DMVPN)   288, 292
Dynamic spectrum sharing (DSS)   255, 272

e

E1   119, 120, 180, 183
E-carrier   118, 119
Edge computing   55, 76, 236, 257, 276–278
8-Phase shift keying (8PSK)   93
Electric energy   4–6, 16, 29, 158, 174
Electricity market   10, 21, 37
Electricity meter   18, 158, 175, 220, 297, 329
Electric Power Research Institute (EPRI)   29, 36–39, 174–176
Electric power system   1, 3–14, 18, 20, 22–25, 29, 30, 34, 36, 37, 76, 149, 165, 168, 174, 282
Electric vehicle (EV)   6, 12, 24, 25, 28, 137, 150, 151, 174
Electric vehicle communication controller (EVCC)   150
Electric vehicle supply equipment (EVSE)   150
Electromagnetic   22, 32, 41, 42, 61, 67, 68, 94, 96, 97, 112–114, 183, 196
Electromagnetic compatibility (EMC)   32, 33, 62, 157, 221
Electromagnetic spectrum   68, 112
Electromechanical   2, 13, 140, 141, 144, 151, 158
Electronic Communications Committee (ECC)   69, 77
Electrostatic discharge   33, 157
Emission   22, 24, 33, 110, 176, 262, 313
Encapsulation   120, 180, 188, 191, 246, 252, 288, 308
Encoder   89, 104–107, 207, 210, 211, 213
Encryption   161, 180, 188, 203, 204, 209, 244, 257, 300, 312, 313
End-device   55, 308, 317, 333
End-point   22, 27, 30, 34, 117, 133, 134, 166, 288
End-to-end   39, 41, 45, 53, 55–57, 117, 124, 129, 208, 239, 250, 256, 308, 317, 329
End-user   10, 15, 21, 24, 45, 70, 133, 167, 185, 246, 255, 256, 272, 276, 284
Energy Conservation and HOmecare NETwork (ECHONET)   167, 177, 297, 334
Energy consumption   8, 17, 159, 168, 229, 271, 314, 318, 320, 335
Energy efficiency   19, 20, 24, 37, 165, 230, 247, 333
Energy management   32, 138, 139, 161, 165–168, 176, 297

Energy management systems (EMS)   29, 32, 138, 139, 162, 166, 168, 170
Energy price   165–167
Energy supplier   159, 161, 329
Enhanced machine type communications unlicensed (eMTC-U)   325, 326, 335
Enhanced mobile broadband (eMBB)   229, 230, 236, 261, 280
Enhanced UMTS terrestrial radio access network (E-UTRAN)   239–241, 244, 246, 247, 250, 251, 316, 321
eNodeB (eNB)   240–244, 246, 247, 250, 264, 266, 268, 271, 280, 316, 321, 322, 326, 327
Environmental   5, 9–12, 17, 19–21, 24, 32, 33, 36, 61, 64, 65, 157, 169, 170, 172, 305, 333
EPS bearer   239, 240, 242, 248, 250, 270, 278
Equalization   97, 98, 246
Equalizer   89, 99, 210
Error control   53, 123
Error correction   58, 62, 73, 104, 106, 312
Error detection   44, 312
Ethernet   33, 46, 109, 118, 121, 132–135, 141, 142, 155, 157, 164, 180, 181, 183, 186–188, 191, 194, 218, 220, 237, 239, 286
Ethernet passive optical network (EPON)   180, 181, 183, 184, 188, 191
European Committee for Electrotechnical Standardization (Comité Européen de Normalisation Electrotechnique) (CENELEC)   28, 38, 74, 177, 192, 198, 206, 207, 209, 221
European Committee for Standardization (Comite Europeen de Normalisation) (CEN)   28, 38, 55, 177, 221
European Conference of Postal and Telecommunications (CEPT)   68, 69, 77
European Telecommunications Standards Institute (ETSI)   28, 29, 31, 33, 38, 76, 177, 221, 227, 235, 278, 290, 291, 297, 302, 334
European Union Agency for Network and Information Security (ENISA)   39
European Utilities Telecom Council (EUTC)   29, 39, 77
E-UTRAN NR Dual Connectivity (EN-DC)   255, 266
Evolved packet core (EPC)   239, 240, 244, 246, 250–255, 316, 321
Evolved packet system (EPS)   76, 239, 240, 242, 248, 250, 270, 278, 317
Experimental   57, 73
Extended access barring (EAB)   321
Extended discontinuous reception (eDRX)   320, 321, 327

f

Fading   115, 265, 304
Fast Fourier transform (FFT)   88, 198, 206, 211–214
Fault   6, 7, 13, 14, 25, 33, 66, 94, 134, 143–145, 147–149, 151, 162, 174
Fault detection isolation and restoration (FDIR)   144
Fault location, isolation, and service restoration (FLISR)   144, 151, 174
Federal Communications Commission (FCC)   192, 206, 207, 209, 281, 326
Feeder   7, 8, 12, 15, 16, 145, 147, 148, 151, 155, 183, 223

FFT-based OFDM   198
Fiber to the home (FTTH)   46, 180, 286
Fiber to the x (FTTx)   179, 180, 220
Field area network (FAN)   47, 297, 300–302, 305–307, 334
Fieldbus   154
Fifth generation (5G)   55, 63, 225, 227–237, 239, 240, 246, 248, 249, 252–257, 261, 271–274, 276–278, 280, 281, 283, 285, 287, 289–292, 326, 327
5G core network (5GC)   236, 239, 247, 248, 252–257, 260, 277, 281
5G QoS identifier (5QI)   278, 279
5G system (5GS)   239, 292
Firmware   329, 333
First generation (1G)   181, 225, 226, 231, 237, 283
Flexible AC transmission system   148, 169, 174, 177
Flexible alternating current transmission system (FACTS)   21, 148, 169, 174, 177
Flicker   5, 14
Fog   55, 76
Forward error correction (FEC)   62, 104, 105, 123, 187, 190, 200, 205, 207, 208, 210, 211, 213, 215, 305, 306, 309, 310, 326
Fourth generation (4G)   55, 217, 225, 227–229, 231, 232, 235, 237, 239, 240, 244, 245, 248, 254–256, 261, 271, 278, 283–285, 290, 296, 314, 330, 334
Frame   15, 109, 110, 119, 125, 126, 132, 134, 157, 158, 173, 180, 184–191, 198, 200, 202–207, 209, 215, 237, 239, 262, 266–268, 270–272, 297, 299, 303, 304, 306, 307, 309, 310, 312, 314, 323, 326
Frame error rate (FER)   200
Free space   43, 44, 60, 96, 192
Frequency allocation   68, 69, 227
Frequency band   68–70, 75, 88, 193, 194, 197, 206, 209, 213, 228, 230–234, 241, 247, 255, 266, 271, 272, 276, 294, 296, 304, 306, 310, 313
Frequency division duplexing (FDD)   107, 228, 240, 247, 261, 271, 273, 315, 318
Frequency division multiple access (FDMA)   108, 225, 226, 262
Frequency division multiplexing (FDM)   99, 100, 108, 192, 194, 218
Frequency hopping (FH)   313, 325, 326
Frequency modulation (FM)   90
Frequency planning   69, 218
Frequency shift keying (FSK)   91, 192, 193, 302, 304–306
Front-end   89, 93, 141, 312
Fronthaul   47
Full service access network (FSAN)   180, 220
Fuel   5, 6, 13, 24
Full-duplex   45, 107, 132, 158, 318
Full function device (FFD)   298, 300
Fuse box   29, 31

g

Gain   9, 35, 113–115, 172
Gas meter   287, 329, 330, 332, 336
Gateway (GW)   131, 135, 164, 196, 214, 220, 223, 246, 250, 251, 308–310, 329
General packet radio service (GPRS)   226–228, 240, 246, 284, 314, 315, 322, 330, 332, 337

General public   47, 192, 225, 229, 282
Generation   1, 3–6, 9–14, 16, 17, 20, 21, 23–25, 29, 36, 121, 135, 137, 138, 148, 149, 159, 164, 165, 194, 208, 220–222, 227, 229, 236, 246–248, 290, 299, 318, 335
Generation plant   6, 14, 138
Generator   5, 6, 10, 13, 21, 24, 106, 168
Generic framing procedure (GFP)   121, 188
Generic object-oriented substation event (GOOSE)   154, 155, 157, 164
Generic routing encapsulation (GRE)   288, 292
Generic substation state event (GSSE)   157
Geographic information system (GIS)   171
G.hn   71, 74, 194, 210, 211, 214, 215, 222
G.hnem   210, 211
Gigabit passive optical network (GPON)   180, 181, 183, 184, 188–190, 222
Global system for mobile (GSM)   226–228, 235, 240, 246, 283, 289, 291, 292, 294, 314, 315, 321, 322, 330, 335
gNodeB (gNB)   246–248, 255, 257, 259–261, 274
G3-PLC   193, 205–210, 222
GPON encapsulation method (GEM)   180, 186–188
GPON transmission convergence (GTC)   189, 190
GPRS Tunneling Protocol (GTP)   246, 248, 250–252, 261
Grid control   23, 25, 151
Grid edge   8, 20, 37, 75, 179
Grid operation   1, 12, 25, 26, 31, 148, 150, 158, 159, 162, 217, 218
Grid topology   7, 194
Grounding   14, 33, 149
GSM Association (GSMA)   75, 256, 337
Guaranteed bit rate (GBR)   278
Guaranteed time slot (GTS)   303, 304
Guideline   9, 64, 67, 155, 168, 172, 217

h

Half-duplex   45, 57, 107, 132, 314, 318
Half duplex frequency division duplexing (HD-FDD)   318, 322
Handover   243, 244, 246, 251, 257, 260, 261, 271, 309, 315
Harmonics   14, 25, 152
Harmonization   29, 70, 231
Head-end   160, 161, 195, 213, 284, 286, 287
Header   51, 53, 57, 58, 133, 142, 187, 189, 198–200, 204, 205, 244, 246, 248, 274, 301, 309, 310
Healing   21, 25, 144, 147, 299, 300,
Heating   17, 167, 169
Hidden Node   110
Hierarchical   4, 48, 119, 158, 161, 196
High availability   222
High data rate (HDR)   69, 74, 75, 110, 123, 193, 218, 219, 227, 307
High speed packet access (HSPA)   227, 284, 330
High voltage (HV)   3–7, 14, 16, 17, 29, 30, 33, 35, 39, 60, 62–64, 72, 73, 75, 149, 170, 191–195, 197, 283
High voltage direct current (HVDC)   169
Home area network (HAN)   3, 47, 65, 116, 242, 248, 250, 251, 260, 274, 280, 297, 305, 309, 314, 315, 321, 330, 331
Home energy management system (HEMS)   167, 176, 177, 297
Home subscriber server (HSS)   250, 251, 254
Hub   48, 124, 125, 133, 288, 329, 336
Hub-and-spoke   48, 133
Hybrid automatic repeat request (HARQ)   244, 265, 271, 273, 274, 319, 322, 324, 325
Hybrid electric vehicle (HEV)   150
Hybrid fiber-coaxial (HFC)   46, 179, 216

i

Idle mode   242, 243, 246, 250, 251, 254, 257, 270, 271, 320–322, 327
Impedance   3, 72, 73, 97, 145, 169, 197, 218
IMT band   231, 232
In-building   30, 161
Incentive   11, 23, 150, 151, 166
Inductive   17, 30, 31, 194, 197
Industrial   4, 6, 8, 22, 23, 47, 74, 139, 141, 158, 160, 166, 175, 193, 219, 222, 229, 234, 278, 280, 283, 289, 290, 293, 294, 296–298, 304, 327, 336, 337
Industrial area network (IAN)   47
Industrial Internet of Things (IIoT)   229
Industrial, Scientific, and Medical (ISM)   293, 304, 310
Information and Communication Technologies (ICT)   1, 19–22, 27–29, 34, 37, 139, 153, 159, 290, 291
Information system   38, 42, 138, 140, 160
Information technology   20–22, 56, 76
In-home   71, 73, 75, 161, 167, 192, 194, 211, 214, 216, 296, 330
In-home display (IHD)   330, 331
Injected   71, 73, 106, 111, 112, 123, 158, 192, 194, 195
Insulated   15, 17, 18, 62, 169
Insulated gate bipolar transistor (IGBT)   169
Insulation   17, 18, 33, 71, 143
Insulators   6, 17, 18, 33
Integrated Services Digital Network (ISDN)   183
Integration of DER   148, 155
Integrity   57, 66, 146, 204, 207, 244, 248, 257, 309, 312, 328
Intelligence   22, 24, 25, 34, 36, 42, 43, 70, 149, 153, 309
Intelligent electronic device (IED)   141, 149, 155
Inter-carrier interference (ICI)   100, 102, 103
Inter-cell interference coordination (ICIC)   244, 246
Inter-control center protocol (ICCP)   141, 143, 174
Interest group   29, 118, 284, 333
Interface   15, 27, 36, 51, 56, 67, 111, 118–121, 123, 124, 126, 129, 130, 132–134, 139–141, 146, 149, 154, 164, 169, 183, 184, 188, 194, 227, 235, 236, 240, 244, 246–248, 250–255, 259–261, 271, 273, 274, 276, 278, 281, 288, 289, 308, 309, 312, 316, 317, 327, 334
Interference   60, 68, 73, 91, 93, 94, 99, 100, 103, 107, 108, 114, 183, 194, 234, 244, 247, 264, 266, 271, 304, 310, 325, 333
Interleaving   106, 197, 200, 209, 210
Intermodal dispersion   111, 112
Internal combustion engine (ICE)   150
International organization for standardization (ISO)   32, 38, 56, 61, 143, 155, 194

International mobile telecommunications (IMT)   32, 227–232, 289, 335
Internet   34, 46, 48, 58, 75, 76, 133, 135, 183, 215, 216, 218, 226, 229, 238, 239, 256, 277, 284, 291, 293, 311, 314, 315, 333–335, 337
Internet access   46, 75, 183, 215, 216, 218, 226, 284
Internet engineering task force (IETF)   58, 133–135, 251, 297, 327, 328, 334
Internet of Things (IoT)   29, 34, 55, 76, 176, 229, 232, 237, 280, 290, 293–296, 307–309, 314–317, 320, 322–329, 331–335, 337
Internet protocol (IP)   31, 33, 46, 57–59, 121, 133, 134, 142, 157, 194, 218, 220, 221, 230, 237, 239–241, 243, 244, 246, 251, 254, 256, 259, 270, 284, 286–288, 291, 309, 311, 315–317, 327
   address   134, 157, 239, 251, 254, 259, 286–288, 317
Interoperability   28, 29, 36, 76, 133, 141, 149, 153, 154, 180, 183, 191, 221, 231, 280, 296, 297, 305, 308
Interruption   33, 144, 151–153, 230, 243
Intramodal dispersion   112
Inter-symbol interference (ISI)   96, 97, 99, 102, 109, 112, 135, 198, 212
Intrusion   173
Inverter   93, 149, 167
Investment   9–12, 19, 22, 27, 62, 70, 150, 164, 169, 170, 216
IoT market   280, 332, 333
IP multimedia subsystem (IMS)   240, 251, 254, 256, 278, 291
IPsec   288, 292
IPv4   134, 239, 327
IPv6   134, 239, 297, 301, 307, 327, 328
IPv6 Low Power Wireless Personal Area Network (6LoWPAN)   297, 301
Isolation   72, 144, 151, 157, 215, 231
ITU Radiocommunication Sector (ITU-R)   28, 35, 67, 68, 75, 88, 227, 231, 289, 290

j
J-carrier   118
Jitter   213, 278

k
kWh   158

l
L2   301
L3   301
Label switched path (LSP)   134
Land Mobile Radio (LMR)   302
Latency   30, 31, 34, 51, 53, 55, 58, 107, 109, 117, 141, 145–147, 164, 170, 218, 219, 229–231, 237, 248, 255, 265, 271, 273, 274, 276, 278, 280, 282, 284, 285, 294, 300, 303, 304, 316, 317, 327
Layer 2   124, 125, 133, 157, 218, 244, 253, 288, 297, 301, 307, 334
Layer 3   253, 288, 297, 301, 307
Layer protocol   58, 194, 198, 327, 328
Legacy   30, 45, 54, 68, 125, 133, 141, 183, 214, 218, 234, 247, 255, 272, 286, 322
Legislation   9, 69
License   31, 69, 235, 293, 294, 326

Licensed   26, 35, 70, 77, 171, 234, 235, 280, 281, 290, 293, 294, 308
Licensed assisted access (LAA)   280, 281
Licensed spectrum   35, 70, 171, 235, 280, 281, 294
License-exempt   235, 293, 294, 326
Light amplification by stimulated emission of radiation (LASER)   110
Light emitting diode (LED)   110, 133, 192, 198, 305, 329
Lightweight machine-to-machine (LwM2M)   328, 329, 335
Line-of-sight (LoS)   70, 113, 115, 233, 306
Line protection   145
Link budget   93, 115, 116, 185, 229, 295, 314
Link capacity adjustment scheme (LCAS)   121
Link layer   56, 304, 312
Listen before talk (LBT)   235, 280, 310, 313, 325, 326
Logical link identification (LLID)   191
Load   8, 12–14, 17, 23, 73, 74, 88, 109, 125, 128, 130, 147, 148, 150, 152, 153, 158, 159, 166–168, 176, 192, 220, 247, 248, 255, 296, 321, 331, 333
Load control   166, 167, 296, 331
Load curve   23
Load management   73, 74, 176, 192, 220
Load profile   150, 159
Local area data network (LADN)   277
Local area network (LAN)   47, 124–126, 133, 134, 141, 154, 157, 164, 194, 280, 281, 291, 292, 306
Location management   243, 257
Logical channel   244, 269, 270
Logical connection   51, 186, 187, 189
Logical device (LD)   155
Logical network   48, 203, 255, 304
Long-term evolution (LTE) cell   240–243, 246, 248, 255, 264, 315–318, 322
Long-term evolution for machines (LTE-M)   261, 285, 294, 314–322, 325–327, 332, 335, 337
Losses   4–6, 11, 14, 61, 111, 114, 115, 147, 148, 185
Low data rate (LDR)   73, 74, 193, 219, 294
Low-density parity check (LDPC)   106, 213, 215, 271
Low power WAN (LPWAN)   294, 307, 314, 331, 333, 334, 337
Low power wide area (LPWA)   229, 337
Low-rate WPAN (LR-WPAN)   294, 334
Low voltage (LV)   3, 4, 7, 8, 14–18, 30, 31, 33, 35, 37, 71–73, 75, 148, 149, 151, 159, 166, 192–197, 216, 217–220, 223, 284
   feeder   15, 16, 18, 159, 220, 223
   grid   7, 15, 17, 18, 30, 31, 72, 73, 75, 193, 194, 197, 217, 219, 223
   panel   15, 18

m
Machine-to-machine (M2M)   229, 284, 287, 292, 328, 329, 332, 335, 336
Machine type communications (MTC)   229, 287, 315–318, 322, 333
Maintenance   11, 12, 14, 17, 48, 64, 67, 131, 141, 148, 149, 151, 170, 183–185, 217, 231, 307
Major event day (MED)   152
Management information base (MIB)   219, 270, 271, 275, 323

Mapping   93, 105, 155, 187, 246–248, 266, 267, 269, 275, 277, 288, 323
Massive machine type communications (mMTC)   229, 230, 261, 294, 314, 327
Master   109, 140, 141, 157, 195, 196, 209, 211, 213–215, 255, 270
Master station   140, 141, 157
Maximum coupling loss (MCL)   295
Medium access control (MAC)   125, 132, 135, 157, 185, 187, 191, 198, 200, 202, 203, 205, 207–211, 213–215, 244, 246, 248, 265, 266, 269, 270, 274, 294, 296, 297, 300–306, 309, 310, 312, 314, 334
Medium voltage (MV)   3, 4, 7, 8, 14–18, 30, 33, 71–73, 75, 148, 149, 151, 166, 194, 195, 197, 211, 217, 218, 222, 283, 284
   grid   7, 17, 18, 30, 151, 166, 194, 211, 217, 218, 283
Mesh   48, 49, 208, 288, 299, 300, 330
Message   43, 45, 46, 53, 57, 59, 89, 91, 97, 104, 105, 107–110, 126, 130–132, 141–143, 150, 154, 156, 157, 160, 161, 166, 167, 188, 191, 203, 205, 207–211, 214, 226, 243, 244, 246, 257–261, 267, 270, 271, 283, 294, 300, 304–306, 308–317, 321, 322, 324, 327, 328, 330
Message queuing telemetry transport (MQTT)   328, 335
Metallic cable   33, 60, 61, 63, 64, 66, 71
Meter   9, 11, 15, 18, 29, 31, 34, 42, 60, 73, 75, 114, 139, 158–162, 167, 175, 192, 193, 195, 196, 198, 217–223, 238, 283–287, 294, 297, 329–332, 336, 337
Meter data management (MDM)   34
Meter data management system (MDMS)   160, 195, 219, 220
Metering   9, 14, 69, 70, 74, 142, 147, 154, 156, 158–161, 175, 192, 193, 196, 198, 217–219, 221, 223, 228, 283–287, 329–332, 336, 337
Meter reading   11, 34, 73, 158–160, 192, 219, 285
Meter room   31, 73, 219
Metro Ethernet Forum (MEF)   133, 135, 183, 220
Metrologically   161, 286
Metropolitan area network (MAN)   47, 93, 133, 139
Microgrid   6, 149, 166, 168, 177
Microgrid energy management system (MEMS)   168
Microprocessor   81, 88, 144
Microwave   70, 71, 113, 118, 171, 173, 283
Mission-critical   38, 70, 77, 231, 256, 283
Mobile network   32, 38, 77, 171, 229, 231, 237–239, 242, 252, 253, 256, 257, 277, 278, 283–285, 332, 337
Mobile network operator (MNO)   238, 283, 285, 286, 288
Mobile radio   69, 302
Mobile telecommunications   29, 172, 227, 228, 289
Mobile virtual network operator (MVNO)   292
Mobility management   243, 246, 247, 250, 251, 253, 254, 257
Mobility management entity (MME)   246, 250, 251, 316, 317
Modbus   149, 154, 175, 301
Modem   43, 217, 226, 228, 237, 283, 284, 286, 287, 292, 314
Modulation   56, 60, 62, 70, 79, 85, 88–91, 93, 94, 96, 100, 110, 193, 199, 205–210, 219, 227, 246, 264, 302, 303, 305–307, 312, 313, 322, 326
Modulation and coding scheme (MCS)   264, 266, 270, 319

Modulator   89, 209
Momentary average interruption event (E) frequency index (MAIFIE)   153
Momentary average interruption frequency index (MAIFI)   153
Momentary interruption   152, 153
Monitoring   20, 23, 30, 32, 59, 74, 123, 129, 138, 140, 148, 154, 156, 162, 164, 167, 170, 175–177, 222, 254, 271, 293, 296–298, 302, 312, 316, 320, 333
Monopoly   9, 11
Multi-antenna transmission and reception   265, 276
Multicarrier   100, 102, 107, 116, 193, 196, 323
Multicarrier modulation   100, 107, 196
Multicast   125, 133, 288, 301, 326, 328
Multi hop delivery service (MHDS)   301, 307
Multipath   94, 96, 97, 304
Multiple access   56, 108, 109, 114, 261, 262,
Multiple-input multiple-output (MIMO)   71, 109, 194, 215, 221, 248, 265, 266, 273, 276
Multiplexer   99, 115, 121, 123, 190
Multiplexing   46, 62, 99, 100, 108, 109, 118, 120–124, 128, 181, 183–187, 190, 244, 265, 266, 270, 276
Multiplex section-shared protection ring (MS-SPRING)   121
Multi-point MAC control protocol (MPCP)   187, 191
Multi-protocol label switching (MPLS)   118, 134, 135, 222
Multi-protocol label switching-transport profile (MPLS-TP)   118, 134, 135
Multiservice   160, 287, 288
MV/LV transformer   15

n

Narrowband   69–71, 73–75, 103, 109, 140, 172, 193, 194, 197, 198, 218, 223, 225, 310, 314, 318, 323, 324, 335
NarrowBand IoT (NB-IoT)   261, 280, 285, 294, 314–317, 320–327, 332, 335, 337
Narrowband PBCH (NPBCH)   323, 326
Narrowband PDCCH (NPDCCH)   324, 325, 327
Narrowband PDSCH (NPDSCH)   324, 325
Narrowband PLC   71, 73, 75, 193, 198, 218, 223
Narrowband PSS (NPSS)   323, 326
Narrowband PUSCH (NPUSCH)   324, 325
Narrowband SSS (NSSS)   323, 326
National Institute of Standards and Technology (USA) (NIST)   29, 38, 76
NB-IoT Unlicensed (NB-IoT-U)   325, 326, 335
Neighborhood area network (NAN)   47, 296
Network (NWK)   300, 305
Network address translation (NAT)   38
Network architecture   75, 194, 211, 225, 300, 308, 311, 315, 316
Network attach   315, 316
Network design   26, 48, 54, 217, 218, 284
Network edge   55
Network element   14, 35, 48, 50, 53, 55, 108, 117, 118, 123, 134, 219
Network function (NF)   252, 255, 280
Network function virtualization (NFV)   238, 248, 280, 290
Networking device   33, 157

Network interface   58, 130, 133, 134, 183, 247, 286
Network interface card (NIC)   286
Network layer   55, 57, 133, 238, 297, 301, 302, 304, 306
Network link   35, 128
Network management   123, 134, 203, 321
Network management system (NMS)   219
Network node   44, 48, 49, 51, 59, 214, 252
Network protocol   142, 227, 252
Network registration   242, 246, 253, 257, 258, 315
Network server (NS)   308, 309, 311
Network slice   253–255, 257, 278–280
Network slicing   231, 236, 252, 255, 278, 280, 282, 291
Network topology   48, 49, 126, 130, 220, 298, 299
New radio (NR)   228, 235–237, 246–248, 255, 256, 261, 266, 271–277, 280, 281, 290, 291, 327
New radio unlicensed (NR-U)   281
Next generation-passive optical network 2 (NG-PON2)   181, 183, 187, 190
Next generation radio access network (NG-RAN)   236, 239, 246–248, 253, 255, 281, 282, 290
Next hop resolution protocol (NHRP)   288, 292
Noise   44, 60, 73, 74, 79, 93–95, 97, 98, 103, 106, 107, 192–194, 196, 203, 207, 211, 214, 215, 218, 219, 272, 310
Non-access stratum (NAS)   246, 250, 254, 255, 257, 316, 317, 321
Non-IP data delivery (NIDD)   316
Non-public network (NPN)   281, 282
Non-stand alone (NSA)   255
Non-terrestrial network (NTN)   327
Normally open point (NOP)   151
Normal Priority Contention Window (NPCW)   207
North American Synchrophasor Initiative (NASPI)   164, 175, 176
Numerology   272, 323

o

Office of Communications (UK) (OFCOM)   77, 336
Open PLC European Research Alliance (OPERA)   74, 194, 217
Open shortest path first (OSPF)   131, 135
Open system interconnection (OSI)   56–60, 124, 142, 143, 154, 157
Operating band   262, 276, 277
Operational   3, 6, 8, 17, 19–22, 25, 137, 138, 147, 149, 169–172, 183, 184, 192, 217, 218, 225, 226, 231, 256, 262, 283, 300, 324, 332
Operational expenditure (OPEX)   238
Operational procedure   8, 25, 149, 171, 184, 217
Operation and maintenance   11, 12, 22, 26, 32, 63, 64, 153
Operations   12–14, 19, 21–23, 31, 34, 35, 69, 70, 77, 89, 102, 104, 137–139, 147, 151, 154, 158, 159, 167, 168, 170–173, 183, 187, 211, 252, 266, 282, 284, 297, 305, 331, 332
Operations, administration, and maintenance (OAM)   134, 188, 189, 217
Operations management   137–139
Optical amplifier   62, 124
Optical channel   121, 123, 124
Optical carrier (OC)   121
Optical data unit (ODU)   124

Optical fiber   14, 27, 33, 35, 41, 43, 60–67, 69, 71, 75, 79, 93, 110–112, 118, 120–123, 132, 146, 157, 170, 171, 173, 179, 180, 183–185, 190, 192, 215–217, 283, 285
Optical fiber cable   14, 27, 35, 61, 63, 65–67, 69, 170, 171, 180, 216
Optical line terminal (OLT)   180, 183–191
Optical network terminal (ONT)   180, 184, 186–191, 222
Optical network unit (ONU)   180, 184, 188
Optical transport   99, 118, 123, 135
Optical transport network (OTN)   118, 123, 124, 135, 183
Orthogonal frequency division multiple access (OFDMA)   108, 109, 228, 261
Orthogonal frequency division multiplexing (OFDM)   99, 100, 102, 103, 107, 108, 116, 193, 194, 196, 198–200, 204–207, 209–215, 220, 222, 246, 248, 261, 262, 265, 267, 268, 272–275, 302, 322, 323, 326, 334
Outage   11, 12, 14, 139, 143, 151, 152, 159, 174
Outage management system   139
Overhead line   17, 18, 65

p

Packet data network (PDN)   239, 250, 251, 315, 316
Packet data unit (PDU)   53, 57, 207, 213, 239, 240, 242, 254, 255, 257, 259–261, 277, 278
   session   239, 240, 254, 255, 257, 259–261, 277, 278
Packet loss   246, 278
Packet-switched   45, 50–52, 226–228, 237, 240, 256
Packet switching   45, 50, 51, 53
Pad-mounted   30
Paging   243, 244, 246, 251, 254, 258, 259, 270, 271, 320, 321, 324, 327
Paging control CHannel (PCCH)   270
Panel   6, 15, 18, 29
Passive optical network (PON)   179–188, 190, 191, 216, 217, 220, 259
Pattern   8, 19, 23, 107, 113, 114, 164–166, 188, 333
Payload   53, 57, 58, 121, 124, 133, 187, 198–200, 204–207, 209, 230, 309, 310, 312, 313
PDN gateway (P-GW)   246, 250–253, 315–317
Personal area network (PAN)   47, 208, 209, 226, 297–301, 303, 304, 306, 307
Peak   23, 31, 94, 100, 103, 116, 148, 150, 159, 165, 166, 176, 187, 192, 230, 231, 262, 266, 318
Performance indicator   229
Performance monitoring   123, 188
Performance requirement   33, 65, 146, 229, 278, 314, 322
Phase   4, 5, 8, 14, 16–18, 25, 53, 61–64, 71, 83, 91, 93, 94, 96, 97, 99, 103, 112, 114, 127, 131, 143, 145, 159, 161–163, 169, 172, 193, 194, 197, 199, 200, 220, 222, 272, 286, 310, 324
Phase modulation (PM)   91
Phase shift keying (PSK)   91, 94–98, 199, 213
Phasor   5, 161–164, 175
Phasor data concentrator (PDC)   163, 164
Phasor measurement unit (PMU)   162–164
Photodiode   111
Physical (PHY)   198, 204, 205, 207, 209–214, 244, 246, 248, 266, 269, 270, 290, 294, 296, 297, 300–302, 304–306, 308, 309, 312, 334

Physical channel   103, 264, 266–269, 272, 323, 324
Physical control formal indicator CHannel (PCFICH)   268, 272, 274, 318
Physical downlink control CHannel (PDCCH)   264, 267, 271, 272, 274, 318, 325
Physical downlink shared CHannel (PDSCH)   265, 267, 268, 270–272, 318, 319, 325
Physical hybrid ARQ Indicator CHannel (PHICH)   268, 272, 274, 318
Physical interface   56, 141
Physical layer   56–58, 60, 120, 188, 190, 211, 238, 302, 303, 310, 313, 325, 326, 335
Physical layer convergence protocol (PLCP)   211
Physical media   22, 59, 60, 79, 214, 215
Physical medium   59, 60, 107
Physical radio block (PRB)   264, 266–269, 272, 318, 319, 322–324
Physical random access CHannel (PRACH)   268–272, 276, 324
Physical signal   266–268, 271, 317, 322, 323
Physical uplink control CHannel (PUCCH)   268, 271, 272, 319, 320
Physical uplink shared CHannel (PUSCH)   265, 268, 270–272
Plain old telephone system (POTS)   216
Planning   10, 12, 13, 32, 75, 150, 162, 164, 168, 169, 218
Plesiochronous digital hierarchy (PDH)   99, 118–121, 183
Plug-in electric vehicle (PEV)   150, 167
Plug-in hybrid electric vehicle (PHEV)   150
Point of common coupling (PCC)   168, 251
Point-to-area   45, 70, 330
Point-to-multipoint   45, 71, 133, 179–181, 187, 191
Point-to-point   45, 46, 48, 49, 69, 71, 72, 113, 117, 118, 133, 134, 181, 183, 191, 216, 218, 252
Pole   15, 30, 77
Policy and charging function (PCF)   253, 254, 259
Power budget   307
Power control   264, 272
Power factor   14, 148
Power generation   4, 5, 20, 169
Power line   3, 4, 6, 7, 12–17, 25, 29, 30, 33, 35, 36, 41, 43, 60, 63–65, 67, 69, 71–73, 140, 143–145, 147, 164, 169–172, 179, 191–194, 197, 200, 217, 218, 220–223, 297, 332
Power line communications (PLC)
  application   73, 75, 191, 192, 218
  architecture   195
  channel   73, 75, 193, 202
  technology   60, 118, 179, 191–198, 211, 215, 217, 286, 287
PoweRline for Intelligent Metering Evolution (PRIME)   193, 198–207, 209, 219–221, 223
Power plant   4–6, 20, 32, 34, 143, 155
Power quality   5, 14, 20, 21, 148, 152, 156, 159, 168
Power restoration   139
Power save mode (PSM)   320, 321
Power spectral density   205, 218, 322
Power supply   30, 31, 33, 35, 115, 144, 179, 217, 218, 327, 332, 333

Power system   1, 3–6, 9–13, 17–20, 22–24, 32, 47, 76, 77, 137, 139, 140, 143–145, 148, 156, 162–166, 168, 169, 172, 174, 175, 191, 194, 332
Preamble   132, 188, 191, 198, 204–206, 209, 271, 276, 309, 324
Prepayment   331
Pricing   166, 331, 332
Primary substation   7, 195
Prioritization   51, 157, 244
Priority   53, 125, 143, 171, 207, 211, 215, 218, 234, 239, 278, 321
Privacy   204, 244, 313
Private   10, 23, 26, 27, 29, 31, 35, 46–48, 54, 55, 69, 134, 159, 167, 170–172, 216, 217, 229, 231, 233–235, 238, 277, 280–284, 288–291, 305, 308, 311, 332, 333
Private (or professional) mobile radio (PMR)   69, 70, 172, 282, 283
Private network   26, 31, 35, 48, 170, 171, 229, 231, 233–235, 238, 277, 281–284, 288, 291, 311, 332, 333
Private telecommunication network   26, 27, 48
Procedure   2, 11, 25, 53, 56, 63, 99, 107, 121, 149, 161, 162, 169, 171, 188, 200, 217, 218, 242–244, 246, 251, 252, 254, 256–261, 270–272, 284, 291, 299, 312, 314, 317, 320, 324–327
Program   57, 139, 151, 159, 166, 167, 176, 227, 229, 231, 254, 287, 308, 329
Propagation   43, 60, 67, 70–72, 79, 93, 94, 97, 103, 110–115, 132, 186, 187, 192, 193, 197, 217, 220, 227, 232, 233, 242, 331
Proprietary   119, 143, 154, 160, 167, 194, 286, 307, 308, 328, 330
Protection relay   42, 144–146, 162
Protection scheme   13, 145, 162
Protection system   144–146, 149, 162
Protocol   21, 53, 55–59, 76, 110, 121, 122, 126, 127, 129, 131–135, 141–143, 149, 154, 155, 157, 160, 161, 164, 167, 174, 180, 187, 188, 191, 208, 211, 219–221, 239, 242, 244–246, 248–254, 256, 280, 288, 291–294, 296, 297, 300, 301, 306–309, 311, 312, 315–317, 327–329, 333–335
Protocol frame   309, 312
Protocol stack   57, 242, 244, 245, 248, 249, 280, 293, 297, 300, 301, 308, 309, 311, 312, 315
Proxy coordinator (Pco)   214
Public   10, 23, 29, 36, 42, 45, 47, 48, 69, 70, 77, 165, 170, 172, 174, 216, 221, 229, 231, 233–235, 238, 240, 277, 278, 280–284, 286–289, 296, 308, 311, 321
Public land mobile network (PLMN)   278, 281, 321
Public network integrated NPN (PNI-NPN)   281, 282
Public networks   48, 229, 231, 233, 281, 282, 288
Public protection and disaster relief (PPDR)   234, 256, 278
Public switched telephone network (PSTN)   45, 183, 216, 240, 282, 283, 286
Push-to-talk (PTT)   107, 172, 256

q

Quadrature amplitude modulation (QAM)   213, 264
Quadrature phase shift keying (QPSK)   91, 93, 213, 264, 302, 304, 324

Quality of service (QoS)   32, 39, 42, 134, 188, 213, 215, 237, 239, 240, 248, 250–257, 259, 264, 270, 277–279, 284, 316
  flow   240, 248, 255, 257, 277, 278
Quantizing   81, 89

r

Radial   7, 8, 112
Radiation   73, 110, 113, 114, 170
Radio access   227, 228, 230, 235, 240, 247, 255, 270, 288, 290, 291, 316, 317, 322, 323
Radio access network (RAN)   227, 236–239, 242, 246, 248, 250, 252, 253, 255, 271, 278, 282, 283, 290
Radio channel   108, 226, 227, 297, 303, 312
Radio frequency (RF)   33, 67, 68, 76, 183, 192, 240–244, 247, 304, 331
Radio interface   227, 237, 240, 243, 244, 246–248, 261, 265, 270, 272, 273, 289, 291, 308, 311, 312, 317
Radio link   244, 271, 299
Radio link control (RLC)   244, 248, 265, 274
Radio network   55, 255, 330
Radio propagation   112, 241
Radio Regulations (RR)   37, 68, 77, 231, 290
Radio resource control (RRC)   242, 243, 246, 248, 255, 257–260, 270, 271, 317, 321, 322, 324
  connection   242, 243, 257–259, 270, 271, 317, 322, 324
Radio resource management (RRM)   243, 246, 280, 327
Radio service   69, 217, 226, 231, 234, 290
Radio spectrum   29, 67–70, 77, 78, 238, 290
Radio technology   69, 308, 332
Radio transmission   67, 88, 237, 238, 291
Random access   109, 244, 257, 259, 268, 270–272, 317, 324, 327
Reactive power   5, 6, 13, 14, 147–149, 158, 162
Ready to send (RTS)   66, 110
Real-time   3, 23, 24, 32, 38, 45, 53, 55, 138, 144, 147, 159, 162, 166, 167, 169–171, 256, 283
Real-time pricing (RTP)   166, 256
Receiver   43, 49, 53, 56, 62, 67, 89, 93, 94, 96–100, 102–104, 107, 108, 110, 113–115, 123, 187, 194, 196, 207, 210, 215, 228, 265, 266, 295, 304, 306, 309, 310, 318, 322, 326
Reduced function device (RFD)   298, 300
Redundancy   7, 33, 35, 48, 54, 157, 210, 213, 218, 247, 312
Reference architecture   27, 177, 195, 196
Reflection   61, 111, 114
Refraction   111, 115
Regeneration   62, 123, 124
Regional Commonwealth in the field of Communications (RCC)   69, 77, 302
Registration   133, 135, 188, 191, 202, 203, 209, 211, 251, 254–259
Regulation   9–11, 14, 19, 26, 36, 68, 70, 161, 171, 217, 286, 290, 310, 314
Regulatory   6, 9, 10, 12, 19, 27, 29, 68, 151, 161, 281, 302, 310
Relay   32, 33, 48, 144, 145, 154, 162, 196, 213, 246, 278, 297, 305, 308, 317

Relaying   73, 192, 203
Reliability   5, 9, 10, 12, 17, 19, 21, 26, 27, 31, 34, 35, 128, 144, 146, 148, 153, 157, 158, 164, 167, 168, 173, 222, 229–231, 237, 248, 278, 284, 285, 294, 304, 332
Remote connection   69, 159
Remote control   2, 73, 138, 140, 143, 195, 220, 287
Remote terminal unit (RTU)   43, 141, 283
Renewable   5, 19, 174, 234
Repeater   44, 123, 192, 195
Repetition   48, 57, 204, 211, 219, 252, 319, 320, 322, 324, 325
REpresentational State Transfer (REST)   3, 10, 19, 33, 48, 60, 67, 100, 103, 108, 109, 115, 117, 121, 131, 142, 144, 167, 170, 180, 200, 211, 212, 217–219, 252, 254, 268, 285–287, 324, 328–330, 333
Request for comments (RFC)   58, 133–135, 244, 251, 328, 334
Requirement   8, 19, 22, 23, 26–28, 31–33, 42, 59, 62, 64, 65, 70, 85, 107, 144, 146, 149–151, 153, 155, 164, 166, 168, 171, 172, 176, 180, 186, 192, 195, 198, 208, 213, 217, 223, 227–231, 234, 235, 248, 251, 255, 274, 282–286, 288, 289, 293, 294, 296–298, 315, 318, 327, 330, 333, 335, 336
Residential   4, 8, 22, 23, 32, 45, 46, 150, 158, 160, 166, 167, 183, 214, 216, 218, 222, 285–287, 330
Resilience   21, 35, 39, 70, 149, 217, 255
Resiliency   21, 22, 34, 35, 171, 218
Resilient   2, 13, 19–21, 34, 48, 64, 70, 148, 174
Resilient AC distribution system (RACDS)   148, 174
Resistance   17, 66, 170
Restoration   25, 123, 144, 151
Retail   45, 46, 331
Retransmission   58, 73, 104, 125, 187, 244, 265, 274, 304, 325
Ring   7, 48, 49, 89, 121, 123, 151, 152, 157, 158, 180
Ripple control   73, 166, 176, 220
Robustness   89, 93, 94, 98, 164, 197, 199, 200, 207,
Roll-out   159, 175, 222, 284, 329
Root mean square (rms)   3, 8
Round-trip time (RTT)   191
Router   127–131, 134, 284, 286–288, 300, 301, 307
Routing   43, 48–50, 54, 58, 59, 76, 117, 123–131, 133–135, 157, 208, 221, 237, 244, 246, 250, 251, 253, 254, 277, 288, 296, 297, 299–301, 306, 307, 317
Routing information protocol (RIP)   130, 131, 135
Routing protocol   129, 131, 133, 208, 288, 301
Routing protocol for low power and lossy networks (RPL)   301, 307
Routing table   126–128, 130, 131, 208, 301
Routing technology   117
Rural   30, 34, 69, 232, 238, 284

s

Safety   6, 13, 14, 17, 22, 25, 27, 33, 60, 62, 77, 99, 143, 149, 159, 170–172, 194, 217, 218, 235, 289
Sampled value (SV)   157, 163, 164
Sampling   66, 80, 81, 89, 100, 162, 198, 212–214, 322
Satellite   30, 46, 113, 172, 283, 327
Scalability   54, 117, 123, 133, 238, 252, 287, 333, 335
Scalable   27, 41, 55, 56, 124, 159, 175, 248, 272, 273

Service-based architecture (SBA)   252, 254, 256
Service capability exposure function (SCEF)   316, 317
Schedule   108, 151, 301, 306, 307, 318, 325
Secondary substation   7, 16, 195, 217
Second generation (2G)   217, 225–229, 231, 232, 235, 237, 240, 283, 284, 330
Security   6, 9, 31, 35, 70, 78, 115, 146, 159, 172–174, 188, 203, 204, 209, 211, 217, 222, 231, 234, 237, 244, 246, 257, 259, 278, 282, 292, 293, 296–298, 300, 301, 303, 305–307, 309, 312, 321, 328, 329
Segmentation   53, 72, 73, 203, 244
Segmentation and reassembling (SAR)   203, 210, 219, 305
Segmentation and reassembly   53
Sensitivity   62, 115, 144, 295, 304, 306, 310
Sensor   20, 148, 153, 156, 170, 173, 175, 214, 287, 297, 298, 305, 335
Serial interface   141, 154, 286
Service framework   256
Service-level agreement (SLA)   31, 32
Service node (SN)   200, 202, 203
Service platform   238, 256, 282
Service provider   55, 133, 166, 297, 329–331, 333
Service requirement   32, 217, 231, 289
Serving gateway (S-GW)   244, 246, 250–253, 315–317
Session management   237, 242, 246, 250, 253, 254, 256, 257
Shared contention period (SCP)   200, 215
Shared spectrum   234, 235, 325
Shield   18, 61, 71
Short message service (SMS)   46, 226, 283, 291, 317, 337
Shutdown   13, 149
Signal coupling   197
Signalling radio bearer (SRB)   244, 269, 270
Signal propagation   73, 110, 216, 218, 287
Signal-to-noise ratio (SNR)   97–99, 201, 208, 211, 215, 265, 310
Simplex   45
Single carrier FDMA (SC-FDMA)   326
Single network slice selection assistance information (S-NSSAI)   255, 257, 259, 278, 280
Single-phase   4, 8, 18, 158, 219
Sixth generation (6G)   229
Slave   109, 141, 195, 196, 213
Slice/service type (SST)   280
Smart city   168, 176, 177, 228, 333
Smart energy   167, 175, 296, 329, 330, 334, 336, 337
Smart energy profile (SEP)   36, 167, 296, 330
Smart grid application   22, 70, 111, 137, 205, 215, 222, 295
Smart grid concept   19, 25
Smart meter   28, 29, 47, 70, 71, 73, 75, 159–161, 166, 167, 175, 193, 195, 196, 198, 200, 217–223, 228, 285–287, 292, 296, 297, 309, 329–334, 336, 337
Smart metering   28, 47, 70, 71, 75, 159, 160, 166, 175, 193, 195, 196, 198, 200, 217–221, 223, 228, 285–287, 296, 309, 329–332, 334, 336, 337
Smart metering WAN (SMWAN)   329, 330
Society of Automotive Engineers (SAE)   150, 290
Software defined networking (SDN)   252, 253

Solar   5, 6, 24, 29, 170
Solid-state   14, 44, 144, 158
Sounding reference signal (SRS)   264, 272
Source   2, 5, 13, 21–24, 42, 45, 47, 53, 62, 68, 69, 79, 89, 91, 99, 103, 110, 111, 114, 117, 119, 125–128, 130, 132, 133, 144, 148–150, 162, 168, 169, 176, 191, 196, 214, 215, 218, 239, 260–262, 277, 300, 301, 307, 330, 331
Specification   29, 55, 58, 63–67, 119, 120, 123, 143, 154, 158, 160, 168, 183, 194, 198, 200, 205, 207, 209–211, 215, 227–229, 235, 236, 248, 252, 255, 256, 262, 273, 278, 280–282, 289, 290, 294, 296, 297, 299–301, 304, 305, 314, 315, 321, 325, 327, 329, 330, 334–336
Spectral efficiency   230
Spectrum   28, 67–71, 73, 77, 78, 85, 91, 92, 94, 96, 97, 99, 100, 102, 103, 110, 113, 135, 177, 181, 183, 184, 192–194, 198, 199, 204, 206, 210, 211, 217, 228, 230–235, 242, 247, 255, 266, 276, 280, 281, 284, 290, 291, 293, 294, 302, 306, 307, 310, 313, 317, 322, 325, 333, 336
Speed   5, 13, 31, 38, 60, 63, 77, 83, 94, 110, 112, 121, 141, 144, 157, 158, 162, 169, 170, 180, 183, 190, 192, 227, 230
Splitter   115, 180, 184–186
Spreading factor (SF)   267, 310, 323
Spread spectrum   302, 307, 310
Stability   6, 13, 21, 25, 148, 162, 164, 169, 293,
Standalone NPN (SNPN)   281, 282
Star   28, 47–49, 75, 89, 94, 140, 157, 160, 211, 222, 228, 229, 257, 259, 260, 298–300, 308, 311
State-of-the-art   26, 73, 122, 140, 158, 192, 198, 287
STATic COMpensator (STATCOM)   169
Station bus   154, 157
Storage   3, 5, 12, 19–21, 24, 33, 55, 66, 150, 160, 168, 254, 312, 327
Subscriber identity module (SIM)   45, 237, 285, 287, 288
Substation   3–6, 12, 14, 16, 25, 26, 30, 32–35, 47, 72, 138, 140, 141, 143–145, 147, 148, 153–155, 157, 158, 162, 163, 170–174, 215, 283, 285, 287
Substation automation   25, 147, 153
Substation configuration language (SCL)   154, 155
Suburban   30, 34, 232
Supervision   9, 123, 138, 140, 156
Supervisory   32, 123, 138, 139, 156
Supervisory control and data acquisition (SCADA)   32, 34, 43, 77, 138–143, 147, 149, 151, 154, 157, 162, 170, 174, 195, 228, 283, 284, 293
Surface wave   36
Surge   14, 33, 157
Surveillance   172, 173, 287
Sustained Interruption   152, 153
Switch   14, 16, 18, 49–51, 115, 121, 123–125, 147, 200, 202, 203, 253, 274, 289, 298, 302, 304, 318
Switchgear   14, 15, 72, 147, 156, 176, 197, 218
Switching   14, 41, 43, 48–51, 53, 54, 75, 76, 117, 118, 124–126, 132, 134, 135, 147, 157, 240, 243, 244, 252, 260, 266, 288, 318
Switching/routing   41, 48–50, 252
Switching technology   117, 118

Synchronization   118, 119, 132, 146, 157, 186, 188, 191, 198, 241, 242, 266, 268, 270, 271, 275, 299, 317, 322, 323, 326
Synchronization signal block (SSB)   275, 276
Synchronous digital hierarchy (SDH)   118–123, 133, 135, 180, 183
Synchronous optical network (SONET)   119–123, 135, 183
Synchronous transport module (STM)   121
Synchronous transport signal (STS)   121
Synchrophasor   155, 161–164, 176
System architecture   184, 237, 252, 254, 291
System average interruption duration index (SAIDI)   152, 153
System average interruption frequency index (SAIFI)   152, 153
System operator   7, 13, 16, 23, 162

t

T1   119, 120, 180, 183
Tail-Biting Convolutional Code (TBCC)   273, 322
Tamper   159
Tariff   159, 330
T-carrier   118, 119
TCP/IP   57–59, 141, 142, 167
Technical specification   149, 168, 176, 205, 227, 229, 235, 330, 334, 336
Telecommunication channel   44, 46, 144, 146
Telecommunication circuit   45, 46
Telecommunications connectivity   20, 22, 23, 161, 171, 173
Telecommunications Industry Association (TIA)   61, 297, 301, 334
Telecommunication line   46
Telecommunication network   14, 21–23, 26, 29, 30, 34, 35, 37, 41–43, 45–47, 50, 51, 54–57, 60, 62, 75, 141, 146, 161, 176, 222, 285, 292, 330
Telecommunication service   14, 22, 26, 29–31, 35, 41, 42, 45, 46, 50, 147, 192, 285, 330
Telecommunication service provider (TSP)   26, 27, 31, 32, 54, 55, 180, 184, 216, 217, 317
Telecommunication system   29, 32, 41–46, 48, 55, 73, 75, 79, 88, 94, 100, 104, 115, 172, 225
Telecontrol   32, 70, 138, 142, 143
Telecontrol application and service element (TASE)   143, 174
Telemetry   69, 70, 328
Telephone   22, 42, 43, 45, 51, 71, 99, 214, 216, 240, 286
Telephone network   42, 43, 216, 240
Teleprotection   32, 139, 142, 145, 146, 193
Temperature   14, 17, 33, 66, 94, 148, 170, 187, 305
10 Gbps Ethernet passive optical network (10G-EPON)   183, 184, 188, 191
Terminal   42, 44, 57, 121, 123, 138, 140, 144, 180, 202, 203, 225, 228, 230, 237, 243, 246, 247, 250, 255–257, 259, 262, 264, 266, 268, 270–272, 274, 284, 285, 293, 315, 318, 319
Territory   9, 11, 13, 14, 26, 29, 30, 41, 46, 69, 170, 282, 283
TERrestrial Trunked RAdio (TETRA)   177, 283

Thermal rating   169, 170
Third generation (3G)   217, 227–229, 231, 232, 235, 237, 240, 256, 271, 283, 284, 291, 314, 330
Third generation (3GPP)   29, 38, 46, 55, 188, 227–229, 235, 236, 240, 246, 248, 251, 252, 254–257, 278, 280–282, 286, 289–292, 314, 325–327, 329, 332, 333, 335
Three-phase   3–5, 8, 18, 158
Throughput   30, 31, 34, 46, 79, 128, 193, 217, 218, 230, 294, 335
Time and wavelength division multiplexing (TWDM)   187, 190
Time division duplexing (TDD)   107, 228, 240, 247, 261, 271, 273, 274, 276, 315, 327
Time division multiplexing (TDM)   46, 99, 118, 120, 146, 180, 183, 186–188
Time division multiple access (TDMA)   108, 187–189, 191, 213, 226, 227
Time-sensitive communications (TSC)   278, 281
Time-sensitive networking (TSN)   281, 291
Time slotted channel hopping (TSCH)   304
Topology   6–8, 48, 54, 72, 126, 128–131, 157, 158, 179, 180, 196, 202, 208, 209, 218, 219, 297–300, 308, 311
Tower   6, 16, 17, 35, 63, 64, 66
Tracking area update (TAU)   320, 321
Traffic   44, 51, 53, 56, 59, 109, 120, 121, 125, 128, 130, 134, 157, 180, 186–189, 202, 203, 218–220, 230, 233, 237–239, 241–244, 247, 250, 251, 253–255, 260, 270, 274, 276, 282, 284, 288, 307, 312, 315–318, 320, 322, 323, 327
Traffic class   187, 189, 239
Traffic engineering   44, 134
Traffic monitoring   188
Traffic pattern   187, 327
Transceiver   103, 118, 195–198, 204, 207, 209, 210, 303, 307
Transformer   5, 6, 8, 14–16, 18, 30, 72, 73, 144, 147, 148, 150, 151, 156, 192, 194, 196, 221
Transient   13, 33, 144, 164, 197, 307
Transit   133, 158
Transmission capacity   6, 11, 12, 62, 121, 128
Transmission channel   45, 115, 204
Transmission control protocol (TCP)   57–59, 131, 141, 142, 157, 167, 239, 246, 272, 301, 327, 328
Transmission grid   4, 6, 11, 17, 169
Transmission line   6, 16, 43, 67, 76, 170
Transmission media   43, 48, 54, 59, 94, 141, 197
Transmission medium   44, 62, 71, 91, 93, 94, 108, 109, 132, 186, 187, 191
Transmission rate   60, 89, 93, 96, 105, 111, 193, 205, 212–215
Transmission speed   61, 92–94, 98, 99, 112, 131, 193, 194, 212
Transmission system   6, 13, 42, 76, 169, 198
Transmission time interval (TTI)   262, 269
Transmit power   60, 73, 235, 241, 264, 295, 304, 310, 313–315, 318, 324
Transmitter   42, 43, 49, 53, 62, 67, 89, 93, 94, 97, 99, 102–104, 107, 108, 113, 115, 117, 123, 196, 198, 200, 205–207, 209–212, 215, 238, 265, 309, 310, 322

Transport   1, 4, 21, 38, 41, 43, 45, 46, 48–50, 54, 57, 58, 117–124, 132–135, 143, 148, 150, 180, 188, 238, 246, 251, 269, 270, 297, 305, 322, 328, 334
Transport capacity   119, 121
Transport layer   57, 58, 297, 328
Transport level   54, 251
Transport network   48, 117, 120, 122, 123, 135, 238, 246, 251
Transport technology   117–119, 121, 123, 133, 134
Tree   7, 48, 49, 72, 125, 126, 133, 143, 180, 184, 187, 196, 202, 298–300, 333
Tunnel   18, 134, 246, 248, 250–252, 259, 261, 288
Turbo coding   326
Twisted pair   60, 107, 183
Type of service   133

u

Ultrahigh frequency (UHF)   67, 69, 70, 113, 330
Ultra narrowband (UNB)   73, 193
Ultra reliable low-latency communications (URLLC)   229, 230, 248, 261, 280, 281, 294
Unavailability   35, 283
Underground   6, 16–18, 29, 30, 63, 64, 71, 217, 218, 287
Unidirectional   25, 45
Unified data management (UDM)   253, 254, 257–259
Universal mobile telecommunications system (UMTS)   227, 239, 240, 246, 256, 284, 289
Unlicensed spectrum   235, 280, 281, 291, 294, 307, 313, 325, 331
Uplink   230, 244, 251, 258–262, 264, 268–274, 276, 309, 311–314, 318, 319, 322–326, 330
Uplink control information (UCI)   176, 268
Upstream   180, 181, 183–191, 307
Urban   17, 30, 34, 47, 176, 217, 232, 238, 284
Use case   70, 76, 117, 145, 150, 217, 229, 280, 282, 285, 307, 327, 328, 332–334
User datagram protocol (UDP)   58, 157, 239, 297, 301, 327, 328
User plane function (UPF)   247, 248, 252–254, 259–261, 277
Utilities Technology Council (UTC)   29, 69, 162, 292
Utility control center (UCC)   140, 171, 282, 283

v

Vehicle-to-everything (V2X)   280, 289
Vendor   22, 27, 117, 118, 194, 198, 286, 296, 307, 331–333
Vertical industry   233, 254, 257
Vertical user   234, 235, 291
Very High Frequency (VHF)   67, 69, 70, 113
Virtual Concatenation (VCAT)   121
Virtual Container (VC)   120
Virtual Private Network (VPN)   134, 288, 292, 311

Viterbi   106
Voice communication   36, 42, 51, 69, 70, 73, 119, 171, 172, 192, 282, 283, 315
Voice over LTE (VoLTE)   256, 290, 315
Voltage   3–8, 12–16, 25, 33, 44, 63, 71, 72, 74, 79, 85, 102, 111, 143–149, 159, 161, 162, 164, 166, 169, 173, 174, 194, 197, 221
Voltage level   3, 4, 7, 13, 14, 63, 102, 144, 146, 148, 173, 194, 197
Voltage regulation   14, 148

w

Water   5, 13, 33, 65, 66, 159, 166, 282, 296, 309, 332, 336, 337
Wavelength   51, 59, 61, 62, 93, 99, 110, 114, 118, 122–124, 181, 183, 185, 186, 190
Wavelength add-drop multiplexer (WADM)   124
Wavelength division multiplexing (WDM)   62, 99, 122–124, 181, 184, 187, 190
Wavelet-based OFDM   211, 212
Wholesale   10, 11, 45, 46
Wide-area measurement/monitoring system (WAMS)   162, 164, 175
Wide area network (WAN)   47, 134, 164, 220, 301, 330
Wideband   227, 256, 287, 322
Wi-Fi   216, 235, 255, 280–282, 291, 294, 307, 325, 333
Wind   5, 6, 24, 29, 156, 170
Window   36, 61, 62, 96, 102, 191, 207, 276, 311, 321
Wireless   35, 36, 43, 44, 47, 55, 60, 67, 69–71, 73, 75, 76, 79, 93, 96, 97, 109, 110, 112–115, 150, 216, 220, 225, 235, 252, 280, 281, 287, 288, 290, 293–298, 301, 307, 329, 333, 334, 337
Wireless access   75, 290, 334
Wireless network   296, 334
Wireless personal area network (WPAN)   47, 294
Wireless technology   55, 293, 307
Wireline   55, 60, 67, 69, 73, 93, 97, 115, 179, 191, 252, 283, 288
Workforce   171
World Radiocommunication Conference (WRC)   28, 232

x

x digital subscriber line (covering various types of DSL) (xDSL)   46, 179, 183, 192, 216, 286
X (10) Gbps passive optical network (XG-PON)   180, 190
X (10) Gbps symmetrical passive optical network (XGS-PON)   180, 184, 190

z

ZigBee   167, 177, 293, 294, 296, 297, 300, 304, 305, 307, 330, 331, 334, 336

CPSIA information can be obtained
at www.ICGtesting.com
Printed in the USA
BVHW020242040523
663563BV00003B/125